Studies in Computational Intelligence

Volume 1214

The series "Studies in Computational Intelligence" (SCI) publishes new developments and advances in the various areas of computational intelligence—quickly and with a high quality. The intent is to cover the theory, applications, and design methods of computational intelligence, as embedded in the fields of engineering, computer science, physics and life sciences, as well as the methodologies behind them. The series contains monographs, lecture notes and edited volumes in computational intelligence spanning the areas of neural networks, connectionist systems, genetic algorithms, evolutionary computation, artificial intelligence, cellular automata, self-organizing systems, soft computing, fuzzy systems, and hybrid intelligent systems. Of particular value to both the contributors and the readership are the short publication timeframe and the world-wide distribution, which enable both wide and rapid dissemination of research output.

Indexed by SCOPUS, DBLP, WTI AG (Switzerland), zbMATH, SCImago.

All books published in the series are submitted for consideration in Web of Science.

Anis Koubaa · Adel Ammar · Lahouari Ghouti ·
Wadii Boulila · Bilel Benjdira
Editors

Generative AI and Large Language Models: Opportunities, Challenges, and Applications

Volume 1

Springer

Editors
Anis Koubaa
Alfaisal University
Riyadh, Saudi Arabia

Adel Ammar
Prince Sultan University
Riyadh, Saudi Arabia

Lahouari Ghouti
Abjad AI
Riyadh, Saudi Arabia

Wadii Boulila
Robotics and Internet-of-Things Lab
Prince Sultan University
Riyadh, Saudi Arabia

Bilel Benjdira
Robotics and Internet-of-Things Lab
Prince Sultan University
Riyadh, Saudi Arabia

ISSN 1860-949X ISSN 1860-9503 (electronic)
Studies in Computational Intelligence
ISBN 978-3-031-90572-8 ISBN 978-3-031-90573-5 (eBook)
https://doi.org/10.1007/978-3-031-90573-5

This Springer imprint is published by the registered company Springer Nature Switzerland AG
The registered company address is: Gewerbestrasse 11, 6330 Cham, Switzerland

Preface

The rapid evolution of Generative Artificial Intelligence (AI) and Large Language Models (LLMs) has brought about a new era of technological innovation, transforming industries, reshaping communication, and redefining the boundaries of human–machine collaboration. From healthcare and natural language processing to software development and environmental sustainability, the applications of these advanced AI systems are vast and far-reaching. This book, *Generative AI and Large Language Models: Opportunities, Challenges, and Applications*, aims to provide a comprehensive exploration of this dynamic field, offering insights into the latest advancements, practical applications, and the challenges that lie ahead.

The journey of AI, from its early statistical models to the sophisticated neural architectures of today, has been marked by significant achievements and paradigm shifts. Central to this transformation are transformer-based models, which have revolutionized natural language processing, computer vision, and beyond. This book examines the architecture of these models, their training processes, and their ethical implications, providing readers with a clear understanding of the mechanisms that power modern AI systems.

One of the key themes of this book is the adaptation of AI technologies to address language-specific challenges, particularly in the context of Arabic. With the development of models like ArabianGPT and frameworks for enhancing semantic understanding in Arabic NLP, this book highlights the importance of tailoring AI solutions to meet the unique linguistic and cultural needs of diverse communities. These advancements not only bridge the gap in language modeling but also pave the way for more inclusive and accessible AI systems.

In addition to language-specific innovations, this book explores the role of generative AI in code development, where AI-driven tools are transforming software engineering practices. From code generation and optimization to testing and refactoring, the potential of AI to enhance productivity and creativity in software development is immense. However, this potential comes with challenges, including issues of contextual understanding, data bias, and the need for explainable AI, all of which are thoroughly examined in this volume.

The book also examines the applications of multimodal AI and diffusion models, showcasing their use in computer vision, audio synthesis, and beyond. These models, which mimic natural processes like diffusion, offer stable and high-quality outputs, opening new possibilities for creative and technical applications. Furthermore, the integration of AI with intellectual property (IP) frameworks is explored, particularly in the context of Saudi Arabia's Vision 2030, where AI is envisioned as a catalyst for sustainable development and economic diversification.

As editors, we are deeply grateful to the contributing authors, whose expertise and dedication have made this book possible. Each chapter reflects the collective effort of researchers and practitioners who are at the forefront of AI innovation. Their work not only advances the field but also inspires future generations to explore the potential of AI.

This book is intended for a broad audience, including researchers, practitioners, students, and policymakers who seek to understand the transformative power of generative AI and LLMs. Whether you are new to the field or an experienced professional, we hope this book will serve as a valuable resource, sparking new ideas and fostering a deeper appreciation for the opportunities and challenges that lie ahead in the ever-evolving landscape of AI.

We invite you to join us on this journey, as we explore the cutting-edge developments, practical applications, and future prospects of generative AI and large language models. Together, let us envision a future where AI not only enhances technological capabilities but also enriches human lives and fosters global progress.

Riyadh, Saudi Arabia
February 2025

Anis Koubaa
Adel Ammar
Lahouari Ghouti
Wadii Boulila
Bilel Benjdira

Contents

Survey of AI-Driven Advancements in Diagnosis and Clinical Decisions for Healthcare

Aymen Abid, Khaoula Tayari, and Omar Cheikhrouhou

Abstract The integration of Artificial Intelligence (AI) technologies is revolutionizing healthcare by driving advancements in disease detection, personalized medicine, and clinical decision support. This chapter provides a comprehensive overview of recent developments in deep learning and machine learning, focusing on generative models for medical data synthesis, deep learning architectures for multimodal data analysis, and advanced techniques for predictive analytics and precision medicine. Key applications discussed include early diagnosis, treatment optimization, and real-time clinical insights, emphasizing the role of AI in addressing challenges such as data scarcity and enhancing diagnostic capabilities. The chapter explores how advanced AI models facilitate the integration of complex and heterogeneous datasets, enabling comprehensive decision-making in healthcare. Furthermore, critical challenges related to data scalability, privacy concerns, model interpretability, and clinical integration are examined. By addressing both opportunities and barriers, this chapter offers valuable insights into the evolving role of AI-driven methodologies in modern healthcare, ultimately aiming to enhance patient outcomes and operational efficiency.

A. Abid (✉) · K. Tayari
CES-LAB National Engineers School of Sfax, Sfax, Tunisia
e-mail: aymen.abid@enis.tn

O. Cheikhrouhou
National School of Electronics and Telecommunications of Sfax, University of Sfax, Sfax, Tunisia
e-mail: omar.cheikhrouhou@enetcom.usf.tn; cheikhrouhou@gmail.com

A. Koubaa et al. (eds.), *Generative AI and Large Language Models: Opportunities, Challenges, and Applications*, Studies in Computational Intelligence 1214,
https://doi.org/10.1007/978-3-031-90573-5_1

1 Introduction

Artificial Intelligence (AI) is revolutionizing healthcare by enhancing efficiency, improving diagnostic accuracy, and enabling personalized treatment. Its applications span various domains of medicine and healthcare delivery, addressing both clinical and operational challenges (Fig. 1). AI technologies are reshaping traditional medical practices by facilitating early disease detection, optimizing healthcare workflows, and driving significant advancements in biomedical research and clinical applications [1, 2].

Recent advancements in AI, including generative models for medical data synthesis, transformer-based deep learning frameworks for multimodal learning, and advanced machine learning (ML) techniques for predictive analytics and precision medicine, have paved the way for groundbreaking innovations in healthcare [3–5]. However, despite these advancements, integrating AI into clinical practice presents challenges such as data privacy concerns, inconsistent data quality, and the need for greater interpretability and trust in AI-driven decision-making.

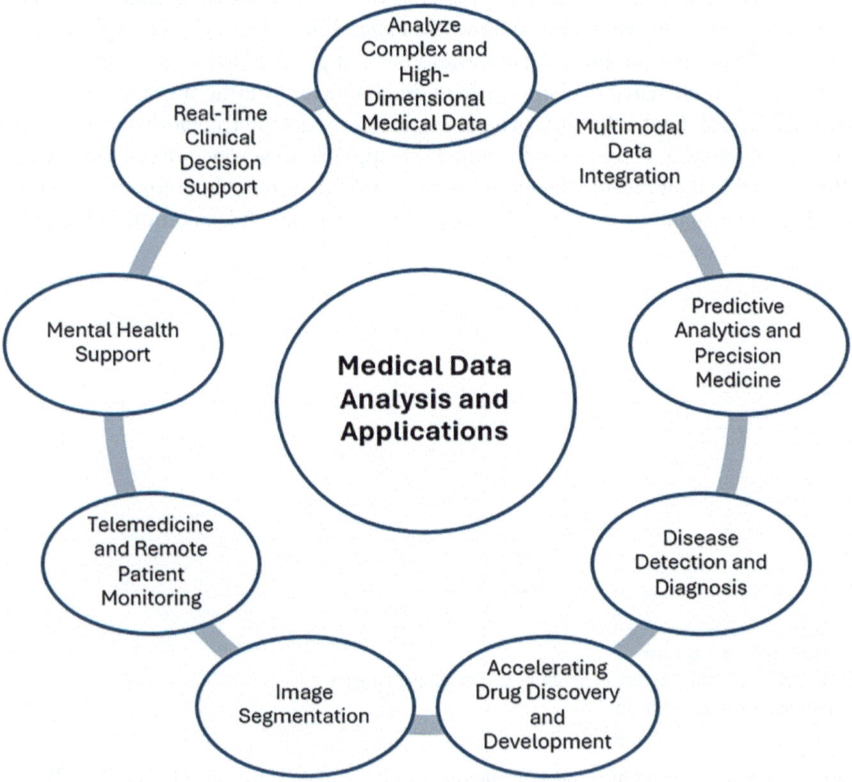

Fig. 1 Overview of AI applications in healthcare

Several studies have investigated AI's transformative potential in healthcare. In [1], the authors explore the application of machine learning and deep learning techniques in analyzing complex medical data to enable accurate predictions and personalized treatment plans. These applications include early disease detection, precision medicine, and resource optimization, while challenges such as data privacy, algorithmic bias, and clinical workflow integration are highlighted. Similarly, [2] reviews recent trends in biomedical AI research, focusing on the application of deep learning frameworks such as transformers for disease prediction, medical image analysis, and multimodal data integration. The study underscores AI's role in accelerating drug discovery and supporting real-time clinical decision-making, while emphasizing challenges related to explainability and data standardization.

Despite AI's vast potential, several barriers hinder its widespread adoption in healthcare settings. These challenges include fragmented data quality, lack of interpretability, algorithmic bias, and complexities in clinical workflow integration (Fig. 2). Additionally, ethical concerns, high computational costs, and regulatory compliance pose further obstacles to the large-scale deployment of AI-driven solutions. Building trust among healthcare professionals and ensuring the ethical use of AI in clinical decision-making remain critical hurdles.

Research Objective This study aims to explore the evolving landscape of AI applications in healthcare, with a particular focus on medical signal and image analysis. The key objectives include:

- Investigating the effectiveness of AI-driven approaches in enhancing disease diagnosis and patient monitoring through the analysis of physiological signals such as ECG, EEG, and PPG.
- Exploring state-of-the-art deep learning techniques, including transformers and generative models, to assess their impact on advancing healthcare applications.
- Identifying key challenges that hinder the scalability and adoption of AI in clinical settings and proposing potential solutions.

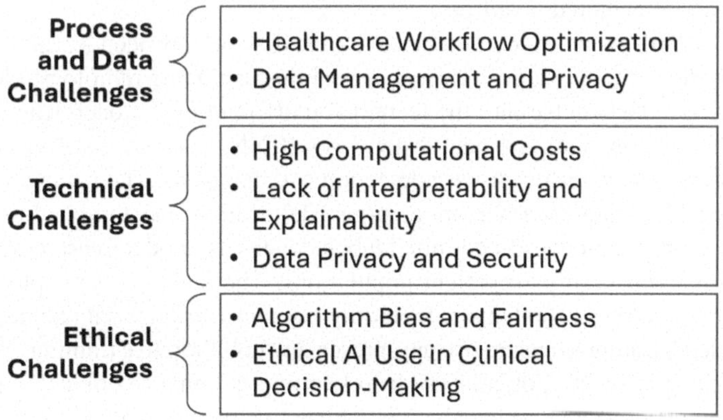

Fig. 2 AI challenges in healthcare

Thereby, this study provides a comprehensive perspective on AI's transformative role in healthcare, with an emphasis on medical signal and image analysis. The subsequent sections will explore specific AI implementations, evaluating their practical applications, benefits, and inherent limitations. By offering a holistic analysis, this work aims to provide valuable insights into how AI-driven methodologies are revolutionizing healthcare practices and improving patient outcomes.

2 Advancing AI-Driven Medical Imaging for Healthcare Applications

The integration of artificial intelligence (AI) into medical domain has revolutionized traditional diagnostic workflows, offering new levels of precision, efficiency, and scalability. In radiology, AI has enhanced disease detection and classification, enabling automated analysis of CT and MRI scans to identify pathologies such as tumors and fractures with unprecedented accuracy. In pathology, deep learning models have revolutionized histopathological image analysis, aiding in cancer detection and reducing diagnostic variability. AI-driven techniques have also made a substantial impact in ophthalmology, where retinal imaging tools support early detection of diabetic retinopathy and other vision-threatening conditions. Similarly, in dermatology, advanced algorithms have improved the classification of skin lesions, bolstering the early diagnosis of melanoma.

This section further explores specialized applications, such as AI in mammography for breast cancer screening and its role in cardiac imaging for personalized cardiovascular care and further using neuroimaging to investigate Alzheimer's disease and neuroimaging in the study of Alzheimer's disease. These advancements illustrate the broad scope of AI's impact across medical disciplines. By examining these advancements, in the following and as summarized by the Table 1 we aim to showcase how AI is not only enhancing diagnostic capabilities but also shaping the future of patient-centered healthcare.

Authors in [6], provide a comprehensive overview of how deep learning is transforming medical imaging across various modalities, including radiology, pathology, and cardiology. They delve into the technical underpinnings of deep learning techniques, such as convolutional neural networks (CNNs), which excel in analyzing complex medical images for disease detection and diagnosis. The review highlights applications like lesion segmentation, organ delineation and tumor classification while addressing challenges like limited labeled datasets, model interpretability and generalization across diverse patient populations. The authors also explore strategies to overcome these challenges, including transfer learning, data augmentation, and federated learning to ensure privacy-preserving collaborative training. The paper underscores the growing potential of deep learning to enhance clinical workflows, reduce diagnostic errors and enable personalized medicine while emphasizing the need for interdisciplinary collaboration and ethical considerations. Thus, the paper

Table 1 Summary of advancing deep learning applications in medical imaging for healthcare

Studies	Area	Technique	Dataset	Limitations
[6] (2023)	Radiology, Pathology, Cardiology	Convolutional Neural Networks (CNNs), Transfer Learning, Data Augmentation, Federated Learning	ImageNet, LIDC-IDRI	Limited labelled data; generalization across different populations; interpretability of deep learning models
[7] (2022)	Ophthalmology (Diabetic Retinopathy and Macular Oedema Detection)	Convolutional Neural Networks (CNNs)	Retinal fundus images (e.g., EyePACS)	Challenges in image quality variability; class imbalance in datasets and limited generalizability across diverse populations
[8] (2018)	Medical Image Reconstruction (Dynamic MRI)	Convolutional Recurrent Neural Networks (CRNNs) for image reconstruction	10 fully sampled short-axis cardiac cine MR scans	Need for high-quality training datasets; challenges in generalization across MRI scanners; real-time applicability not fully optimized
[9] (2024)	Pulmonology (Lung Cancer Detection)	• Deep learning-based system (CNNs and other advanced ML models) for: AI-Rad Companion (Siemens Healthineers), • AVIEW Lungscreen (Coreline Soft), • ClearRead (Riverain Technologies), • InferRead CT Lung (Infervision), • VUNO Med LungCT AI (VUNO), • Veolity (MeVis)	CT scan images (MEDLINE, Embase, CRD, INAHTA)	Variability in study methodologies; inconsistent data quality and limited generalizability across diverse populations and imaging protocols
[10] (2023)	Dermatology (Skin Lesion Classification)	• CNNs (Convolutional Neural Networks), • Transfer Learning, • Ensemble Learning	• ISIC, • DermNet, • MSK Skin Lesion Dataset, • Atlas of Dermoscopy	Limited availability of high-quality labeled data; Bias in dataset composition; Variability in image quality and acquisition methods
[11] (2023)	Medical Imaging	• CNNs (Convolutional Neural Networks), • RNNs (Recurrent Neural Networks), • GANs (Generative Adversarial Networks), • Transformers	• CT Scan, • MRI, • X-ray, • Public medical imaging: Chest X-ray 14, LUNA16 (Lung Nodule Analysis 2016), NIH Chest X-ray Dataset, ISLES (Ischemic Stroke Lesion Segmentation)	Limited dataset diversity; reliance on large annotated datasets; challenges with overfitting; interpretability of deep learning models

(continued)

Table 1 (continued)

Studies	Area	Technique	Dataset	Limitations
[12] (2024)	Biomedicine (Alzheimer…)	• CNNs, • Attention-based Models, • Early, Intermediate, • Late Fusion	• ADNI (Alzheimer's Disease Neuroimaging Initiative), • BraTS (Brain Tumor Segmentation Challenge), • TCIA (The Cancer Imaging Archive), • OASIS (Open Access Series of Imaging Studies), • SPC (Skin Lesion Classification Dataset), • TCGA (The Cancer Genome Atlas), • ABIDE (Autism Brain Imaging Data Exchange), • ADHD-200 (Attention Deficit Hyperactivity Disorder 200), • COBRE (Center for Biomedical Research Excellence), • GAMMA (Glaucoma Dataset), • CPM-RadPath (Cancer Pathology Dataset), • ISIT-UMR (Gastrointestinal Lesions Dataset), • MRNet (Knee MRI Dataset), • CTU-UHB (Fetal Distress Diagnosis Dataset)	Data Availability and Labeling; Data Integration and Alignment; Generalization Across Datasets; Interpretability of Multimodal Models; Computational Complexity; Transferability and Transfer Learning; Scalability
[21] (2023)	Medical Imaging	Deep Residual Learning (ResNet50)	28,378 multi-modal medical images	Potential challenges in generalization across diverse imaging modalities and limited interpretability of deep learning models
[24] (2024)	Diagnosis using imaging	LLM + CNN integration	MRI, CT, X-ray datasets from various medical institutions	Limited data diversity and potential bias in training datasets
[25] (2023)	Image-based diagnosis	LLM + visual data processing	Mixed hospital image data	Difficulty in distinguishing between subtle image anomalies
[26] (2024)	Clinical report generation	LLMs for report generation	Image datasets (X-ray, CT scans)	Dependence on high-quality images for accurate output

provides a broad overview of the techniques applied in medical imaging, addressing common challenges like dataset limitations and model interpretability, as summarized in the table. It explores the use of CNNs and various training techniques such as transfer learning and federated learning while also pointing out the need for better generalization across diverse clinical settings and patient populations.

In [7], authors focus on the development and application of deep learning models for the automated screening of diabetic retinopathy (DR) and diabetic macular edema (DME), two major complications of diabetes that can lead to vision loss if untreated. The study leverages convolutional neural networks (CNNs) to analyze retinal fundus photographs, aiming for accurate detection and classification of these conditions. The proposed approach demonstrates the potential to significantly improve early diagnosis by providing fast and reliable assessments, which are crucial in regions with limited access to ophthalmologists. The paper discusses training the models using large, annotated datasets and evaluates their performance against human experts, showcasing competitive accuracy, sensitivity, and specificity. Challenges such as variability in image quality, class imbalance and model generalizability are also addressed. The work highlights the promise of AI in enhancing DR and DME screening, reducing diagnostic delays, and improving patient outcomes. A systematic review of the test accuracy of AI software in the context of lung cancer detection using CT scans is presented in [8]. This study critically evaluates various AI technologies, including deep learning models, designed to assist in the detection of lung nodules and potential cancers during screening. By analyzing multiple studies, the authors assess how AI tools, such as convolutional neural networks (CNNs), improve diagnostic accuracy and reduce human error in the interpretation of CT images. The review also examines the variability in performance across different studies, highlighting the challenges associated with dataset quality, model generalizability and clinical implementation. Despite these limitations, the study underscores the growing potential of AI-driven solutions in early lung cancer detection, demonstrating their capability to enhance early diagnosis, optimize clinical workflows, and support clinicians in making more informed decisions.

The paper [9] provides a systematic review of the test accuracy of AI software in the context of lung cancer detection using CT scans. It critically evaluates various AI technologies, including deep learning models, designed to assist in the detection of lung nodules and potential cancers during screening. By analyzing multiple studies, the authors assess how AI tools, such as convolutional neural networks (CNNs), improve diagnostic accuracy and reduce human error in the interpretation of CT images. The review also examines the variability in performance across different studies, highlighting the challenges associated with dataset quality, model generalizability and clinical implementation. Despite these limitations, the study underscores the growing potential of AI-driven solutions in early lung cancer detection, demonstrating their capability to enhance early diagnosis, optimize clinical workflows and support clinicians in making more informed decisions. The growing role of deep learning techniques, particularly convolutional neural networks (CNNs), in the automated classification of skin lesions is explored in [10]. The study highlights various applications of CNNs for skin cancer detection, focusing on melanoma and other common skin lesions. By utilizing large image datasets, the paper discusses how deep learning models can outperform traditional methods in terms of diagnostic accuracy and efficiency, enabling earlier detection and improving patient outcomes. It also evaluates the challenges in acquiring labeled datasets, the need for high-quality images and the need for further fine-tuning of models to enhance generalization across diverse populations. The paper underlines the importance of such advancements for dermatology, enabling non-specialist medical practitioners to provide faster, reliable diagnoses while minimizing human error.

Authors of [12] explore the emerging field of Multimodal Deep Learning (MDL) in several biomedical domains, including Alzheimer's disease diagnosis, breast cancer screening via mammography and personalized cardiovascular care. They focus on integrating diverse data types such as medical images, electronic health records (EHR) and genomics to enhance diagnostic accuracy. The paper discusses key deep learning techniques that are designed to handle complex, high-dimensional medical data. It also examines the application of these methods in various medical tasks, such as disease diagnosis, prognosis prediction and personalized treatment recommendations. Additionally, the paper addresses challenges associated with multimodal data fusion, the need for large and diverse datasets and limitations of current models, including overfitting and interpretability issues. The authors highlight future directions in multimodal learning, emphasizing better integration of diverse data sources and the development of more robust, generalized models. However, they note limitations related to data availability, manual annotation requirements and the scalability of these systems in real-world medical applications.

To improve the classification accuracy of multi-modal medical images, the authors in [21] utilized deep residual learning through the ResNet50 architecture. Their dataset comprised 28,378 multi-modal medical images used for training and validation. The proposed model demonstrated superior performance compared to other state-of-the-art methods across key metrics, including accuracy, precision, recall and F1 score. Furthermore, the study highlights the application of deep learning techniques in medical image classification and explores the role of machine learning in risk prediction.

Large Language Models (LLMs) are increasingly being integrated with medical imaging systems to assist in image interpretation and diagnosis. By processing images such as X-rays, CT scans, and MRIs, LLMs can support the detection of various conditions, including tumors, fractures, and organ abnormalities. Wang et al., in [24], explore the transformative role of LLMs in healthcare, highlighting their applications in medical question-answering, clinical dialogue summarization, electronic health record generation, scientific research, medical education and medical imaging analysis. LLMs enhance diagnostic reasoning, streamline administrative tasks and support clinical decision-making. Despite their potential, challenges persist, including data security risks, biases, inaccuracies, ethical concerns and high computational costs. The paper also discusses solutions such as de-identification frameworks, fairness prompting techniques and standardized guidelines to address these issues, offering a comprehensive overview of LLMs' impact and future prospects in healthcare.

The paper [25] highlights the potential of LLMs to enhance diagnostic accuracy by providing context and insights into image features, particularly when used in combination with deep learning models. LLMs can also generate reports based on the findings from medical images, streamlining workflows and aiding clinicians in making more informed decisions. However, challenges remain in terms of ensuring model reliability, handling the variability of image quality, and ensuring interpretability of model predictions in a clinical setting.

In [26], authors systematically examine the transformative role of Large Language Models in healthcare across various domains. The study explores their applications in medical writing, diagnostics, education, research management and patient communication. It highlights LLMs' capabilities in generating medical documents, aiding clinical decision-making, summarizing patient records, enhancing medical education through simulations and improving doctor-patient communication. However, the paper also identifies significant challenges, including contextual limitations, ethical concerns, data privacy issues and risks of over-reliance on AI. It emphasizes the need for rigorous validation, integration of multimodal LLMs and collaboration between medical experts and developers to ensure responsible and effective adoption in healthcare.

Then, Fig. 3 provides an overview of key aspects driving AI-powered medical imaging applications. It categorizes the landscape into three main areas: AI applications, deep learning approaches, and challenges. The AI applications section highlights various medical fields, such as radiology, pathology, and cardiology, where

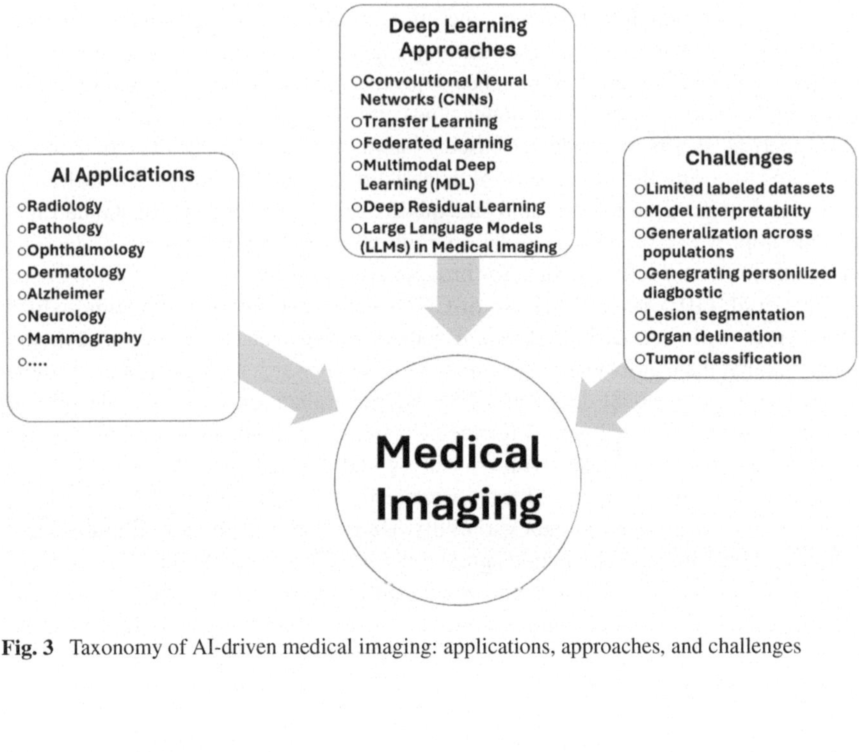

Fig. 3 Taxonomy of AI-driven medical imaging: applications, approaches, and challenges

AI technologies have been effectively deployed to enhance diagnostic accuracy and workflow efficiency. The deep learning approaches section outlines advanced techniques, including convolutional neural networks (CNNs), transfer learning, and large language models (LLMs), which have been pivotal in improving image analysis and interpretation. Lastly, the challenges section addresses critical obstacles such as limited labeled datasets, model interpretability, and generalization across populations, emphasizing the ongoing need for robust solutions to ensure reliable and ethical deployment of AI in healthcare. This comprehensive representation underscores the multifaceted nature of AI integration in medical imaging and the opportunities it presents to improve patient care.

In summary, several studies exploring AI-driven medical imaging for healthcare applications have been discussed in this section. The work by Zhang and Qie [6] provides a comprehensive review of deep learning techniques applied to medical imaging, highlighting the promise of AI in radiology, pathology, and cardiology. However, the review acknowledges limitations such as the lack of labeled data and challenges in generalizing models across diverse populations. Future work could focus on enhancing dataset diversity and improving model interpretability.

Li et al. [7] applied CNNs for detecting diabetic retinopathy and macular oedema in retinal fundus images. While the technique demonstrated effectiveness, challenges such as image quality variability and class imbalance were identified. Future directions could involve refining model robustness against image quality inconsistencies and exploring the integration of generative models to augment data.

Qin et al. [8] focused on the application of Convolutional Recurrent Neural Networks (CRNNs) for dynamic MRI reconstruction, offering a novel approach for improving image quality. However, the study's reliance on high-quality datasets and the challenges in real-time applicability suggest the need for more efficient and adaptive models. Generative approaches or leveraging transfer learning could further optimize the model for diverse scanner types and settings.

Geppert et al. [9] reviewed AI-based systems for lung cancer detection in CT scans, underscoring the importance of accurate detection algorithms. Nevertheless, the variability in methodologies and data quality across studies calls for standardization and improved cross-study validation. Future research should aim to create more unified datasets and refine models for broader applicability.

Silva et al. [10] explored the use of CNNs for skin lesion classification, which showed promise in dermatology. Despite challenges like dataset bias and variability in image quality, the results were promising. Moving forward, incorporating ensemble learning and multi-modal data could improve generalization. Li et al. [11] applied various deep learning techniques for medical image analysis, including GANs and transformers, but faced issues with overfitting and data reliance. Their work could be enhanced by using larger and more diverse datasets, along with fine-tuning the architecture to better handle data variability. Duan et al. [12] reviewed multimodal biomedical data fusion, which holds great potential for integrating data from various sources, such as MRI, CT scans, and genetic information. Future work could focus on more efficient fusion techniques and seamless integration of heterogeneous data.

Lastly, Yang et al. [25] and Meng et al. [26] explored the use of large language models (LLMs) in healthcare, with promising applications in diagnosis and clinical report generation. However, the heavy reliance on high-quality labeled datasets and challenges related to model interpretability and bias remain. In the future, efforts could focus on enhancing the accuracy and reliability of LLMs, as well as addressing ethical concerns such as data privacy and bias in AI models.

Subsequently, while these studies demonstrate significant advances in AI-based medical imaging, future work should focus on improving data diversity, enhancing model robustness, and exploring generative and multimodal approaches to address the limitations identified in current research. One promising direction for future development is the integration of generative deep learning techniques, such as Generative Adversarial Networks (GANs), to improve synthetic augmentation of medical images, particularly for underrepresented conditions or rare pathologies. These techniques can help tackle the challenges of data scarcity and variability while producing more diverse and realistic images for training robust models.

Another key area of focus is the adoption of multimodal learning approaches, combining different imaging modalities (e.g., MRI, CT, X-rays) and integrating them with clinical data. This would enable AI models to gain deeper insights into complex medical conditions, leading to more accurate and comprehensive diagnoses. Additionally, leveraging advancements in edge computing and the Internet of Things (IoT) for real-time image acquisition and analysis can provide dynamic, personalized healthcare solutions. The integration of real-time medical imaging with cloud-based platforms will facilitate continuous learning and classification, offering the potential for immediate diagnosis and decision support.

Future developments could also explore better handling of large-scale datasets, leveraging distributed computing to improve the efficiency and accuracy of AI models. By utilizing vast amounts of medical imaging data and computational resources, these systems can become more scalable, providing more accurate diagnoses in a timely manner. However, the integration of these technologies into clinical workflows will require addressing challenges related to data privacy, system interoperability, and ensuring real-time data transmission.

Ultimately, by embracing cutting-edge technologies and expanding the scope of AI-driven medical imaging, the future of healthcare can benefit from even more accurate, efficient, and scalable imaging solutions, improving patient outcomes and transforming clinical practices.

3 Advancing AI-Driven Medical Signal Analysis for Healthcare Applications

Artificial intelligence (AI) continues to transform healthcare by enabling advanced analysis of physiological signals such as electrocardiograms (ECGs), electroencephalograms (EEGs), photoplethysmograms (PPGs), and respiratory signals. These technological advances, as we will see in the following and the summarized Table 2 facilitate early disease and anomaly detection, accurate diagnostics, and real-time health monitoring, ushering in a new era of personalized and predictive medicine.

Authors in [13], introduce an approach for electrocardiogram (ECG) classification by leveraging transformer-based architectures, which have gained significant attention in natural language processing and computer vision. In this study, the authors apply transformers to ECG signal data for improved classification accuracy, particularly in detecting arrhythmias and other heart conditions. The proposed model benefits from the self-attention mechanism of transformers, which allows the model to focus on important time-dependent patterns in the ECG signals, enhancing its ability to distinguish between various heart abnormalities. The study compares transformer-based methods with traditional machine learning techniques, demonstrating superior performance in terms of accuracy and robustness. Furthermore, the paper emphasizes the scalability of transformer models, making them suitable for large-scale ECG datasets. The research underscores the potential of transformer-based models to revolutionize ECG classification by providing faster, more accurate, and automated heart disease diagnosis. Challenges such as the model's computational complexity and the need for large labeled datasets for training are also discussed.

For electroencephalogram (EEG) signals classification, the application of convolutional neural networks (CNNs) is explored in [14], with a focus on enhancing the accuracy of brain-computer interfaces (BCIs) for tasks such as cognitive state monitoring and disease detection. The authors propose a CNN architecture specifically designed to process and classify EEG data by automatically learning hierarchical features from raw EEG signals, which traditionally required handcrafted feature extraction. Through extensive experimentation, the study demonstrates the effectiveness of CNNs in accurately classifying EEG signals across different cognitive states, highlighting their superior performance over conventional machine learning techniques. The CNN model is trained on a publicly available dataset and optimized to capture both spatial and temporal patterns in EEG signals. This approach addresses challenges such as signal variability and noise inherent in EEG data. Additionally, the paper discusses the potential of CNN-based EEG classification systems in real-time applications, offering promise for non-invasive, accurate monitoring of neurological disorders like epilepsy and sleep disorders. The limitations identified include the need for large datasets and the computational complexity of training deep CNN models.

Moreover, Aymen et al., in [16], introduce a novel machine learning-based approach to improve the accuracy and reliability of EEG signal analysis. The study proposes an enhanced K-Nearest Neighbor (KNN) algorithm with an integrated outlier detection mechanism designed to address common issues in EEG data, such as noise and signal artifacts. By leveraging a smart adaptive mechanism, the model dynamically identifies and removes outliers, ensuring cleaner data for classification tasks. Extensive experiments on benchmark EEG datasets demonstrate significant performance improvements in terms of classification accuracy and robustness compared to conventional KNN and related machine learning models. The research

Table 2 Summary of advancing deep learning applications in signal-based healthcare for disease detection

Studies	Area	Technique	Dataset	Limitations
[13] (2023)	ECG Classification	Transformers	MIT-BIH Arrhythmia [Link], PTB Diagnostic ECG [Link]	Limited to specific arrhythmias
[14] (2023)	EEG Signal Analysis	CNN	BCI Competition IV-2a Link, EEG Motor Movement/Imagery [Link]	Potential overfitting on small datasets
[16] (2024)	EEG Signal Analysis	Smart K-Nearest Neighbor (KNN) with Outlier Detection	BCI Competition IV-2a [Link]	Susceptible to incorrect identifications differentiating between typical and unusual signals due to unclear patient replies
[15] (2025)	ECG Classification	Vision Transformer with Data Augmentation	Multi-label ECG dataset	Computational complexity and need for diverse training data
[17] (2022)	PPG Signal Analysis	Deep Learning (CNN-based)	MIMIC-III [Link], PPG-DaLiA Link	Limited validation on diverse populations
[18] (2022)	Respiratory Signal Analysis	RNN	CAPTURE [Link], PhysioNet Respiratory Sound [Link]	Challenges in noisy environments
[19] (2023)	Multi-disease Diagnosis (Alzheimer's, Cancer, Chronic Diseases, Heart Diseases, Tuberculosis)	Various AI/ML: CNN, RNN, SVM, Decision Trees	Alzheimer's Disease Neuroimaging: ADNI [Link], Cardiology and Respiratory: PhysioNet [Link], Cancer Imaging Archive: TCIA [Link]	Limited dataset availability in specific diseases, data heterogeneity, and integration challenges across multiple domains

(continued)

Table 2 (continued)

Studies	Area	Technique	Dataset	Limitations
[20] (2024)	Disease Diagnosis with AI	• Convolutional Neural Networks (CNN), • Recurrent Neural Networks (RNN), • Generative Adversarial Networks (GANs), • Transfer Learning, • Deep Reinforcement Learning.	Real datasets for Alzheimer's, cancer, heart disease, chronic diseases, and tuberculosis	Dataset limitations, need for more real-world validation, challenges with interpretability and ethical concerns in AI applications
[23] (2020)	Healthcare, Epilepsy Prediction	SVM and Random Forest [link1, link2]	EEG signals	Small dataset; signal noise; variations in seizure patterns across patients
[25] (2023)	Signal-based treatment recommendations	LLMs for biosignal analysis	EEG, ECG signal datasets	Lack of interpretability and training data consistency
[28] (2023)	Signal-based diagnosis	LLMs for bio signal report generation	EEG, ECG signal datasets	Noise sensitivity; model complexity; challenges in real-time processing and interpretability issues
[29] (2024)	Signal-based diagnosis	AI-LLM integration for healthcare with EEG and BCI	EEG datasets, electronic health records (EHR)	Computational complexity; large dataset requirements; real-time processing challenges; integration of environmental/lifestyle data

highlights the potential of adaptive outlier detection in advancing EEG-based applications, including cognitive monitoring and brain-computer interfaces (BCIs).

In addition to transformer-based and CNN approaches, recent work has demonstrated the potential of combining data augmentation with vision transformer architectures for robust multi-label ECG arrhythmia classification. The paper "A Data-Augmented Vision Transformer Model for Robust Multi-Label ECG Arrhythmia Classification" [15] presents a model that leverages data augmentation techniques to enhance the diversity of ECG inputs, thereby addressing the challenge of limited labeled data. By utilizing a vision transformer (ViT) framework, the model employs self-attention mechanisms to capture critical temporal and spatial patterns inherent in 12-lead ECG recordings, enabling the accurate classification of multiple arrhythmia types. The study reports superior performance over traditional methods, demonstrating improved generalization and robustness in detecting subtle cardiac abnormalities. However, the authors also note limitations, such as the computational complexity associated with transformer architectures and the need for further validation on larger, more heterogeneous ECG datasets to fully establish its clinical utility.

In [17], a deep learning-based method for analyzing photoplethysmogram (PPG) signals, focusing on the application of Convolutional Neural Networks (CNNs) to enhance the extraction of features from PPG data, is presented. The study aims to improve the accuracy and efficiency of cardiovascular health monitoring using PPG signals, which are commonly used for heart rate and blood oxygen saturation level measurement. By employing CNNs, the authors demonstrate how these models can automatically learn and extract high-level features from the raw PPG signals, eliminating the need for traditional manual feature extraction. The paper explores the potential of CNNs in detecting arrhythmias, classifying heartbeats, and monitoring other vital signs from PPG data. The proposed CNN architecture is optimized through extensive training on a labeled PPG dataset, demonstrating its superior performance in comparison to traditional machine learning techniques. The results indicate that CNN-based models can significantly enhance the reliability of PPG-based health monitoring systems. The limitations discussed include challenges related to the quality of PPG signals in real-world settings and the need for larger, more diverse datasets to generalize the model effectively.

The study, in [18], proposes a deep learning-based approach for identifying respiratory diseases using raw lung auscultation sounds. It explores the application of Recurrent Neural Networks (RNNs) to analyze respiratory sounds and classify them into different pathology categories, such as wheezing, crackles, and normal breathing patterns. The authors emphasize the use of raw, unprocessed lung sounds, which eliminates the need for complex feature engineering, offering a more scalable and efficient approach to disease detection. The proposed model demonstrated high accuracy in detecting various respiratory conditions, highlighting the potential of deep learning in real-time clinical diagnostics. The paper also discusses the challenges of noise interference in respiratory sound recordings and the need for robust models capable of handling such noise to ensure reliable performance in diverse clinical settings. Overall, the research suggests that deep learning models, particularly RNN-based architectures, can significantly enhance the automation of respiratory disease diagnosis, aiding healthcare professionals in early and accurate identification of pathologies.

Physiological signal analysis extends beyond traditional applications such as ECG classification, EEG classification, PPG signal analysis, and respiratory signal analysis. Kumar et al. explore a broader spectrum of healthcare domains where artificial intelligence (AI) has transformative potential [19]. Their study covers critical areas like Alzheimer's disease, cancer detection, chronic disease management, heart disease diagnostics, and tuberculosis screening. The authors systematically review AI-driven diagnostic tools, emphasizing their underlying machine learning (ML) and deep learning (DL) methodologies. They present a comprehensive framework synthesizing key AI applications, their associated datasets, and performance metrics. The study highlights both opportunities and challenges in applying AI for disease diagnosis, stressing the need for robust datasets, ethical considerations, and real-world deployment strategies. This work serves as a crucial reference for advancing AI-based healthcare solutions across diverse medical contexts. Building upon the broader applications of physiological signal analysis, the paper [20] expands the scope from domains such as Alzheimer's disease, cancer detection, and heart disease diagnostics to include chronic disease management and tuberculosis screening. It provides a comprehensive review of AI-driven diagnostic tools, focusing on deep learning techniques and their associated datasets. The study highlights key challenges, ethical considerations, and the need for robust real-world deployment strategies to advance AI in healthcare diagnostics.

In [23], Savadkoohi et al. investigate the use of ECG signals (a type of medical signal) in combination with machine learning techniques like Random Forest and Support Vector Machine (SVM) to predict the risk of cardiovascular events. The research focuses on extracting features from ECG signals and using them as structured data inputs for predictive modeling.

In the realm of signal analysis, LLMs have shown significant promise in interpreting complex biosignals, including ECG, EEG and other physiological data. The integration of LLMs allows for more accurate classifications and predictions based on the patterns observed in the signals, which are often noisy or irregular. In the paper [25], the authors emphasize the use of LLMs for diagnosing conditions such as arrhythmias from ECG data or epilepsy from EEG data, where the models can recognize subtle, clinically relevant features that might otherwise be missed. Despite the potential, the application of LLMs in signal processing faces challenges, including the need for large, high-quality datasets, real-time processing limitations, and difficulties in interpreting the model's decision-making process.

The application of Large Language Models for generating reports derived from biosignal data is explored in [28].The authors propose the integration of LLMs with biosignal data to generate structured, human-readable reports that can be used for diagnosis and treatment recommendations. However, the paper also highlights several challenges, including ensuring the accuracy of generated reports, handling noisy or incomplete signals, and the need for large, high-quality datasets to train these models. The complexities of real-time processing and model interpretability remain key concerns that need to be addressed for clinical adoption.

The paper [29] presents an integrated healthcare system that leverages artificial intelligence, including large language models, deep learning neural networks and machine learning platforms such as TensorFlow. The system incorporates diverse data types, such as genetic, lifestyle and environmental information, alongside traditional and advanced three-dimensional multilayer EEG systems. This AI-driven approach facilitates early detection, diagnosis, treatment, management and prevention of various health conditions, including diabetes, cardiovascular diseases, stroke, autism and epilepsy. The system aims to enhance clinical decision support, improve health outcomes and provide actionable insights, while also exploring the potential of brain-computer interfaces (BCIs) for human-machine interactions and brain-to-brain communication. Despite its promising capabilities, challenges such as computational complexity, large dataset requirements and the integration of diverse data types remain significant hurdles.

In summary, Fig. 4 provides a comprehensive overview of the key components driving AI-powered medical signal analysis. It categorizes the landscape into three primary areas: AI applications, deep learning approaches, and challenges. The AI applications section emphasizes diverse healthcare domains, including cardiology, neurology, respiratory health, and general health monitoring, where AI technologies are being applied to enhance the detection and diagnosis of diseases through various physiological signals like ECG, EEG, PPG, and respiratory sounds. The deep

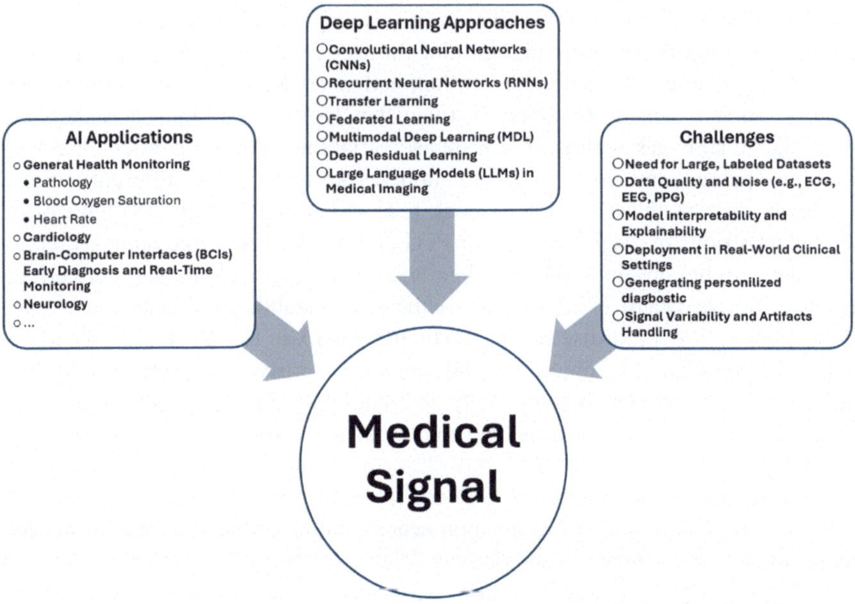

Fig. 4 Taxonomy of AI-driven medical signal: applications, approaches, and challenges

learning approaches section highlights advanced techniques such as convolutional neural networks (CNNs), recurrent neural networks (RNNs), and transformer-based models, which are critical in improving signal classification, feature extraction, and predictive modeling. Finally, the challenges section identifies key obstacles such as limited labeled datasets, model interpretability, noise handling, and computational complexity, underscoring the need for robust, efficient, and scalable solutions for AI to be reliably integrated into medical signal analysis. This comprehensive taxonomy highlights the multifaceted role of AI in transforming healthcare through improved diagnostics, monitoring, and personalized medicine.

Thus, the reviewed studies demonstrate significant advancements in AI-driven medical signal analysis, particularly in disease detection and classification across various biosignals. Leveraging deep learning techniques such as Transformers, CNNs, and RNNs, alongside traditional and emerging machine learning approaches like Smart K-Nearest Neighbors (KNN) and Generative Adversarial Networks (GANs), these works enhance signal processing and diagnostic accuracy. The datasets span multiple physiological domains, including ECG, EEG, PPG, and respiratory signals, providing a solid foundation for medical applications. However, challenges such as limited labeled data, signal noise, and model interpretability persist, posing obstacles to the generalizability and scalability of these solutions.

For example, Zhang et al. [13] successfully applied Swin Transformer-based GANs for cardiac segmentation, offering a promising cross-modality adaptation. Yet, the narrow focus on specific arrhythmias restricts broader applicability, suggesting future research could expand the model's scope to include diverse conditions and enhance its robustness across various patient populations. Wang et al. [14] introduced a self-supervised model for EEG emotion recognition but identified overfitting concerns with smaller datasets, highlighting the need for larger, more diverse datasets to address this issue. Aymen et al. [16] applied smart KNN with outlier detection to EEG analysis, which effectively distinguishes typical from unusual signals but encounters challenges with patient response variability. This emphasizes the need for refined patient interaction models to improve robustness across different settings.

Studies such as those by Kotzen et al. [17] and Alqudah et al. [18] have made significant strides in deep learning for PPG and respiratory signal analysis. However, these studies are limited by smaller, homogeneous populations. Future efforts should prioritize data diversity to improve the generalizability and clinical relevance of the models. The multi-disease diagnostic framework in Kumar et al. [19] demonstrated the potential of combining AI/ML models for cross-disease predictions. Yet, challenges related to data heterogeneity and dataset availability persist, suggesting the need for standardized, interoperable datasets across diseases to improve model versatility.

Looking ahead, integrating AI with IoT for real-time, continuous monitoring, and leveraging multimodal learning approaches, could significantly enhance disease diagnosis and treatment recommendations. The exploration of generative models and transfer learning techniques could address data scarcity, especially for rare diseases. Additionally, improving model interpretability, addressing data privacy issues, and enhancing explainability will be vital for ensuring widespread clinical adoption.

Incorporating environmental and lifestyle data, as suggested by Liu et al. [29], could lead to more personalized and effective healthcare solutions, transforming the field of AI-driven medical signal analysis.

Future directions also include integrating generative deep learning techniques, such as GANs, to further improve synthetic augmentation of ECG signals and address challenges like data scarcity and variability. Furthermore, leveraging IoT for real-time ECG monitoring and data collection could create dynamic, personalized systems that provide scalable, real-time classification for immediate diagnosis and continuous learning. The future of ECG classification may involve advanced cloud-based platforms and distributed computing to enhance deep learning models by utilizing extensive patient data and computational resources. However, overcoming challenges in data privacy, system interoperability, and real-time data transmission will be crucial. By integrating these advanced technologies, the development of robust, efficient, and scalable ECG classification systems could revolutionize healthcare practices.

4 Discussion

4.1 Multimodal Data Integration

Integrating multimodal data-such as imaging (e.g., MRI, CT scans) with signal data (e.g., EEG, ECG, wearable sensors)-provides a holistic approach to disease diagnosis and monitoring, enhancing diagnostic accuracy, personalized treatment, and early detection. For instance, combining brain imaging with EEG in epilepsy reveals structural and functional abnormalities [40], while integrating echocardiography with ECG in cardiovascular diseases correlates structural and electrical heart issues [41]. However, challenges like data heterogeneity, complexity, lack of standardized protocols, privacy concerns, and ethical issues such as algorithmic bias and patient consent hinder integration. Overcoming these requires unified data platforms, advanced AI for multimodal fusion, real-time integration systems, and collaborative research with explainable AI (XAI) to ensure transparency and clinician trust. Establishing ethical guidelines and regulatory frameworks to address data ownership and bias is also crucial. By tackling these challenges, healthcare can fully leverage multimodal data for accurate diagnoses, personalized treatments, and proactive monitoring, ultimately improving patient outcomes and driving innovation. Integrating AI and multimodal features to create a more user-friendly and effective healthcare support system was the purpose of the project in [42]. The paper focuses on leveraging Artificial Intelligence (AI), Natural Language Processing (NLP), and deep learning techniques to create a multimodal chatbot capable of interacting with users through text, voice, and other modalities. Also, authors, in [43–45], focus on the transformative role of multimodal technology in clinical practice, particularly its integration with AI to enhance diagnostic accuracy and redefine patient-clinician interactions. LLMs demonstrate capability to integrate diverse data types to improve diagnostic accuracy, suggesting potential for signal processing integration [46]. LLMs, with their advanced multimodal capabilities, can be adapted to interpret both text-based and biosignal data, enabling a more holistic analysis of patient information. Traditional signal processing techniques, such as Fourier transforms and wavelet analysis, can preprocess raw medical signals like ECG and EEG, extracting critical features before feeding them into LLMs for higher-order reasoning and decision-making [47]. Furthermore, inference acceleration technologies, including model compression and KV-cache optimization, ensure that LLMs can process large-scale biosignal data efficiently in real time [48]. Storing preprocessed signal features in vector databases allows LLMs to query structured medical data, reducing computational overhead and enhancing interpretability [49]. However, as LLMs still face contextual limitations and potential biases, human oversight and hybrid AI-expert models are essential to ensure accuracy and reliability in clinical applications. By combining the strengths of signal processing with LLM-driven analysis, healthcare AI systems can achieve faster, more efficient, and ethically sound diagnostic solutions.

4.2 Generalizability of AI Models in Healthcare

The generalizability of AI models is a cornerstone of their successful deployment, particularly in healthcare, where diverse patient populations and clinical settings demand robust and adaptable solutions. Generalizability ensures that AI systems perform consistently across different environments, making them reliable tools for diagnosis, prognosis, and treatment planning. In medicine, where decisions directly impact patient outcomes, the inability of AI models to generalize can lead to biased predictions, inequitable care, and reduced trust in AI-driven technologies.

This challenge has been widely recognized in recent literature, including studies such as [33], which highlights the need for diverse datasets and rigorous validation frameworks, and *"Generalizability of AI models across healthcare systems"* specially for COVID case [32], which underscores the limitations of models trained on narrow datasets. Despite these advancements, significant barriers remain. Dataset bias, where training data over represent certain demographics or clinical settings, often leads to poor performance in underrepresented groups. Dataset shift, caused by differences in data distributions between training and deployment environments, further exacerbates this issue. Overfitting to specific datasets, lack of external validation, and the complexity of clinical workflows also hinder generalizability. In addition, regulatory and ethical challenges, such as ensuring fairness and equity among populations, complicate the deployment of AI models in real-world healthcare settings.

Recent studies, such as [6, 11], emphasize the importance of addressing these challenges through advanced techniques like transfer learning and multimodal data fusion. For instance, [6] highlights the role of convolutional neural networks (CNNs) and federated learning in improving model adaptability across diverse datasets, while [11] discusses the potential of transformers and generative adversarial networks (GANs) to enhance generalizability in medical imaging. To address these limitations, future research must focus on developing standardized benchmarks to evaluate generalizability consistently across studies. Other promising directions include leveraging synthetic data to augment underrepresented populations, advancing explainable AI (XAI) to improve model transparency, and fostering interdisciplinary collaboration between AI researchers, clinicians, and policymakers [34].

By tackling the points mentioned, the medical sector can advance toward developing AI models that are not only precise but also fair and reliable in various healthcare environments.

4.3 *Practical Challenges of Deploying AI Models in Real-World Clinical Settings*

The deployment of AI models in real-world clinical settings presents significant practical challenges that extend beyond technical performance. One major barrier is the integration of AI into existing clinical workflows, which often requires substantial modifications to accommodate new technologies. As highlighted in [35], the misalignment between AI systems and clinical workflows can disrupt routine practices, leading to resistance from healthcare providers. In addition, acceptance by clinicians remains a critical hurdle, as trust in AI systems is often hampered by lack of transparency and interpretability. A systematic review by [36] found that factors such as usability, perceived usefulness, and alignment with clinical goals significantly influence clinician adoption and acceptance of AI.

Furthermore, data quality and interoperability issues pose substantial challenges, as AI models trained on high-quality, curated datasets may struggle to generalize to real-world data with missing or inconsistent entries [37]. Recent work by [19, 20] highlights the importance of addressing these issues through multimodal data integration and advanced AI techniques. For example, [19] discusses the use of CNNs and recurrent neural networks (RNNs) for multi-disease diagnosis, while [20] explores the role of generative adversarial networks (GANs) and transfer learning in improving data quality and model robustness.

Regulatory and ethical concerns, including patient privacy and algorithmic bias, further complicate deployment efforts. For instance, [38] emphasizes the importance of addressing these concerns to ensure compliance with healthcare regulations and maintain public trust for wide-scale artificial intelligence (AI) deployment in healthcare practices. Finally, the lack of robust validation frameworks for real-world settings often results in models that perform well in controlled environments but fail in practice. Then, lessons learned from early deployments underscore the need for continuous monitoring and iterative improvements to adapt AI systems to dynamic clinical [39]. Addressing these challenges requires a multidisciplinary approach, involving collaboration between AI researchers, clinicians, and policy makers to ensure that AI solutions are not only accurate, but also practical, ethical and aligned with the needs of healthcare systems.

4.4 Ethical and Regulatory Challenges

The integration of AI in healthcare introduces significant ethical and regulatory challenges, particularly in terms of algorithmic bias, data privacy, and compliance with regulatory frameworks. AI models trained on non-representative datasets may produce skewed outcomes, leading to disparities in diagnosis and treatment across diverse patient populations [33]. Addressing these issues requires bias-aware training, fairness audits, and the incorporation of diverse data to ensure equitable healthcare delivery. Furthermore, as AI systems rely on large-scale patient data, data privacy and security concerns are paramount [34, 38]. Advanced technologies such as blockchain, federated learning, and homomorphic encryption have emerged as promising solutions by enabling decentralized and secure approaches to data sharing and model training [35]. Despite these innovations, significant challenges remain in establishing comprehensive regulatory frameworks, ensuring interoperability, and mitigating bias—challenges that are further compounded by limited generalizability across clinical environments [32, 33].

Recent literature further examines these ethical dimensions. For example, Santosh et al. [50] explore the privacy, security, and ethical dilemmas posed by AI in healthcare, emphasizing the need to build trust in AI-driven systems. Similarly, Sidra et al. [51] provide an in-depth analysis of ethical and regulatory challenges, highlighting issues such as the lack of explainability and the potential for discrimination when biased datasets are used. Akter et al. [52] discuss algorithmic biases in the context of data-driven innovation, underlining implications for fairness and justice, while authors of the paper [53] propose a framework for responsible AI adoption through improved governance. Collectively, these studies underscore the necessity for a multidisciplinary approach—integrating technical, clinical, and policy expertise—to develop AI systems that are both ethically responsible and practically effective in real-world healthcare settings.

5 Conclusion and Future Work

The integration of AI technologies into healthcare has brought transformative advancements across various domains, including medical imaging, biomedical signal analysis, and clinical decision-making. AI-driven innovations have significantly enhanced diagnostic accuracy, enabled real-time health monitoring, and facilitated personalized medicine through deep learning frameworks such as Convolutional Neural Networks (CNNs), Recurrent Neural Networks (RNNs), and transformer-based architectures. Generative models, including Generative Adversarial Networks (GANs), have further contributed by addressing data scarcity and enhancing model robustness through synthetic data generation.

In medical imaging, AI has revolutionized diagnostic workflows by improving disease detection and classification in radiology, pathology, dermatology, ophthalmology, and cardiology. AI-powered models have demonstrated their potential in tasks such as automated analysis of CT and MRI scans, early detection of diabetic retinopathy, and accurate classification of skin lesions. Similarly, in medical signal analysis, deep learning techniques have advanced the processing of physiological signals such as electrocardiograms (ECGs), electroencephalograms (EEGs), and photoplethysmograms (PPGs), supporting early disease detection and continuous health monitoring.

Despite these remarkable advancements, several challenges remain. These include ensuring data privacy and security, addressing biases in training datasets, developing explainable AI models, and overcoming the complexities associated with integrating AI solutions into clinical workflows. Standardizing data formats, improving model interpretability, and enhancing computational efficiency will be crucial to overcoming these challenges and fostering the widespread adoption of AI in healthcare.

Future Work

Building on the discussed advancements and challenges, future research directions should focus on the following key areas:

- **Enhancing Data Diversity and Quality:** Future efforts should prioritize the development of diverse and high-quality datasets to improve model generalizability across diverse populations. This includes the curation of large-scale multi-institutional datasets and the adoption of data augmentation techniques to mitigate class imbalances.
- **Advancements in Generative AI:** The integration of generative deep learning techniques, such as GANs and diffusion models, can further enhance the augmentation of underrepresented medical conditions and rare pathologies, ensuring more robust model training.
- **Multimodal and Fusion Techniques:** Combining multiple data modalities (e.g., imaging, physiological signals, and clinical data) using advanced fusion techniques can provide a more comprehensive understanding of patient conditions, leading to more accurate diagnoses and treatment plans.
- **Edge Computing and IoT Integration:** Leveraging edge AI and Internet of Things (IoT) technologies can enable real-time health monitoring and on-device analysis, offering personalized healthcare solutions while reducing latency and enhancing privacy.
- **Explainability and Transparency:** Developing interpretable AI models that provide insights into their decision-making processes will be critical for gaining trust from clinicians and patients. Explainable AI (XAI) methods should be further explored to ensure accountability and transparency.
- **Ethical and Regulatory Considerations:** Establishing robust regulatory frameworks and ethical guidelines will be essential to address concerns related to bias, fairness, and data privacy in AI-driven healthcare solutions.

- **Scalability and Deployment:** Efforts should focus on making AI models more scalable and computationally efficient to facilitate their deployment in resource-constrained settings and ensure widespread accessibility.

In conclusion, AI continues to reshape the healthcare landscape by enabling precise diagnostics, personalized treatment strategies, and efficient clinical workflows. By addressing current challenges and focusing on future opportunities, AI-driven healthcare solutions can achieve greater impact, ultimately improving patient outcomes and advancing precision medicine.

Competing Interests The authors have no conflicts of interest to declare that are relevant to the content of this chapter.

Acknowledgements The authors thank the Ministry of Higher Education and Research of Tunisia for supporting this work under Project Number 22PEJC-D3P3.

References

1. B. Rawat, Y. Joshi, A. Kumar, AI in healthcare: opportunities and challenges for personalized medicine and disease diagnosis, in *2023 5th International Conference on Inventive Research in Computing Applications (ICIRCA)*. IEEE (2023)
2. S. Pandian, R. Murugan, Unveiling the recent trends in biomedical artificial intelligence research. i-manager's J. Softw. Eng. **18**(4) (2024)
3. B. Rawat, Y. Joshi, A. Kumar, AI in healthcare: opportunities and challenges for personalized medicine and disease diagnosis, in *2023 5th International Conference on Inventive Research in Computing Applications (ICIRCA)*. IEEE (2023)
4. X. Yi, E. Walia, P. Babyn, Generative adversarial network in medical imaging: a review. Med. Image Anal. **58**, 101552 (2019)
5. T. Lai, Interpretable medical imagery diagnosis with self-attentive transformers: a review of explainable AI for health care. BioMedInformatics **4**(1), 113–126 (2024)
6. H. Zhang, Y. Qie, Applying deep learning to medical imaging: a review. Appl. Sci. **13**(18), 10521 (2023)
7. F. Li et al., Deep learning-based automated detection for diabetic retinopathy and diabetic macular oedema in retinal fundus photographs. Eye **36**(7), 1433–1441 (2022)
8. C. Qin, et al., Convolutional recurrent neural networks for dynamic MR image reconstruction. IEEE Trans. Med. Imaging **38**(1), 280–290 (2018)
9. J. Geppert, et al., Software using artificial intelligence for nodule and cancer detection in CT lung cancer screening: systematic review of test accuracy studies. thorax **79**(11), 1040–1049 (2024)
10. G.M. Silva, A.E. Lazzaretti, F.C. Monteiro, Deep learning techniques applied to skin lesion classification, in *2023 2nd International Conference on Machine Learning, Control, and Robotics (MLCR)*. IEEE (2023)
11. M. Li, et al., Medical image analysis using deep learning algorithms. Front. Public Health **11**, 1273253 (2023)
12. J. Duan, et al., Deep learning based multimodal biomedical data fusion: an overview and comparative review. Inf. Fusion 102536 (2024)
13. Y. Zhang, et al., ST-GAN: a swin transformer-based generative adversarial network for unsupervised domain adaptation of cross-modality cardiac segmentation. IEEE J. Biomed. Health Inform. (2023)

14. X. Wang, et al., Self-supervised EEG emotion recognition models based on CNN. IEEE Trans. Neural Syst. Rehabil. Eng. **31**, 1952–1962 (2023)
15. A. Abid, O. Cheikhrouhou, A data-augmented vision transformer model for robust multi-label ECG arrhythmia classification. Int. J. Inf. Technol. 1–7 (2025). Springer
16. A. Aymen, et al., Catalyzing EEG signal analysis: unveiling the potential of machine learning-enabled smart K nearest neighbor outlier detection. Int. J. Inf. Technol. 1–12 (2024)
17. K. Kotzen, et al., SleepPPG-Net: a deep learning algorithm for robust sleep staging from continuous photoplethysmography. IEEE J. Biomed. Health Inform. **27**(2), 924–932 (2022)
18. A.M. Alqudah, S. Qazan, Y.M. Obeidat, Deep learning models for detecting respiratory pathologies from raw lung auscultation sounds. Soft Comput. **26**(24), 13405–13429 (2022)
19. Y. Kumar, et al., Artificial intelligence in disease diagnosis: a systematic literature review, synthesizing framework and future research agenda. J. Ambient Intell. Humanized Comput. **14**(7), 8459–8486 (2023)
20. T. Islam, P. Washington, Non-invasive biosensing for healthcare using artificial intelligence: a semi-systematic review. Biosensors **14**(4), 183 (2024)
21. M.H. Abid, et al., Multi-modal medical image classification using deep residual network and genetic algorithm. Plos One **18**(6), e0287786 (2023)
22. H. Wang, et al., Using machine learning to predict cardiovascular risk using self-reported questionnaires: findings from the 45 and Up Study. Int. J. Cardiol. **386**, 149–156 (2023)
23. M. Savadkoohi, T. Oladunni, L. Thompson, A machine learning approach to epileptic seizure prediction using Electroencephalogram (EEG) Signal. Biocybern. Biomed. Eng. **40**(3), 1328–1341 (2020)
24. D. Wang, S. Zhang, Large language models in medical and healthcare fields: applications, advances, and challenges. Artif. Intell. Rev. **57**(11), 299 (2024)
25. R. Yang, et al., Large language models in health care: development, applications, and challenges. Health Care Sci. **2**(4), 255–263 (2023)
26. X. Meng, et al., The application of large language models in medicine: a scoping review. Iscience **27**(5) (2024)
27. Z.A. Nazi, W. Peng, Large language models in healthcare and medical domain: a review. Informatics **11**(3). MDPI (2024)
28. C. Liu, et al., BioSignal Copilot: leveraging the power of LLMs in drafting reports for biomedical signals. medRxiv 2023-06 (2023)
29. F.E. Ekpar, A comprehensive artificial intelligence-driven healthcare system. Eur. J. Electr. Eng. Comput. Sci. **8**(3), 1–6 (2024)
30. O. Ronneberger, P. Fischer, T. Brox, U-net: convolutional networks for biomedical image segmentation, in *Medical Image Computing and Computer-Assisted Intervention-MICCAI 2015: 18th International Conference*, Munich, Germany, Oct 5–9, 2015, Proceedings, Part III 18. Springer (2015)
31. D. Shen, G. Wu, H.-I. Suk, Deep learning in medical image analysis. Ann. Rev. Biomed. Eng. **19**(1), 221–248 (2017)
32. J. Yang, A.A.S. Soltan, D.A. Clifton, Machine learning generalizability across healthcare settings: insights from multi-site COVID-19 screening. NPJ Digit. Med. **5**(1), 69 (2022). Nature Publishing Group UK London
33. H. Singh, V. Mhasawade, R. Chunara, Generalizability challenges of mortality risk prediction models: a retrospective analysis on a multi-center database. PLOS Digit. Health **1**(4), e0000023 (2022) Public Library of Science San Francisco, CAUSA
34. W. Samek et al., Explainable AI for healthcare: from black box to interpretable models. Nat. Mach. Intell. **3**(1), 1–10 (2021). https://doi.org/10.1038/s42256-020-00286-8
35. B. Lainjo, Integrating artificial intelligence into healthcare systems: opportunities and challenges. Acad. Med. **1** (2024)
36. S.I. Lambert, M. Madi, S. Sopka, A. Lenes, H. Stange, C.-P. Buszello, A. Stephan, An integrative review on the acceptance of artificial intelligence among healthcare professionals in hospitals. NPJ Digit. Med. **6**(1), 111 (2023)

37. A.L. Beam, A.K. Manrai, M. Ghassemi, Challenges to the reproducibility of machine learning models in health care. JAMA **323**(4), 305–306 (2020)
38. P. Esmaeilzadeh, Challenges and strategies for wide-scale artificial intelligence (AI) deployment in healthcare practices: a perspective for healthcare organizations. Artif. Intell. Med. **151**, 102861 (2024)
39. J. Yin, K.Y. Ngiam, H.H. Teo, Role of artificial intelligence applications in real-life clinical practice: systematic review. J. Med. Internet Res. **23**(4), e25759 (2021)
40. P. Babaeeghazvini, et al., Brain structural and functional connectivity: a review of combined works of diffusion magnetic resonance imaging and electro-encephalography. Front. Hum. Neurosci. **15**, 721206 (2021)
41. M.B. Abubaker, B. Babayiğit, Detection of cardiovascular diseases in ECG images using machine learning and deep learning methods. IEEE Trans. Artif. Intell. **4**(2), 373–382 (2022)
42. B.S. Garimella, et al., Advancing healthcare accessibility: development of an AI-driven multimodal Chatbot, in *2023 4th International Conference on Intelligent Technologies (CONIT)*. IEEE (2024)
43. Y. Artsi, et al., Advancing clinical practice: the potential of multimodal technology in modern medicine. J. Clin. Med. **13**(20), 6246 (2024)
44. X. Chen, et al., Artificial intelligence and multimodal data fusion for smart healthcare: topic modeling and bibliometrics. Artif. Intell. Rev. **57**(4), 91 (2024)
45. M. Usman, et al., Advancing metaverse-based healthcare with multimodal neuroimaging fusion via multi-task adversarial variational autoencoder for brain age estimation. IEEE J. Biomed. Health Inform. (2024)
46. M. Yuan, et al., Large language models illuminate a progressive pathway to artificial intelligent healthcare assistant. Med. Plus 100030 (2024)
47. X. Xu, et al., A comprehensive review on synergy of multi-modal data and AI technologies in medical diagnosis. Bioengineering **11**(3), 219 (2024)
48. G. Qu, et al., Mobile edge intelligence for large language models: a contemporary survey. IEEE Commun. Surv. Tutor. (2025)
49. Y. Zheng, et al., Large language models for medicine: a survey. Int. J. Mach. Learn. Cybern. 1–26 (2024)
50. K.C. Santosh et al., Privacy, security, and ethical issues. *Artificial Intelligence and Machine Learning in Public Healthcare, Opportunities and Societal Impact* (2021), pp. 65–74
51. N. Sidra, R.A. Khan, S. Bai, Ethical framework for harnessing the power of AI in healthcare and beyond. IEEE Access **12**, 31014–31035 (2024)
52. S. Akter et al., Algorithmic bias in data-driven innovation in the age of AI. Int. J. Inf. Manage. **60**, 102387 (2021)
53. H. Siala, Y. Wang, SHIFTing artificial intelligence to be responsible in healthcare: a systematic review. Soc. Sci. Med. **296**, 114782 (2022)

ArabianGPT: An Arabic Language Adaptation of GPT-2

Anis Koubaa, Adel Ammar, Lahouari Ghouti, Omer Nacar, and Serry Sibaee

Abstract The dominance of English and Latin-based LLMs has created a gap in native Arabic models. These models are often hindered by the inclusion of English tokens, hampering effectiveness due to Arabic's complex morphology and syntax. To address this deficiency, we introduce ArabianGPT, a series of transformer-based models specifically designed for Arabic. This series includes three variants of increasing size: ArabianGPT-0.1B, ArabianGPT-0.3B, and ArabianGPT-0.8B. A key contribution of these models is the development and integration of a purely Arabic tokenizer (AraNizer), which ensures precise and dedicated processing of Arabic text by excluding all non-Arabic tokens. Comparative analysis shows that fine-tuning greatly enhances performance in tasks such as question answering, and summarization. This highlights the efficacy of tailored transformer architectures in advancing Arabic NLP. Evaluating ArabianGPT against AraGPT2, GPT2, Bloom-7B1, and Llama-7B on benchmarks for scientific reasoning, common-sense understanding, cross-domain knowledge, and truthfulness, revealed promising results. ArabianGPT-0.8B achieved the highest score in HellaSwag (43.3) and the highest overall score (36.2), while ArabianGPT-0.3B excelled in TruthfulQA with a score of 52.5. Furthermore, fine-tuning these models on tasks like sentiment analysis and summarization led to significant improvements, with ArabianGPT-0.1B achieving 95% accuracy in sentiment analysis, despite its tiny size, and ArabianGPT-0.3B achieving 100%.

A. Koubaa · A. Ammar · L. Ghouti · O. Nacar (✉) · S. Sibaee
Robotics and Internet-of-Things Lab, Prince Sultan University, Riyadh, Saudi Arabia
e-mail: onajar@psu.edu.sa
URL: http://www.psu.edu.sa

A. Koubaa
e-mail: akoubaa@psu.edu.sa

A. Ammar
e-mail: aammar@psu.edu.sa

L. Ghouti
e-mail: lghouti@psu.edu.sa

S. Sibaee
e-mail: sibaee@psu.edu.sa

A. Koubaa et al. (eds.), *Generative AI and Large Language Models: Opportunities, Challenges, and Applications*, Studies in Computational Intelligence 1214,
https://doi.org/10.1007/978-3-031-90573-5_2

Keywords Large language models · Natural language processing · Transformers · Arabic language · Deep learning

1 Introduction

A notable theoretical and empirical challenge in natural language processing (NLP) is the underrepresentation of non-English languages in developing large language models (LLMs). This issue is particularly acute for languages with intricate linguistic structures, such as Arabic. The complexity of Arabic, with its rich morphology and diverse dialects, demands specialized attention in LLM development, a necessity often overlooked in the predominantly English-centric NLP landscape. This disparity limits the scope of NLP applications and hinders progress in understanding and processing multilingual contexts. While GPT-1 and GPT-2 [1, 2] have laid the groundwork to demonstrate the capabilities of generative pre-trained transformers, their primary focus has been English, creating a gap in the Arabic NLP landscape. This gap becomes more evident when considering the success of models like Hulmona [3] and ARABERT [4], which, despite their advances in Arabic NLP, still grapple with the challenges of adapting models trained predominantly on English data.

Addressing this gap, ArabianGPT, within the broader ArabianLLM initiative, emerges as a targeted response. Unlike conventional LLMs, which are primarily designed to focus on English and exhibit a propensity to include a significant proportion of English tokens even in their multilingual models, ArabianGPT-Base is explicitly developed for the Arabic language. This model diverges from the norm by prioritizing Arabic's linguistic peculiarities in its design and implementation.

The methodology underpinning ArabianGPT-Base involves three critical components: an architectural adaptation, a rigorous training regimen, and a novel tokenization strategy. Firstly, ArabianGPT-Base adapts the proven GPT-2 architecture, tailoring it to accommodate Arabic's syntactic and morphological idiosyncrasies. This adaptation ensures that the model's structure is inherently aligned with the linguistic realities of Arabic, unlike conventional models that retrofit a predominantly English-centric architecture.

Secondly, the training of ArabianGPT-Base is conducted on a carefully curated corpus, specifically chosen to cover a wide range of Arabic text, thus enabling the model to learn from and adapt to the language's complexity and diversity. This aspect of the methodology is particularly critical, as it ensures that the model is exposed to and learns from a representative sample of the language rather than being limited to a subset that might bias its learning.

Lastly, ArabianGPT-Base introduces AraNizer, an innovative tokenizer developed to address the unique challenges of Arabic text segmentation specifically. This tokenizer represents a significant departure from standard tokenization methods, often inadequate for Arabic due to its complex script and morphological richness.

AraNizer's design allows for more precise and contextually aware segmentation, a critical factor in enhancing the model's performance on downstream NLP tasks.

By integrating these methodological innovations, ArabianGPT-Base addresses the gap in Arabic NLP and sets a new standard for developing language-specific models. This approach marks a significant shift in the field, moving away from the one-size-fits-all strategy of previous LLMs towards a more nuanced and inclusive future for NLP.

This chapter presents several key contributions to the field of Arabic natural language processing:

- **Development of ArabianGPT:** A suite of transformer-based models (0.1B, 0.3B, and 0.8B parameters) specifically engineered for Arabic language tasks.
- **Introduction of AraNizer:** A new tokenizer designed to address the unique challenges of Arabic text processing.
- **Extensive evaluation:** Comprehensive benchmarking of ArabianGPT models against existing language models, demonstrating their competitive performance.
- **Fine-tuning analysis:** In-depth exploration of fine-tuning impacts on ArabianGPT models for targeted NLP tasks including summarization, sentiment analysis, and question answering.

The paper is structured as follows: Section 2 surveys related work in Arabic language modeling. Section 3 outlines the datasets employed for training and evaluation. Section 4 elaborates on the AraNizer tokenizer. Sections 5 to 7 detail the architecture and training methodologies for ArabianGPT-0.1B, 0.3B, and 0.8B models, respectively. Section 8 presents results from few-shot evaluations across various benchmarks. Section 9 investigates the effects of fine-tuning on specific NLP tasks. Finally, Sect. 10 concludes the paper and discusses future research directions in Arabic language model development.

2 Related Works

Trends

The evolution of Generative Pre-trained Transformers (GPTs) represents a significant milestone in Natural Language Processing (NLP), marking a departure from traditional models towards more sophisticated, contextually aware systems. Initiated by GPT-1 [1], the series showcased the efficacy of Causal Language Modeling (CLM) in understanding and generating human-like text. CLM, by predicting subsequent tokens based on previous context, enables deep learning models to capture nuanced linguistic patterns, facilitating both comprehension and production of complex language structures.

The evolution of Generative Pre-trained Transformers (GPTs) represents a significant milestone in Natural Language Processing (NLP), marking a departure from traditional models towards more sophisticated, contextually aware systems. Initiated by GPT-1 [1], the series showcased the efficacy of Causal Language Modeling (CLM) in understanding and generating human-like text. CLM, by predicting subsequent tokens based on previous context, enables deep learning models to capture nuanced linguistic patterns, facilitating both comprehension and production of complex language structures.

GPT-2 [2] further advanced the capabilities of these models, demonstrating that language models could begin to learn various tasks without explicit supervision when trained on extensive datasets like WebText [5], which comprises millions of webpages. The largest GPT-2 model variant, featuring 1.5 billion parameters, set new benchmarks by achieving state-of-the-art results on seven out of eight tested language modeling datasets in a zero-shot setting, despite still underfitting WebText. These advancements illustrated the potential of large-scale, unsupervised language models to generate coherent paragraphs of text and suggested a promising path towards creating language processing systems capable of learning tasks from naturally occurring demonstrations.

Progressing to GPT-3 [6], the field witnessed an exponential increase in model size and the diversification of training corpora. GPT-3, with its 175 billion parameters, leveraged an extensive dataset comprising vast swathes of internet text, setting new benchmarks across a multitude of NLP tasks. This leap in scale introduced advancements in few-shot learning, empowering models to perform tasks with minimal prior examples, a testament to their evolving contextual understanding.

Building on these advancements, GPT-4 [7] introduced a large-scale, multimodal approach, capable of processing both image and text inputs to generate text outputs. While GPT-4's performance in real-world scenarios still falls short of human capabilities, it exhibits human-level performance on various professional and academic benchmarks, such as scoring in the top 10% on a simulated bar exam. GPT-4 benefits from an enhanced post-training alignment process, resulting in improved factuality and adherence to desired behavior. The development of GPT-4 also involved optimizing infrastructure and methods to predict its performance reliably, even with models trained on a fraction of the compute resources. This progression highlights the growing sophistication and versatility of GPT models in advancing language processing and understanding.

Simultaneously, the advent of transfer learning emerged as a critical strategy for broadening the applicability of GPT models beyond English, addressing the challenges of linguistic diversity and dataset scarcity in languages like Arabic. The complexity of Arabic, with its rich morphological structure and dialectal variations, underscores the need for specialized approaches in data curation and model tuning.

This trajectory of GPT development not only underscores the importance of scale and data diversity in achieving linguistic comprehension but also highlights the ongoing challenges and opportunities in adapting these models for multilingual NLP, particularly for languages with unique syntactic and morphological features.

Existing Arabic LLMs

Bilingual Focus: Models such as Jais & Jais-Chat [8] and AceGPT [9] have explored bilingual training regimes to enhance Arabic NLP capabilities. These approaches, while innovative, often struggle to fully capture the nuanced morphological and syntactic features of Arabic due to the inherent complexities of balancing bilingual corpora.

Dialectal Emphasis: Efforts by GigaBERT, ARBERT, and MARBERT [10, 11] (inspired by the original BERT model [12]) have been directed towards accommodating the rich dialectal diversity within the Arabic language. Despite their successes, these models may not fully address the intricacies of Modern Standard Arabic (MSA) and classical Arabic, limiting their applicability across all Arabic linguistic tasks.

Understanding/Extraction Focus: ARBERT, AraELECTRA, and CAMeLBERT [4, 13, 14] have set new benchmarks in Arabic NLP for tasks centered around understanding and information extraction. However, their generative capabilities remain limited, underscoring a gap that ArabianGPT seeks to fill.

Text-to-Text Focus: Models like AraT5, QARiB, and JABER & SABIR [15–17] have demonstrated proficiency in text-to-text transformations. Nevertheless, their performance in generative tasks requiring deep linguistic comprehension of Arabic remains an area for further exploration.

Advanced Arabic Generation: The AraGPT2 [18] series represents a significant advancement in Arabic language generation. Trained from scratch on a comprehensive corpus of Arabic internet text and news articles, it includes AraGPT2-mega, with 1.46 billion parameters. AraGPT2-mega has shown remarkable success in generating coherent and human-like news articles, achieving a perplexity of 29.8 on held-out Wikipedia articles. It has also demonstrated high efficacy in zero-shot question answering. An automatic discriminator model developed alongside AraGPT2 has achieved a 98% accuracy in distinguishing model-generated text from human-written content. These developments mark a notable progress in Arabic NLP, offering new opportunities for research and application in the field.

AraNizer's Advantage

The AraNizer tokenizer stands as a testament to ArabianGPT's commitment to addressing the nuanced challenges of Arabic morphology. By suitably processing the unique morphological features of Arabic, AraNizer directly enhances ArabianGPT's ability to engage in more accurate and linguistically faithful text generation and processing tasks, setting it apart from the limitations noted in previous models.

ArabianGPT's Positioning and Impact

ArabianGPT distinguishes itself by meticulously addressing the gaps left by existing Arabic LLMs. Through a deliberate design that emphasizes Arabic fluency and morphological understanding, ArabianGPT offers unparalleled capabilities in generative tasks. Its architecture, informed by the shortcomings of prior models, is specifically

Table 1 Comparative analysis of ArabianGPT and related Arabic language models

Feature	ArabianGPT	Jais/AceGPT	GigaBERT/ARBERT	AraELECTRA	AraT5/QARiB	AraGPT2
Arabic-specific focus	✓	Partial	✓	✓	✓	✓
Bilingual capabilities	–	✓	–	–	–	–
Understanding/Extraction	✓	✓	✓	✓	✓	✓
Advanced generation	✓	Limited	Limited	Limited	Partial	✓
Morphological processing	✓	Limited	Partial	Partial	Partial	Partial
Text-to-text tasks	✓	Partial	Limited	Partial	✓	Partial
Zero-shot capabilities	✓	Limited	Limited	Limited	Limited	✓

engineered to navigate the complexities of Arabic syntax and morphology, thereby facilitating advanced generative applications in native Arabic contexts, a leap forward in the pursuit of linguistic inclusivity and technological advancement in Arabic NLP.

To elucidate the distinctive features and capabilities of ArabianGPT in relation to other recent Arabic language models, we present Table 1. This comparative analysis encapsulates the key attributes and focus areas of ArabianGPT alongside models such as Jais/AceGPT, GigaBERT/ARBERT, AraELECTRA, AraT5/QARiB, and AraGPT2. The table highlights various aspects crucial to Arabic NLP, including language-specific focus, advanced generation capabilities, and morphological processing. By juxtaposing these models, we aim to illustrate ArabianGPT's comprehensive approach to Arabic language modeling, emphasizing its strengths in areas where other models may have limitations. This comparison underscores ArabianGPT's potential to address the multifaceted challenges in Arabic NLP more holistically than its predecessors.

ArabianGPT represents a significant stride towards addressing the multifaceted challenges of Arabic NLP. By harnessing specialized techniques such as the AraNizer tokenizer and focusing on the generative aspects of language processing, ArabianGPT not only bridges existing gaps but also paves the way for future innovations in the field. Its impact on the research community is already evident, as demonstrated by recent studies leveraging its capabilities. For instance, Haouhat et al. [19] deployed ArabianGPT embeddings alongside the Segment Anything Model (SAM) and CLIP for image representation in their work on multimodal propagandistic memes classification. Their system achieved competitive performance, scoring a 0.7290 macro F1 and ranking 6th among participants in the ArAIEval Shared Task [20]. This application underscores ArabianGPT's versatility and its potential to enhance performance in complex, multimodal NLP tasks within the Arabic language domain.

3 Datasets

3.1 Dataset D1

Dataset D1 represents a carefully curated compilation of contemporary Arabic newspaper articles, embodying an expansive collection of 5.37 million texts. This voluminous dataset, which aggregates to a substantial 15.5 GB, encompasses an extensive lexicon tallying 1.8 billion tokens, thereby providing a robust and comprehensive linguistic foundation for the ArabianGPT's training regimen. This corpus encompasses a diverse array of news articles, each contributing to a broad spectrum of linguistic and cultural narratives spanning various chronological timelines. This diversity is paramount in ensuring that the model is exposed to a wide range of dialectical nuances, idiomatic expressions, and syntactic variations, all of which are intrinsic to modern Arabic vernacular and discourse.

The D1 Corpus is not simply a collection of texts; it is a systematic aggregation aimed at capturing the multifaceted nature of Arabic language and culture as represented in contemporary media. The corpus's vastness and variety provide a rich, contextual environment conducive to training sophisticated models capable of understanding and generating Arabic text with a high degree of linguistic fidelity. By covering a broad temporal range, from the early 2000s to 2014, the dataset facilitates ArabianGPT's comprehension of temporal linguistic shifts and trends, thereby enhancing its capability to produce pertinent and context-aware text.

This corpus has been strategically segmented and organized to facilitate efficient and effective training processes. Each article within the corpus has been preprocessed and structured to ensure optimal compatibility with the ArabianGPT's learning algorithms. The segmentation of the corpus into discrete articles allows for a granular approach to model training, where the nuances of each piece can be individually learned and integrated into the broader linguistic model.

The D1 Corpus, and consequently the ArabianGPT model, is limited by its knowledge cutoff in 2014. This restricts the model's understanding to the language, culture, and world events only up to that year, omitting subsequent developments and modern lexicon. This temporal limitation can impact the model's relevance and effectiveness in contemporary contexts, as it may produce outputs that are outdated or misaligned with recent trends and events. Users and developers must be aware of this constraint, especially for tasks requiring current knowledge.

3.2 Dataset D2: Composite Linguistic Resource

Dataset D2, employed for the training of ArabianGPT-0.3B, amalgamates three distinct yet complementary data sources. Collectively, these datasets constitute a comprehensive linguistic corpus with a substantial size of 23 GB and an impressive lexicon totaling approximately 3.34 billion (3,348,610,363) tokens. The strategic integration of these datasets is predicated on the premise of fostering a multifaceted and robust linguistic environment that is imperative for training a versatile and nuanced Arabic language model.

Collectively, these datasets create a rich tapestry of linguistic scenarios, spanning from formal to colloquial and from historical to contemporary, thus equipping the ArabianGPT-0.3B model with the versatility needed to understand and generate Arabic text across a myriad of contexts and styles. It is this comprehensive and diverse linguistic environment that lays the groundwork for a robust and adaptable Arabic language model, capable of addressing a broad spectrum of language processing tasks with high efficacy.

3.3 Dataset D3: Arabic Web

Dataset D3 (Table 2), utilized in the training of ArabianGPT-0.8B, comprises two substantial and complementary web-based corpora. This dataset represents a significant leap in scale and diversity, boasting a combined size of 117 GB and an impressive token count of 14.69 billion. The strategic selection and integration of these web-based sources aim to capture the dynamic and evolving nature of Arabic language usage in digital spaces.

The ArabicWeb22-A corpus, with its 67 GB of data and over 8.27 billion tokens, provides a contemporary snapshot of Arabic web content. This is complemented by the ArabicWeb16 corpus, contributing an additional 50 GB and 6.41 billion tokens, which offers a slightly earlier perspective on Arabic web language. Together, these datasets create a comprehensive linguistic landscape that spans recent years of Arabic internet usage.

The inclusion of web-based corpora is particularly valuable for training a state-of-the-art language model. It exposes the model to a wide array of linguistic styles, from formal to colloquial, and covers diverse topics reflecting the breadth of online discourse in the Arab world. This rich tapestry of real-world language use equips our

Table 2 Data sources for the dataset D3 with size and token counts

Data ID	Source	Size (GB)	Tokens
1	ArabicWeb22-A	67	8,278,145,557
2	ArabicWeb16	50	6,412,078,236

model with the ability to understand and generate Arabic text that is both current and contextually relevant across various digital platforms and communication styles.

By leveraging this extensive and diverse dataset, we aim to develop a language model that not only comprehends the nuances of modern Arabic usage but also adapts to the rapidly evolving nature of language in digital spaces. This approach positions our model to excel in a wide range of contemporary NLP tasks, from content generation to sentiment analysis, while maintaining cultural and linguistic authenticity.

4 Tokenization

4.1 Overview

Effective natural language processing (NLP) hinges on accurate and efficient text tokenization, particularly for morphologically rich languages like Arabic. Standard tokenization methods often falter when confronted with the unique script, complex morphology, and extensive diacritics inherent in Arabic. This can lead to misinterpretations and hinder the performance of large language models (LLMs) trained on Arabic data.

To address these challenges, we introduced AraNizer, a custom-trained SentencePiece tokenizer specifically designed for Arabic NLP tasks. Leveraging the strengths of the SentencePiece model and incorporating domain-specific knowledge of Arabic linguistics, AraNizer delivers precise and semantically meaningful tokenization, paving the way for improved LLM performance.

Traditional tokenization techniques for Arabic often struggle with:

- **Word root identification:** Separating prefixes, suffixes, and inflections from the core root morpheme poses challenges for standard tokenizers.
- **Diacritic handling:** Diacritics play a crucial role in Arabic morphology and semantics, yet their inconsistent representation and omission in some datasets can lead to ambiguities.
- **Ligature recognition:** The presence of ligatures, where multiple individual characters combine into a single glyph, further complicates word segmentation.

These limitations can adversely affect LLM performance on Arabic tasks, leading to inaccuracies in sentiment analysis, text classification, and other downstream applications. AraNizer aims to overcome these hurdles by:

- **Fine-grained tokenization:** Utilizing the subword-level capabilities of SentencePiece, AraNizer achieves a deeper granularity than simple word segmentation, preserving morphological and semantic information within tokens.
- **Diacritic-aware processing:** AraNizer incorporates diacritics into its tokenization process, ensuring their contribution to word meaning is preserved and reflected in LLM representations.

– **Ligature handling:** AraNizer is trained to recognize and appropriately segment ligatures, avoiding misinterpretations and maintaining accurate morpheme boundaries.

4.2 Description

AraNizer [21] is a custom-trained SentencePiece tokenizer [22] specifically developed for the ArabianLMM series. It utilizes the SentencePiece model, empowering ArabianLMM to parse and generate coherent, contextually relevant Arabic text. The Aranizer tokenizer addresses the complexities of Arabic text with remarkable effectiveness. It leverages SentencePiece and Byte Pair Encoding (BPE) [23] methodologies to optimize tokenization accuracy and efficiency specifically for the Arabic language. This tailored approach ensures optimal handling of the language's morphological richness and unique script challenges. To address the varying complexities and nuances of Arabic text, Aranizer is available in four distinct versions with vocabulary sizes of 32K, 50K, 64K, and 86K. This flexibility allows users to select the version that best suits their specific needs and application domains. For training ArabianGPT models, we opted for the 64K version, striking a balance between accuracy and efficiency.

5 Tiny-Scale Model: ArabianGPT-0.1B

5.1 Model Overview

The key objective of the unsupervised pre-training process is to train a language model that excels at predicting upcoming words in a text sequence, without requiring any labeled training data. The dataset D1, described in Sect. 3.1, is used as the unsupervised corpus. ArabianGPT model architecture employs a multi-layer transformer decoder, a powerful neural network architecture adept at capturing long-range dependencies within text. The Transformer architecture has proven highly effective in various language modeling tasks due to its ability to capture long-range dependencies. The transformer decoder applies multi-headed self-attention to focus on relevant relationships between words within the context. The embedding layer transforms the input tokens U = into numerical representations using token and position embeddings. The objective function consists in maximizing the likelihood of predicting the next word in each sequence, given its preceding words. The objective function is expressed as:

$$L(U) = \sum_i \log P(U_i|U_{i-k}, \ldots, U_{i-1}; \theta) \tag{1}$$

where k is the context window size, and θ represents the model parameters (neural network's weights) that are iteratively updated using stochastic gradient descent. In the ArabianGPT model architecture, the employed loss function characterizes the maximization of the conditional likelihood for sequential word prediction, thereby facilitating the model's proficiency in synthesizing contextually coherent and linguistically accurate Arabic text, a quintessential feature for advanced language understanding and generation. The trained model will ultimately generate a probability distribution over potential target tokens using a softmax function. The model's ability to predict upcoming words stems from its learned understanding of language patterns and relationships between words. Pre-training on a massive, diverse corpus is crucial for the model to capture a broad range of linguistic phenomena.

5.2 Description

ArabianGPT-0.1 follows the architecture of GPT-2-small [2], which is basically equivalent to GPT1 [1] (Fig. 1). We retain the underlying principles of the GPT model while incorporating modifications (especially the tokenizer) to better suit Arabic text

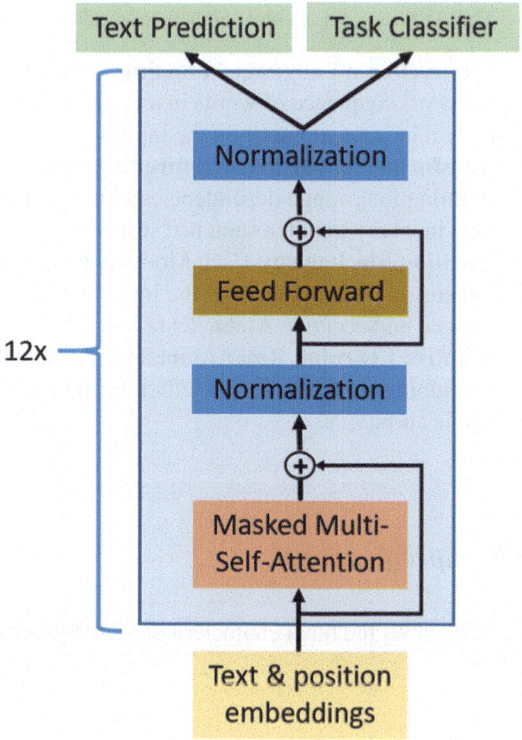

Fig. 1 Architecture of ArabianGPT-0.1 model based on GPT2-small.

processing. The ArabianGPT-0.1 model represents a significant advancement in the field of Natural Language Processing (NLP) for Arabic language. Leveraging the GPT-2 architecture, ArabianGPT is custom-tailored to grasp and generate pure Arabic text. Figure 1 shows the main components of this GPT model. It consists of an embedding layer followed by 12 identical layers, each made up of three sub-layers: masked multi-head self-attention, a feed-forward network, and layer normalization. This stacked architecture allows the model to learn complex relationships between words in a sentence:

- **Text and Position Embedding Layer:** Adds information about the individual words and their positions in the sentence, which helps the model to better understand the meaning of the text.
- **Masked Multi-Head Self-Attention:** The core component of the GPT-1 architecture. It allows the model to attend to different parts of the input sentence simultaneously, using multiple heads that focus on different aspects of the words. This gives the model the ability to capture long-range dependencies and context in the text.
- **Feed-Forward Sub-Layer:** Adds non-linearity to the model, allowing it to learn more complex relationships between words.
- **Normalization Layer:** Helps to stabilize the training process and improve the model's overall performance.

The ArabianGPT model inherits several key features from the GPT-2 model:

- **Unidirectional Language Modeling:** ArabianGPT, like GPT-2, predicts the probability of a sequence of words in a text, learning to generate coherent and contextually relevant text based on the input it receives.
- **Transformer-based Architecture** [24]: The use of 12 Transformer layers aids in capturing long-range dependencies in text, a crucial aspect for Arabic text which often involves complex sentence structures.
- **Attention Mechanism:** Each Model Attention Layer in ArabianGPT is designed to focus on different parts of the input text, allowing the model to generate more nuanced and accurate Arabic text.
- **Adaptive Learning Rate:** ArabianGPT leverages an adaptive learning rate during training, which facilitates efficient and effective learning from the large-scale Arabic corpus.

5.3 Specifications

Table 3 shows the main characteristics and hyperparameters of the ArabianGPT-0.1 model.

Table 3 Characteristics and hyperparameters of the ArabianGPT-0.1 model

Attribute	Value
Model name	ArabianGPT-0.1
Architecture	GPT-2 (GPT2LMHeadModel)
Number of layers	12
Number of model attention layers (MALs)	12
Number of heads	12
Vocabulary size	64,000
Tokenizer	Aranizer
Model size	134M parameters
Initial learning rate	5×10^{-4}
Context window size	768 tokens
Dropout ratio for attention (attn_pdrop)	0.1
Dropout ratio for embeddings (embd_pdrop)	0.1
Dropout probability for all fully connected layers (resid_pdrop)	0.1
Training duration	3 days

5.4 *Training Procedure*

ArabianGPT-0.1B was trained on dataset D1 (see Sect. 3.1) using 2 NVIDIA A100 GPUs with 80GB each. The A100 GPUs are known for their efficiency and capability in handling large-scale deep learning tasks. The model underwent training with a dataset comprising 7.5 million sequences, each with a sequence length of 768 tokens. The training parameters included a batch size of 512, a total of 313,500 steps, and a duration of 3 days, resulting in a final loss of 3.97. The training was focused on adapting the model to recognize and interpret the nuances and complexities inherent in Arabic. Table 4 summarizes the conditions and parameters of the training, and Fig. 2 shows the training loss convergence through the steps of the training.

5.5 *Testing Examples*

Table 5 showcases several prompt and completion pairs generated by the ArabianGPT-0.1B model after training. The examples demonstrate the model's ability to produce natural and elaborate completions despite its relatively small size. This performance is noteworthy, as it suggests that even with limited parameters, ArabianGPT-0.1B can effectively understand and generate coherent text in response to a variety of prompts. However, there are occasional punctuation issues, indicating that while the model's

Table 4 Details of the training of ArabianGPT-0.1B

Parameter	Value
Model	ArabianGPT-0.1B (Base Model)
Hardware	2 NVIDIA A100 GPU (80GB each)
Number of examples	7.5 million sequences (sequence length = 768 tokens)
Batch size	512
Number of steps	313,500
Number of epochs	6
Training duration	3 days
GPU hours	144
Final loss	3.97

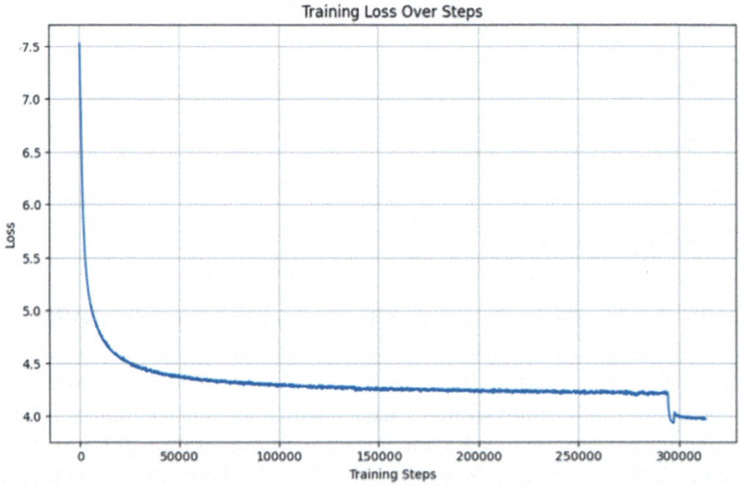

Fig. 2 Evolution of the training loss for the ArabianGPT-0.1 model.

syntactic abilities are strong, there is still room for improvement in its handling of finer grammatical details. Additionally, this small model has a reduced context window of 768 tokens, which limits the length of its outputs. These observations highlight the model's potential and suggest areas for further refinement.

Table 5 Sample output of the ArabianGPT-0.1B model

Completion	Prompt
أعلنت وزاره الحج في المملكه العربيه السعوديه ، انها ستطلق قبل نهاية العام الميلادي 2008 م خدمة " " اركود " " ' الالكترونية ، و " " يسر " " عبر موقعها الالكتروني الجديد ، فيما ستبدأ الوزارة في تشغيل خدمة الهاتف الصوتي المجانية عبر الجوال (الجوال). ويأتي الاعلان انفاذا لتوجيهات وزير الحج الدكتور عبدالواحد بن محمد بن بكر القاضية بايقاف الخدمة عن طريق الجوال اعتبارا من أول إبريل المقبل ، والتي تهدف الى إتاحة المكالمات الصوتية المجانية للحجاج الراغبين في أداء مناسك الحج ، وذلك في سبيل تيسير اداء فريضة الحج بكل يسر وسهولة ودون أي جهد	أعلنت وزاره الحج في المملكه العربيه السعوديه
يبدو اليوم جميلا ، سأقوم بتحضير طعام العشاء لهذا العام ، مع دعوة الجميع إلى العودة إلى المنزل أو الاستراحة لقضاء بعض الوقت مع الأطفال على الشاطئ. أما خلال ايام قليلة ، فإن الجو البارد سيستمر ما يقارب الساعة. كما أن الجو البارد سيكون لطيفا ، وهذا ما سيجعل الأولاد في حال مزاج بارد للغاية. لكن قبل ذلك هناك إمكانية للتخلص من البرد بتناول بعض الاغذية المثلجة ، او الكشخات والعصائر المعلبة والعصائر الساخنة ، أما الوجبات الساخنة ، فهي لذيذة جدا ، حيث يزداد تركيز الطفل في هذه الفترات على الفاكهة والخضار والحبوب الطازجة ، والخضار الطازجة.. والتي تحتوي على سكر السكراب ، وزيت النخيل ، والبرتقال ، وزيت السمسم ، والبرغر ، وعصير الليمون ، وبعض المكسرات اللبنانية ، وعصير الليمون ، والعنب ، والزبيب ، والبطاطس. كما أن تناول الطعام مع الأصدقاء في فترة المساء او تناول بعض الخضروات مثل : الخس والملفوف ، والسردين ، والشاي ، والجرجير ، أو الشاي ، وكذلك بعض العصائر الطازجة	يبدو اليوم جميلا ، سأقوم بتحضير
إن التقنيات الحديثة والمتطورة ساهمت في زيادة انتشار استخدام الانترنت في نشر الاخبار ، حيث تعد اليوم اكبر سوق لنشر الأخبار في الشرق اوسط وذلك في غضون السنوات القادمة. حيث تشهد نشر ما يصل الى 20 % من الأخبار المنشورة على الانترنت زيادة في الوقت الذي زاد استخدام الهاتف المحمول ، كما ان سرعة نقل الأخبار عبر الانترنت قد ساعد على زيادة النمو في هذا المجال. وذلك يرجع بشكل أساسي إلى الجهود المتعددة المبذولة من قبل أجهزة الاعلام الرقمية في كل من قطر والكويت والمملكة العربية السعودية	إن التقنيات الحديثة
عاصمة المملكة العربية السعودية هي في وقتنا الحاضر ملتقى لجميع المسلمين في العالم, وهي تحتضن عاصمة الإسلام والمسلمين, في مكة المكرمة والمدينة المنورة وبمتابعة من من الله, وتشهد دائما ولله الحمد والمنة الكثير من المشاريع التنموية العملاقة التي تشهد فيها المنطقة مزيدا من المشاريع المستقبلية. وهذه المنطقة الجميلة والتي تضم أكبر عدد من الحجاج والمعتمرين, تعتبر أكثر مدينة عربية في آسيا من حيث السكان, كما تتميز بتوافر جميع السلع والخدمات الاجتماعية والتعليمية والترفيهية, مما يساهم في تطور كل منطقة من المناطق, والتي تشكل في مجملها نسبة 90? من مجمل مساحة المملكة.. وهذا العدد غير في الواقع, إلا أن المملكة ليست الوحيدة في الغرب, بل يوجد فيها أكبر عدد من الحجاج والمعتمرين, ولا شك أن هنالك مواطنين متميزين هم بين الأشقاء, مما يعزز مفهوم الترابط والتكامل بينهم, وهم يحظون بكافة الحقوق, إضافة إلى تمتعهم بكافة الحقوق في كل بقعة من بقاع الأرض, مما يعد أكبر دليل على وجود الإسلام في هذه القارة العظيمة وما تتمتع به من مكانة رائدة بين دول العالم, ناهيك عن المكانة التي يحتلها خادم الحرمين الشريفين لدى المسلمين في مختلف أرجاء المعمورة.. والمملكة العربية السعودية تمثل نموذجا رائعا في هذا المجال.. وهي أرض الإسلام, التي تحتضن الكثير من الأديان, وتقوم على التسامح والاحترام, وهذه نعمة على كل مسلم ومسلمة, بفضل الله, بفضل من الله 'سبحانه وتعالى	عاصمة المملكة العربية السعودية هي

6 Small-Scale Model: ArabianGPT-0.3B

6.1 Model Overview

Compared to its predecessor, ArabianGPT-0.1B, ArabianGPT-0.3B is three times larger and more advanced, trained on an increased dataset of examples. This allows it to capture the intricacies and nuances of Arabic across various domains and topics with greater accuracy and effectiveness.

6.2 Description

ArabianGPT-0.3B is a variant of the GPT-2 model [2] with a medium size, specifically pre-trained for the Arabic language. It retains the underlying principles of GPT-2 while incorporating modifications to better suit Arabic text processing. ArabianGPT-0.3B is a more advanced and larger version than ArabianGPT-0.1. It handles a significantly larger number of examples, making it more adept at capturing the nuances and complexities of the Arabic language across various topics.

6.3 Specifications

Table 6 shows the main characteristics and hyperparameters of the ArabianGPT-0.3 model. It has a double number of layers compared to ArabianGPT-0.1, with a larger number of Model Attention Layers (15 compared to 12) and heads (15 compared to 12).

6.4 Training Procedure

Table 7 summarizes the conditions and parameters of the training of the ArabianGPT-0.3B model. Utilizing the power of 4 NVIDIA A100 GPUs, the model underwent an extensive training process, leveraging a dataset consisting of 70 million sequences with a sequence length of 1024. With a batch size of 512 and 1,009,500 steps over 4.97 epochs, the training duration spanned 4.1 days. The culmination of this training journey resulted in a final loss of 3.82, as illustrated in Fig. 3, depicting the convergence of the training loss throughout the steps of the training process.

Table 6 Characteristics and hyperparameters of the ArabianGPT-0.3B model

Attribute	Value
Model name	ArabianGPT-MID
Architecture	GPT-2 (GPT2LMHeadModel)
Number of layers	24
Number of model attention layers (MALs)	16
Number of heads	16
Vocabulary size	64,000
Model size	345M parameters
Initial learning rate	5×10^{-4}
Context window size	1024 tokens
Dropout ratio for attention (attn_pdrop)	0.1
Dropout ratio for embeddings (embd_pdrop)	0.1
Dropout probability for all fully connected layers (resid_pdrop)	0.1

Table 7 Details of the training of ArabianGPT-0.3B

Parameter	Value
Model	ArabianGPT-0.3B (Medium Model)
Hardware	4 NVIDIA A100 GPU (80GB each)
Number of examples	70 million sequences (sequence length = 1024)
Batch size	512
Number of steps	1,009,500
Number of epochs	4.97
Training duration	4.1 days
GPU hours	394
Final loss	3.82

6.5 Testing Examples

Table 8 displays several prompt and completion pairs generated by the ArabianGPT-0.3B model during testing. The completions produced by this model appear more natural and exhibit improved punctuation compared to its predecessors. However, despite these advancements, the generated text still contains several inaccuracies. This indicates that while ArabianGPT-0.3B has made progress in terms of syntactic correctness and coherence, there are still challenges to be addressed in ensuring the accuracy and relevance of the content generated in response to specific prompts.

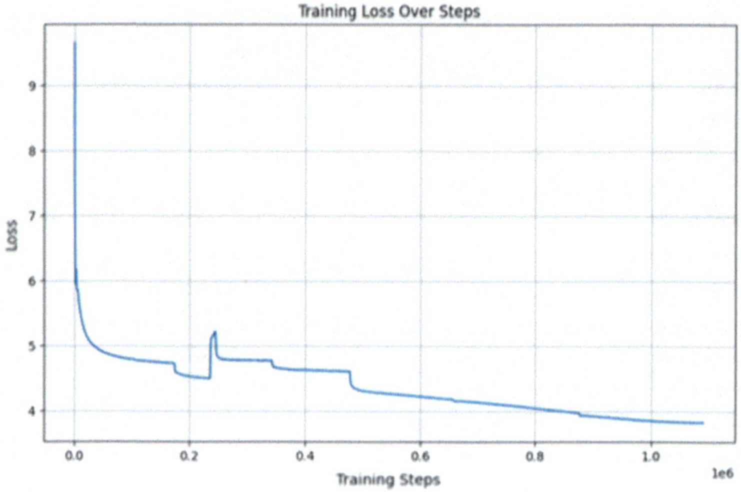

Fig. 3 Evolution of the training loss for the ArabianGPT-0.3 model

7 Medium-Scale Model: ArabianGPT-0.8B

7.1 Model Overview

The development and training of ArabianGPT-0.8B, also based on GPT-2 architecture, leveraged a compilation of web-based data, totaling 117 GB and 15 billion tokens, offering a broad and diverse linguistic foundation that enriches the model's learning. This model's training on such a vast and varied dataset underscores its ability to grasp and generate Arabic text, addressing a wide spectrum of applications from translation to content generation.

7.2 Specifications

Key to the design of ArabianGPT-0.8B is the use of the `gelu_new` activation function, enhancing the model's ability to learn complex patterns with greater efficiency. The model architecture is encapsulated within the `GPT2LMHeadModel`, which is specifically engineered for language modeling, enabling it to predict subsequent tokens in a sequence with high accuracy.

Dropout mechanisms play a critical role in the model's design, with attention dropout (`attn_pdrop`) and embedding dropout (`embd_pdrop`) both set at 0.1. This setup helps prevent overfitting by randomly skipping a proportion of the features during training, thus promoting model generalization. Additionally, the model employs a specific initializer range of 0.02, which is crucial for setting up the initial

Table 8 Sample output of the ArabianGPT-0.3B model

Completion	Prompt
أعلنت وزاره الحج في المملكه العربيه السعوديه انها تنوي تنفيذ عدة مشاريع كبري لخدمة حجاج بيت الله الحرام خلال موسم حج هذا العام من ضمنها مشروعات تطوير الفنادق والشقق المفروشة عبر عدة مشاريع ضخمة في كل من مكة المكرمة والمدينة المنورة والمدينة المنورة. هذا المشروع يعتبر من أكبر المشاريع التي أعلنت وزاره الحج عن نتائجها مؤخرا. المشروع يشمل عدة مشروعات فندقية وسياحيه والتي تم الانتهاء منها في العام 2017 م. مشاريع أخرى سيتم إعلانها مستقبلا خلال هذا العام وسيتم الإعلان عنها بمشيئة الله تعالي مع بداية موسم حج هذا العام. هذا. المشروع هو مشروع سياحي كبير ذو رؤية استراتيجية بعيدة المدى وهو موجه لجميع زوار مدينة بيت الله الحرام وذلك من خلال عدة مشروعات سياحية وسياحية وتجارية وترفيهية وسياحية تستهدف الزوار والمعتمرين القادمين لأداء فريضة الحج من جميع أنحاء العالم	أعلنت وزاره الحج في المملكه العربيه السعوديه
يبدو اليوم جميلا ، سأقوم بتحضير بعض أطباق الطبخ الشهية لكي تتناسب مع هذا البرج ، وذلك بعد قراءة هذا المقال قبل موعد الزيارة ، والذي سنتحدث فيه عن أفضل برج على الإطلاق. شاهدي هذا المقال واكتشفي ما يناسبك ويجعلك تزورين برج باجي مختار في ماليزيا. برج باجي مختار هو برج جميل محاط بمجموعة من الحقول الخضراء الشاسعة ، ويعد هذا البرج أكثر الأبراج جمالا في ماليزيا ، ويتميز بألوانه الزاهية ورداءه الأبيض الناصع البياض ، بالإضافة إلى ذلك ، فهو واحد من أضخم أبراج العالم. البرج هو برج من صنع الإنسان ، ويتميز برج باجي مختار بلونه الأحمر الفاتح المائل إلى الأبيض ، ويعتبر برج باجي مختار هو أطول أبراج العالم ، حيث يبلغ طوله 17 مترا	يبدو اليوم جميلا ، سأقوم بتحضير
إن التقنيات الحديثة المستخدمة في معالجة الصور الحديثة تستخدم طرقا متطورة لمعالجة الصور بهدف تحسين جودة الصورة ، وهذا ما يفسر سهولة الحصول على الصورة المطلوبة سواء باستخدام برنامج فوتوشوب أو باستخدام برامج الرسم المختلفة التي توفر لك إمكانية التعديل على الصور مثل برنامج " بيكاش	إن التقنيات الحديثة
عاصمة المملكة العربية السعودية هي مدينة الرياض حيث تقع مدينة الرياض على ساحل البحر الأحمر غرب مدينة مكة المكرمة وتقع فوق منطقة وسط السعودية وتحديدا على خط العرض 22 و 24 شمالا وتبعد عن مدينة جدة حوالي 80 كم ، وعلى بعد 130 كم عن مدينة جدة ، وتتميز مدينة الرياض بكونها المدينة الساحلية التي تقع على ساحل البحر الأحمر وتبتعد عن العاصمة السعودية الرياض حوالي 38 كم. ويبلغ عدد سكان المدينة حوالي 2. 5 مليون نسمة في سنة 2011 م	عاصمة المملكة العربية السعودية هي

weights to facilitate efficient convergence during training. At the heart of its configuration are the model's dimensions and capacities, including the number of attention heads (`n_head`), the size of embeddings (`n_embd`), and the depth of the network (`n_layer`). These parameters are carefully chosen to balance the model's computational demands with its ability to capture and generate complex language structures. Table 9 summarizes the key technical specifications of ArabianGPT-0.8B.

ArabianGPT-0.8B is configured to dynamically sample from the distribution of possible next tokens, with a predefined maximum length of 50 tokens for generated sequences. This approach ensures that the model can produce coherent and contextually relevant text outputs.

Table 9 Key technical specifications of ArabianGPT-0.8B

Specification	Value
Model type	GPT2
Architectures	GPT2LMHeadModel
Activation function	gelu_new
Attention dropout (attn_pdrop)	0.1
Embedding dropout (embd_pdrop)	0.1
Initializer range	0.02
Layer norm epsilon	10^{-5}
Maximum context size (n_ctx)	1024
Embedding size (n_embd)	1280
Number of attention heads (n_head)	20
Number of layers (n_layer)	36
Residual dropout (resid_pdrop)	0.1
Vocabulary size	64K

7.3 Training Procedure

The training architecture of ArabianGPT-0.8B was expanded to utilize three NVIDIA A100 GPUs, enhancing the parallelization and efficiency of the training process. This expansion allowed for a reduction in time-to-convergence without compromising the model's performance fidelity. By distributing the computational load across multiple GPUs, the setup facilitated increased batch sizes and accelerated iteration cycles, critical for efficient model optimization in large-scale language model training.

To manage memory constraints associated with training larger models, the configuration included gradient accumulation. The parameter `gradient_accumulation_steps` was utilized, effectively enlarging the batch size for gradient updates while keeping within the memory limits of individual GPUs. This method accumulates gradients over multiple forward passes before executing a backward pass, simulating larger batch sizes. This strategy stabilizes the training process and enhances the model's robustness by averaging out noise across multiple mini-batches.

Additionally, the training regime employed mixed-precision training, specifically FP16, rather than the standard FP32. This approach leverages the capabilities of the A100 GPUs, which are optimized for 16-bit floating-point operations, thereby accelerating training operations and reducing memory usage. Mixed-precision training aids in achieving quicker convergence and reduces resource consumption, beneficial for scaling training operations efficiently.

The learning rate was configured with a cosine scheduler, starting with a learning rate of 5×10^{-3} and including a warm-up period of 1,000 steps. This scheduling was chosen to prevent premature convergence to suboptimal minima by gradually increasing the learning rate at the beginning of training. The learning rate follows a

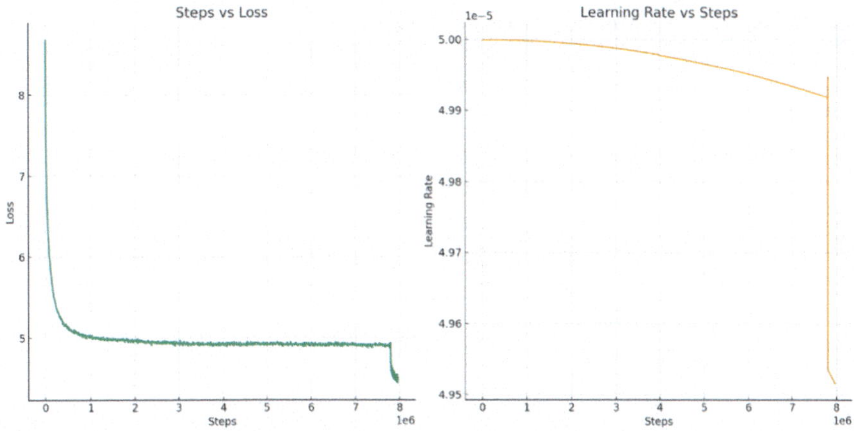

Fig. 4 Evolution of the training loss for the ArabianGPT-0.8 model

Table 10 Training details of ArabianGPT-0.8B

Parameter	Value
Model	ArabianGPT-0.8B
Number of examples	304,019,997
Number of steps	7,954,500
Training duration	60 days
GPU hours	4320
Final loss	4.5

cosine decay pattern after the warm-up, gradually decreasing to nearly zero towards the training's conclusion. This method facilitates effective navigation through the optimization landscape, allowing for smoother transitions during parameter updates and enhancing the model's generalization capabilities.

Figure 4 illustrates the loss trajectory of the ArabianGPT-0.8B model throughout its training phases, reflecting the effectiveness of the optimization strategies employed. Table 10 provides a detailed summary of the training parameters and configurations, elucidating the strategic decisions underlying the training regimen of the ArabianGPT-0.8B model.

7.4 Testing Examples

Table 11 provides examples that demonstrate the ArabianGPT-0.8B model's ability to generate Arabic text based on various prompts.

While the completions generated by the ArabianGPT-0.8B model are generally syntactically correct, they sometimes diverge from the specificity or context of the

Table 11 Sample output of the ArabianGPT-0.8B model

Prompt	Generated Text
كانت ليالي الشتاء طويلة ومليئة بالأسرار،	كانت ليالي الشتاء طويلة ومليئة بالأسرار، وهي من أجمل ليالي الشتاء، كما أن الرياح كانت شديدة الحرارة والرياح ستكون غزيرة وردة إلى معتدلة، والرياح جنوبية شرقية إلى جنوبية غربية.
الذكاء الاصطناعي يغير طريقة استخدامنا ل	الذكاء الاصطناعي يغير طريقة استخدامنا ل" الذكاء الاصطناعي"، والذي يسعى بشكل دائم إلى مساعدة الآخرين، وليس هذا فقط وإنما أيضا، فهو يعلمنا كيفية التعامل مع الطفل، وهو يعرف أن الطفل يتعلم أن يقوم بأجازة إلى غير رجعة، بل وهو يعرف.
في عهد الحضارة الإسلامية، ازدهرت العلوم والفنون، حيث	في عهد الحضارة الإسلامية، ازدهرت العلوم والفنون، حيث أصبح الطب غاية في جميع المجالات، وأعملة الطب هي الآنة، وهذا بالطبع في كل ما يتعلق بهذه العلوم، من علوم العلوم الطبية هي الآنة في مختلف فروع.
في عالم الاقتصاد، تلعب العولمة دوراً هاماً في	في عالم الاقتصاد، تلعب العولمة دوراً هاماً في التنمية الاقتصادية. حيث تعتبر تلك الدول من أوائل الدول التي يمكن تصنيفها تحت مظلة العولمة، والتي تعتبر من بين الدول التي انضمت إلى مجلس التعاون. وفي عام 2007، كان هذا البلد قد دخل في مرحلة.
كرة القدم هي الرياضة الأكثر شعبية في العالم، حيث	كرة القدم هي الرياضة الأكثر شعبية في العالم، حيث تتراوح أعمار الأطفال بين 4 سنين وما قبل عامين، وتتنوع هذه الرياضة بين كرة القدم والماركت والكاراتيه والكاراتيه والشطرنج وغيرها الكثير من أنواع كرة السلة. تعد هذه الرياضة أهم الرياضة وأكثرها استخداما.
تعتبر مدينة مراكش من المدن الساحرة بمغرب، وتتميز ب	تعتبر مدينة مراكش من المدن الساحرة بمغرب، وتتميز بعد جغرافي كبير، نظرا لدورها التاريخي الذي يجعلها واحدة من المدن الرئيسية في المغرب، وتمكنتا من إنشاء عدة جبال، مثل الجبال، والتلال الخضراء.

given prompts. This behavior is not uncommon in pretrained language models, which often require further fine-tuning to adapt to the nuances of specific datasets or task requirements. In the subsequent section, strategies for fine-tuning ArabianGPT-0.8B will be discussed, which are aimed at enhancing the alignment of model outputs with more precisely defined prompt contexts. This fine-tuning process is crucial for optimizing the model's applicability to specialized tasks and improving its overall performance in generating contextually relevant text completions.

Moreover, the qualitative comparisons between the outputs of ArabianGPT-0.8B and its predecessor, ArabianGPT-0.3B (Table 8), reveal no substantial improvement. Table 12 will present more detailed quantitative evaluations to further assess these aspects.

8 Few-Shot Evaluation

This section presents a comparative performance analysis of ArabianGPT-0.1B, ArabianGPT-0.3B, and ArabianGPT-0.8B against benchmarks set by AraGPT2-Base, AraGPT2-Medium, GPT2 Base, GPT2 Medium, Bloom-7B1, and Llama-7B models across diverse benchmarks within a few-shot language model evaluation framework. This analysis offers insights into each model's capabilities in domains such as scientific reasoning, common-sense understanding, cross-domain knowledge, and truthfulness, reflecting their utility in practical applications. The selected benchmarks are multilingual and tailored to assess varying aspects of language model performance. We applied the models only on the Arabic components of these benchmarks.

- **ARC (25-shot) normalized:** This benchmark assesses the scientific reasoning abilities of language models using a set of normalized metrics for comparability. It consists of a selection of multiple-choice questions drawn from science examinations administered to students in grades 3 through 9. These questions test the models' understanding and application of scientific concepts across a broad spectrum of disciplines, thereby providing a robust measure of their reasoning capabilities in academic contexts [25].
- **HellaSwag (10-shot):** Designed to evaluate a model's common-sense reasoning within complex narrative contexts, HellaSwag presents language models with scenarios that require the prediction of logically coherent endings. Despite the task's apparent simplicity and high human performance levels (average accuracy of 95.6%), it poses significant challenges to language models due to its reliance on nuanced understanding of everyday activities and logical progression in narratives [26].
- **MMLU (5-shot):** The Massive Multitask Language Understanding (MMLU) benchmark measures a model's text comprehension abilities across a diverse array of subjects. It includes 57 distinct tasks covering various fields such as elementary mathematics, US history, computer science, and law. Each task is designed to test the model's ability to apply domain-specific knowledge and understanding in a textual analysis context, making it a comprehensive test of cross-domain knowledge [27].
- **TruthfulQA (0-shot):** This benchmark evaluates the capacity of language models to generate accurate and truthful responses, specifically focusing on their ability to avoid replicating common falsehoods found online. TruthfulQA comprises 817 questions spanning 38 categories including health, law, finance, and politics. It challenges models not just on the accuracy of the information provided, but also on the ethical implications of their responses, emphasizing the importance of truthfulness in AI-generated content [28].

Table 12 illustrates the performance of the ten models on these benchmarks. Notably, Bloom-7B1 demonstrates robust performance across ARC and MMLU,

Table 12 Zero and few-shot evaluation scores

Model	Language	Average	ARC (25-shot)	HellaSwag (10-shot)	MMLU (5-shot)	TruthfulQA (0-shot)
Bloom-7B1	Multilingual	34.6	**36.2**	31.4	**43.3**	27.5
Llama-7B	Multilingual	28.9	32.1	24.6	30.9	28.0
AraGPT2-Base	Arabic	31.7	24.6	27.5	25.1	49.5
AraGPT2-Medium	Arabic	32.2	23.9	28.5	26.3	50.0
GPT2 Base	Arabic	30.7	22.5	29.6	26.4	44.1
GPT2 Medium	Arabic	30.3	22.3	29.3	26.0	43.5
ArabianGPT-0.1B	Arabic	31.9	24.0	26.6	25.4	51.8
ArabianGPT-0.3B	Arabic	32.7	24.3	28.4	25.7	**52.5**
ArabianGPT-0.8B	Arabic	**36.2**	31.4	**43.3**	27.5	42.6

achieving the highest scores in these categories yet underperforming in TruthfulQA, suggesting a potential discrepancy in its capability to handle nuanced truth discernment without specific tuning.

In contrast, ArabianGPT-0.3B leads in TruthfulQA with a score of 52.5, outperforming larger models such as Bloom-7B1 and Llama-7B. This superior performance in truthfulness tasks suggests that ArabianGPT-0.3B may benefit from a training regime or architectural nuances that favor accurate information synthesis. Similarly, ArabianGPT-0.1B, despite its smaller size, shows competitive results in ARC and MMLU, indicating effective parameter utilization and potential optimizations that maintain performance with fewer resources.

ArabianGPT-0.8B exhibits the best overall scores among the Arabian models, particularly excelling in HellaSwag with a score of 43.3. However, its lower performance in TruthfulQA suggests challenges in generating responses that require high fidelity to factual accuracy, highlighting a common issue in scaling models-increased capacity does not invariably translate to improved performance across all types of tasks.

The modest results from GPT2-Base and GPT2-Medium, with lower overall scores than the ArabianGPT models, might indicate limitations in their training datasets or fewer optimizations for the evaluated tasks. This disparity underscores the importance of tailored training and the potential benefits of model-specific optimizations.

The analysis underscores that while larger models generally capture complex patterns more effectively, the translation of increased computational capacity into real-world task performance is not always linear. To optimize these larger models' performance further, extending training durations and enlarging datasets could be pivotal. Moreover, this suggests the need for ongoing adjustments in training strategies to better harness the models' capacities for specific task requirements.

9 Fine Tuning

Fine-tuning is a critical procedure for leveraging the extensive capabilities of foundation models and transformer-based architectures to achieve high performance in domain-specific tasks. Foundation models, pre-trained on vast, diverse corpora, develop a generalized understanding by encapsulating a wide range of linguistic features and contextual nuances. To effectively deploy these models in niche applications, it is often imperative to tailor them to the intricacies of specific tasks. Fine-tuning addresses this requirement by adjusting the model's parameters through supervised learning on task-specific datasets, allowing the model to fine-tune its learned representations to the particularities of the target domain.

In this section, we present a detailed analysis of the fine-tuning process applied to ArabianGPT-0.1B, ArabianGPT-0.3B, and ArabianGPT-0.8B models across three core Natural Language Processing (NLP) tasks: summarization, sentiment analysis, and question answering. The fine-tuning effectiveness is quantitatively evaluated through task-specific metrics, demonstrating the models' capabilities to adapt their pre-trained knowledge to generate coherent summaries, accurately gauge sentiments, and provide precise answers. This empirical evaluation serves to underscore the transformative impact of fine-tuning on enhancing the practical utility of pre-trained language models in delivering tailored NLP solutions.

9.1 Summarization

The task of summarization in natural language processing (NLP) involves generating concise and meaningful summaries from a larger text. It is one the most challenging tasks in NLP since it requires a larger context window.

In this project, we have fine-tuned the previously pre-trained ArabianGPT 0.1B, 0.3B, and 0.8B Models for the summarization task. These models have been fine-tuned (FT) on a summarization dataset, with the aim of improving their ability to generate accurate and relevant summaries. The performance of these models has been evaluated using the F1 score, a metric that combines precision and recall.

The dataset used for fine-tuning and evaluation is a subset of xlsum [29], specifically xlsum[1000]. This dataset comprises a range of texts suitable for testing summarization capabilities.

The fine-tuning process involves adjusting the weights of the pretrained model specifically for the summarization task. This phase is crucial as it tailors the model to perform better on the desired task, here summarization.

9.1.1 Hyperparameters

In the fine-tuning process of the ArabianGPT models, the cross-entropy loss function was employed alongside the ADAM optimizer. The Cross-Entropy Loss function is widely used in classification tasks, including tasks like summarization and question answering, which can be formulated as predicting the next token in a sequence.

Cross-Entropy Loss: This loss function measures the performance of a classification model whose output is a probability value between 0 and 1. It increases as the predicted probability diverges from the actual label, making it suitable for training models in tasks like summarization and question answering.

Optimizer: ADAM, an algorithm for first-order gradient-based optimization, is utilized for its efficiency in handling sparse gradients and its adaptability to large-scale problems. It's a popular choice in fine-tuning language models due to its effective handling of large datasets and complex neural network architectures.

Figure 5 depicts the fine-tuning loss for the ArabianGPT-0.1B, ArabianGPT-0.3B, and ArabianGPT-0.8B models. The smooth decrease in loss values from the initial to final stages indicates successful learning and adaptation during fine-tuning. The initial loss values provide a benchmark to gauge the effectiveness of the fine-tuning process. The significant reduction in these values by the end of fine-tuning reflects the models' improved ability to predict accurate tokens for the tasks at hand. The difference in loss values between the Summarization and Question Answering tasks

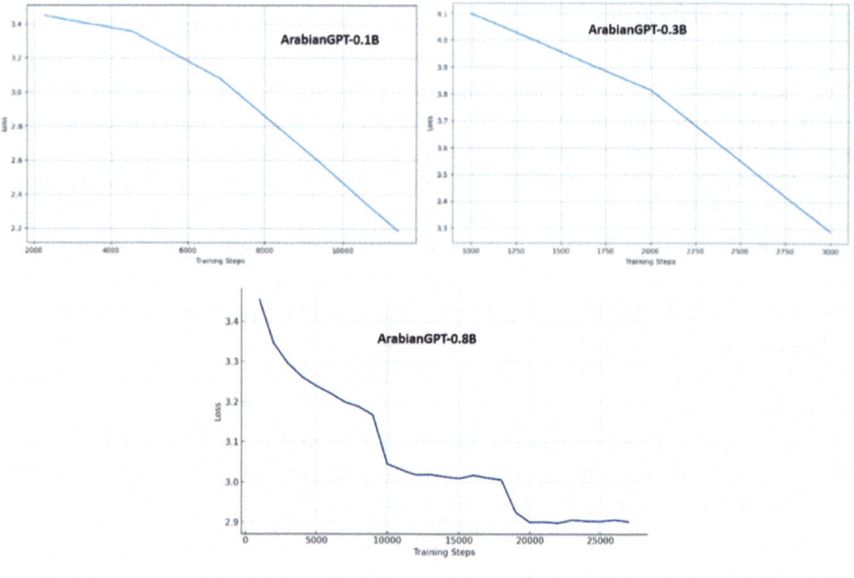

Fig. 5 Fine-tuning loss for the ArabianGPT-0.1B, ArabianGPT-0.3B, and ArabianGPT-0.8B models, on the summarization task.

suggests that these tasks have varying levels of complexity or that the models respond differently to the intricacies of each task.

The ArabianGPT-0.3B models start with a higher initial loss in both tasks compared to both the 0.1B and 0.8B models, potentially indicating a more complex model capacity which, after fine-tuning, also shows a substantial reduction in loss.

9.1.2 Results

The performance of both the original and fine-tuned models was evaluated using the F1 score. The F1 score is a harmonic mean of precision and recall, providing a balance between the completeness and accuracy of the summaries generated by the models. They are calculated using BLUE and ROUGE Scores using the following equation:

$$F1 = \frac{2 \cdot \text{BLUE} \cdot \text{ROUGE}}{\text{BLUE} + \text{ROUGE}}$$

Figure 6 illustrates the F1 scores for the ArabianGPT models (0.1B, 0.3B, and 0.8B) on the summarization dataset, both before and after fine-tuning. The results reveal a nuanced picture of model performance and the impact of fine-tuning across different model sizes.

Across all model sizes, fine-tuning led to substantial improvements in F1 scores, underscoring its effectiveness in enhancing summarization capabilities. This consistent pattern suggests that task-specific adaptation through fine-tuning is crucial for optimizing performance, regardless of model size. The observed improvements likely stem from the models learning domain-specific vocabulary, linguistic

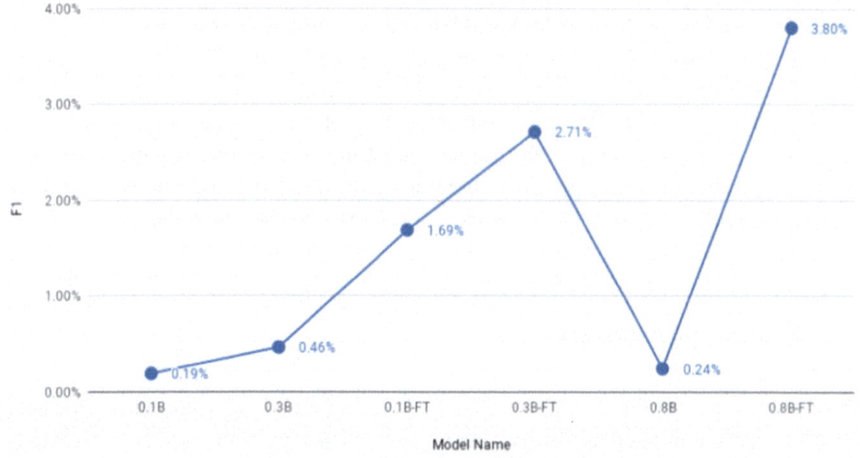

Fig. 6 F1 score for ArabianGPT-0.1B, ArabianGPT-0.3B models, and ArabianGPT-0.8B, before and after fine-tuning on the summarization dataset.

patterns, and summarization strategies present in the fine-tuning dataset. Interestingly, the relationship between model size and performance is not strictly linear. The 0.3B models outperformed their 0.1B counterparts in both fine-tuned and non-fine-tuned scenarios, indicating that the increased capacity allows for better capture of linguistic nuances and summarization techniques. However, the non-fine-tuned 0.8B model unexpectedly underperformed compared to the 0.3B version. This counter-intuitive result might be attributed to overfitting on the pre-training data, a lesser quality dataset, or the challenge of effectively utilizing the larger parameter space without task-specific guidance.

Post fine-tuning, the 0.8B model achieved the highest F1 score, suggesting that with appropriate task-specific training, larger models can indeed leverage their increased capacity more effectively. This highlights the importance of balancing model size with targeted fine-tuning to achieve optimal performance.

Despite these improvements, the overall F1 scores remain relatively low, indicating significant room for enhancement. This could be addressed through several strategies: (1) Extended fine-tuning with larger and more diverse datasets to expose the models to a wider range of summarization scenarios; (2) Experimenting with advanced fine-tuning techniques like gradual unfreezing or discriminative fine-tuning; (3) Exploring alternative model architectures or pre-training approaches that might be more conducive to summarization tasks; and (4) Incorporating human feedback or reinforcement learning to refine the models' summarization strategies.

Tables 13, 14, and 15 show a sample output of the fine-tuned ArabianGPT-0.1B, ArabianGPT-0.3B, and ArabianGPT-0.8B models, respectively, on the summarization task. We notice that the three models understand the task, capture the main points effectively, and summarize the main idea, but the output sometimes include additional irrelevant words. Nevertheless, this drawback is less pronounced for the ArabianGPT-0.8B model, although its output tends to include redundant expressions and sometimes rephrases the same idea multiple times.

The fine-tuning of ArabianGPT models on the summarization task has demonstrated improvements in their performance as indicated by the F1 scores. The results suggest a positive correlation between model size and summarization ability, as well as the effectiveness of fine-tuning in enhancing model performance for specific tasks. Future work could focus on exploring advanced fine-tuning techniques, larger and more diverse datasets, and possibly integrating other AI methodologies to further enhance the summarization capabilities of transformer-based models.

9.2 Sentiment Analysis

We fine-tuned the Arabic Jordanian General Tweets (AJGT) Corpus, which consists of 1,800 tweets annotated as positive and negative. They are expressed either in Modern Standard Arabic (MSA) or Jordanian dialect. Table 16 details the specifications for the sentiment analysis fine-tuning task.

Table 13 Sample output of the fine-tuned ArabianGPT-0.1B model on the summarization task

Completion (ArabianGPT 0.1B)	Prompt
تطورات تكنولوجيا الذكاء الاصطناعي تؤثر على ' الصناعات المختلفة مثل الرعاية الصحية استعراضات".التقدمات جارية".والتصنيع	"لخص النص التالي: يستعرض النص التطورات الأخيرة في مجال الذكاء الاصطناعي وأثرها على الصناعات المختلفة، بما في ذلك الرعاية الصحية والتصنيع. التلخيص:"
التلخيص: استخدام الوسائط المتعددة في تحسين تجربة التعلم للطلاب	"لخص النص التالي: تناول النص أهمية التكنولوجيا في تعزيز التعليم الحديث، مشيراً إلى كيفية استخدام الوسائط المتعددة والأدوات الذكية في تحسين تجربة التعلم لدى الطلاب. التلخيص:"
ن للنص ملخص باللغة العربية: التحديات البيئية" العالمية، التغيرات المناخية والتلوث، وحلول مستدامة للجوانب المعنية	"لخص النص التالي: يتناول النص التحديات التي تواجه البيئة العالمية، مثل التغير المناخي والتلوث، ويقترح حلولاً مستدامة لهذه القضايا. التلخيص:"
التلخيص: دور الإعلام الاجتماعي يؤثر في تشكيل الرأي العام والانتخابات والسياسات العامة. يؤثر أيضا على الانتخابات والاقتصاد. الإعلام الاجتماعي يؤثر	"لخص النص التالي: يناقش النص دور الإعلام الاجتماعي في تشكيل الرأي العام، وكيف يمكن أن يؤثر على الانتخابات والسياسات العامة. التلخيص:"

Here's an updated commentary that includes analytics about the 0.3B and 0.8B models:

Table 17 compares the performance of the ArabianGPT models (0.1B, 0.3B, and 0.8B) on the sentiment analysis task, before and after fine-tuning. We observe a consistent pattern of improvement across all model sizes after fine-tuning.

The 0.1B model showed the most dramatic improvement, with the base model giving almost random responses (56% accuracy) and achieving a marked increase to 95% accuracy after fine-tuning. This significant jump occurred after only 3 epochs on a relatively small fine-tuning dataset.

The 0.3B model demonstrated strong performance even in its base form, with 77.9% accuracy. After fine-tuning, it achieved perfect accuracy (100%), showcasing the potential of mid-sized models when properly tuned.

The largest model, 0.8B, exhibited the best base performance at 86.5% accuracy, indicating that increased model size correlates with better initial task performance.

Table 14 Sample output of the fine-tuned ArabianGPT-0.3B model on the summarization task

Completion (ArabianGPT 0.3B)	Prompt
'لخص النص التالي: يتناول النص التحديات التي تواجه البيئة العالمية، مثل التغير المناخي والتلوث، ويقترح حلولاً مستدامة لهذه القضايا. التلخيص: تلخيص :تلخيص :تلخيص :تلخيص :تلخيص	'لخص النص التالي: يتناول النص التحديات التي تواجه البيئة العالمية، مثل التغير المناخي والتلوث، ويقترح حلولاً مستدامة لهذه القضايا. التلخيص:'
التلخيص: يتحدث النص عن التطورات التكنولوجية وأثرها على مختلف القطاعات	'لخص النص التالي: يتناول هذا النص بالتفصيل التطورات الحديثة في مجال الذكاء الاصطناعي وتأثيرها على مختلف القطاعات. يبدأ النص بتعريف الذكاء الاصطناعي وأصوله التاريخية، موضحًا كيف تطورت هذه التكنولوجيا عبر العقود. التلخيص:'
التلخيص: دراسة الذكاء الاصطناعي: المتغيرات المستمرة والتقدم في التعلم. تلخيص: تلخيص: تلخيص	'لخص النص التالي: يستعرض النص تطور الذكاء الاصطناعي وتأثيره المتزايد على مختلف القطاعات. يبدأ بتاريخ الذكاء الاصطناعي ويتبع تطوره، مع التركيز على التقدم في التعلم الآلي. التلخيص:'

After fine-tuning, it reached a 98.5% accuracy, slightly lower than the fine-tuned 0.3B model but still a substantial improvement over its base version.

Interestingly, while larger models generally performed better in their base form, the fine-tuned 0.3B model outperformed the fine-tuned 0.8B model by a small margin. This suggests that model size is not the only factor in determining performance, and that effective fine-tuning can sometimes allow smaller models to compete with or even surpass larger ones on specific tasks.

Tables 18, 19, and 20 show sample outputs of the three ArabianGPT models (0.1B, 0.3B, and 0.8B, respectively) before and after fine-tuning on the sentiment analysis task. For the 0.1B model, pre-fine-tuning output scores are often close to 0.5, indicating near-random responses, while post-fine-tuning scores approach 1 with mostly correct outputs. The 0.3B model demonstrates even more dramatic improvement,

Table 15 Sample output of the fine-tuned ArabianGPT-0.8B model on the summarization task

Completion (ArabianGPT-0.8B)	Prompt
يناقش التأثير العالمي للكلمات المحفزة للتغيير التكنولوجي على سوق العمل العالمي. يستعرض تأثير التطورات التكنولوجية على الوظائف والمهارات المطلوبة، بما في ذلك الذكاء الاصطناعي وتحليل البيانات، ويقترح حلولا استراتيجية للتكيف. يعتبر النص بديلا استراتيجيا للعمالة وتقنيات المستقبل. يقترح حلولا استراتيجية للتكيف مع هذه التحديات. 1588 يذكر النص بأن هناك	"لخص النص التالي: يناقش النص التأثيرات العميقة للتغيرات التكنولوجية على سوق العمل العالمي. يبدأ النص بمراجعة تاريخية لكيفية تأثير الابتكارات التكنولوجية على الوظائف والمهارات المطلوبة، ويستعرض كيف ساهمت الثورة الرقمية في ظهور مجالات جديدة مثل الذكاء الاصطناعي وتحليل البيانات. بعد ذلك، يناقش التحديات التي يواجهها العمال في ظل الأتمتة والذكاء الاصطناعي، بما في ذلك فقدان الوظائف التقليدية والمهارات القديمة. كما يقترح النص حلولاً استراتيجية للتكيف مع هذه التغيرات، مثل تطوير برامج تدريبية جديدة وتعليم المهارات الرقمية الحديثة، ويستعرض أمثلة من الدول التي نجحت في التعامل مع هذه التحديات. التلخيص:"
يستعرض النص تطور العلاقات الدولية منذ بداية القرن العشرين وحتى الوقت الحالي، ويستعرض أحداث الحروب العالمية والتغيرات الجيوسياسية وتأثيرها على الاقتصاد العالمي. يناقش النص تأثير الأزمات الدولية على الاستراتيجيات والسياسات الوطنية للدول الكبرى، وأثر الأزمات الدولية على الاستراتيجيات والسياسات الوطنية للدول الكبرى.	"لخص النص التالي: يستعرض النص تطور العلاقات الدولية بين القوى الكبرى منذ بداية القرن العشرين حتى الوقت الحالي. يركز النص على الأحداث البارزة مثل الحروب العالمية، الحروب الباردة، وتغيرات التحالفات الجيوسياسية. يعرض كيف شكلت هذه الصراعات والتحالفات السياسات العالمية وأسست التوازنات الدولية الحالية. كما يناقش النص تأثير هذه العلاقات على الاقتصاد العالمي، الثقافة، والأمن الدولي، مشيراً إلى كيف أثرت الأزمات الدولية على الاستراتيجيات والسياسات الوطنية للدول الكبرى. التلخيص:"
ويتين يتناول النص أحدث التطورات في الطب والعلاج، وكيف أحدثت التطورات في تقنيات التشخيص والعلاج تأثيرا كبيرا في الرعاية الصحية. يستعرض النص التحديات الحالية في هذا المجال، وتقدم اقتراحات لتحسين الوصول إلى العلاج وتخفيض التكاليف. ينتهي النص بتقديم اقتراحات لتحسين الوصول إلى العلاج وتخفيض التكاليف. ينتهي النص بتقديم اقتراحات لتحسين الوصول إلى العلاج وتخفيض التكاليف. ينتهي	"لخص النص التالي: يتناول النص أحدث التطورات في مجال الطب والعلاج، مستعرضاً كيف أن التقدم التكنولوجي والابتكارات الطبية قد أحدثت تحولاً كبيراً في الرعاية الصحية. يبدأ النص ببحث التطورات في تقنيات التشخيص والعلاج، مثل الجراحة الروبوتية والعلاج الجيني، ويستعرض كيف أسهمت هذه التقنيات في تحسين نتائج العلاج وجودة الحياة للمرضى. ثم يتناول النص التحديات الحالية في هذا المجال، بما في ذلك التكاليف الباهظة للتقنيات الجديدة وعدم المساواة في الوصول إلى الرعاية الصحية بين مختلف الفئات الاجتماعية والاقتصادية. ينتهي النص بتقديم اقتراحات لتحسين الوصول إلى العلاج وتخفيض التكاليف، مع تقديم أمثلة على المبادرات التي تسعى لتحقيق هذه الأهداف. التلخيص

with fine-tuning increasing confidence scores consistently to 1 and achieving nearly perfect predictions. The 0.8B model also shows significant enhancement after fine-tuning, with increased confidence scores in most cases (barring rare exceptions like the last input sample) and a substantial reduction in misclassifications.

Table 16 Training specifications for the sentiment analysis fine-tuning task

Parameter	Value
Epochs	3
Learning rate	1.4×10^{-5} to 1.7×10^{-6}
Batch	8
GPU	A100
Global step	1500
Train loss	0.203
Training samples	375
Testing samples	375
Training percentage	70%
Testing percentage	30%
Tokenizer	64K aranizer

Table 17 Model Performance on the sentiment analysis task, before and after fine tuning

Model	Accuracy
LLM-0.1B-Base	56%
LLM-0.1B (fine-tuned)	95%
LLM-0.3B-Base	77.9%
LLM-0.3B (fine-tuned)	100.0%
LLM-0.8B-Base	86.5%
LLM-0.8B (fine-tuned)	98.5%

Table 18 Sample output of the ArabianGPT-0.1B model before and after fine tuning on the sentiment analysis task

Sample	Ground-truth label	Fine-tuned Prediction	Fine-tuned confidence	Base Prediction	Base confidence
المطعم جميل جدا	Positive	Positive	0.994	Negative	0.820
المطعم سيئ جدا	Negative	Negative	0.996	Negative	0.856
هذا الكتاب مثير للاهتمام ومفيد	Positive	Positive	0.964	Positive	0.662
الطعام في هذا المطعم كان فظيعا	Negative	Negative	0.994	Negative	0.510
الخدمة ممتازة وسريعة	Positive	Positive	0.999	Positive	0.850
أشعر بخيبة أمل من جودة المنتج	Negative	Negative	0.801	Negative	0.578
منظر البحر من الفندق خلاب	Positive	Negative	0.983	Negative	0.765
لم يعجبني الفيلم، كان مملا	Negative	Negative	1.000	Positive	0.607
المعلم كان ودودًا ومتعاونًا	Positive	Positive	0.946	Negative	0.679
الغرفة كانت قذرة وغير مرتبة	Negative	Negative	0.991	Negative	0.718

Table 19 Sample output of the ArabianGPT-0.3B model before and after fine tuning on the sentiment analysis task. Correct predictions are marked in green, while misclassifications are marked in orange

Sample	Ground-truth label	Fine-tuned Prediction	Fine-tuned confidence	Base Prediction	Base confidence
المطعم جميل جدا	Positive	Positive	1	Negative	0.635
المطعم سيئ جدا	Negative	Negative	1	Negative	0.867
هذا الكتاب مثير للاهتمام ومفيد	Positive	Positive	1	Negative	0.802
الطعام في هذا المطعم كان فظيعا	Negative	Negative	1	Positive	0.843
الخدمة ممتازة وسريعة	Positive	Positive	1	Positive	0.513
أشعر بخيبة أمل من جودة المنتج	Negative	Negative	1	Positive	0.848
منظر البحر من الفندق خلاب	Positive	Positive	1	Positive	0.679
لم يعجبني الفيلم، كان مملا	Negative	Negative	1	Positive	0.649
المعلم كان ودودًا ومتعاونًا	Positive	Positive	1	Negative	0.992
الغرفة كانت قذرة وغير مرتبة	Negative	Negative	1	Positive	0.961

Table 20 Sample output of the ArabianGPT-0.8B model before and after fine tuning on the sentiment analysis task. Correct predictions are marked in green, while misclassifications are marked in orange

Sample	Ground-truth label	Fine-tuned Prediction	Fine-tuned confidence	Base Prediction	Base confidence
المطعم جميل جدا	Positive	Positive	0.999	Negative	0.703
المطعم سيئ جدا	Negative	Negative	0.913	Negative	0.776
هذا الكتاب مثير للاهتمام ومفيد	Positive	Positive	0.982	Positive	0.611
الطعام في هذا المطعم كان فظيعا	Negative	Negative	0.852	Negative	0.850
الخدمة ممتازة وسريعة	Positive	Positive	0.991	Negative	0.979
أشعر بخيبة أمل من جودة المنتج	Negative	Negative	0.999	Negative	0.998
منظر البحر من الفندق خلاب	Positive	Positive	0.998	Positive	0.819
لم يعجبني الفيلم، كان مملا	Negative	Negative	0.938	Negative	0.978
المعلم كان ودودًا ومتعاونًا	Positive	Positive	1.000	Negative	0.990
الغرفة كانت قذرة وغير مرتبة	Negative	Negative	0.940	Negative	1.000

9.3 Question Answering

The task of question-answering in natural language processing (NLP) involves generating accurate and contextually relevant answers to given queries. Since the capabilities of the base models stop at text generation, it is important to expand these capabilities to be able to correctly deal with question/answering.

We have fine-tuned the previously pre-trained ArabianGPT-0.1B, 0.3B, and 0.8B Models for the question-answering task. These models have undergone fine-tuning (FT) on a question-answering dataset, specifically the mkqa-1000 dataset, with the goal of enhancing their capability to provide precise and informative answers.

The mkqa-1000 dataset, derived from the Google Natural Questions dataset, serves as the basis for fine-tuning and evaluation. This dataset encompasses a diverse set of queries and passage-independent answers, which facilitates testing the models' question-answering capabilities across various contexts.

The fine-tuning process involves adjusting the weights of the pretrained model, tailoring it specifically for the question-answering task. This crucial phase aims to optimize the model's performance and enhance its proficiency in generating accurate responses to a wide range of queries in Arabic language.

Table 21 presents the evaluation scores for the fine-tuned ArabianGPT models (0.1B, 0.3B, and 0.8B) on the benchmarks introduced in Sect. 8. The results reveal interesting patterns across model sizes and benchmark tasks.

Interestingly, all three models achieve similar average scores (31.2 for 0.1B, 31.4 for both 0.3B and 0.8B), suggesting that raw model size does not necessarily correlate with overall performance improvement across these diverse tasks. However, a closer examination of individual benchmarks reveals nuanced differences in model capabilities.

The 0.8B model shows superior performance on the ARC (27.2) and HellaSwag (29.46) benchmarks, indicating that the larger model size benefits tasks requiring broader knowledge and common-sense reasoning. This suggests that increased parameter count allows for better capture and utilization of general knowledge.

Surprisingly, the 0.1B model outperforms its larger counterparts on the MMLU (24.9, tied with 0.3B) and TruthfulQA (49.6) benchmarks. This unexpected result,

Table 21 Model evaluation scores for ArabianGPT-0.1B, ArabianGPT-0.3B, and ArabianGPT-0.8B after fine tuning on the Question/Answering dataset

Model	Language	Average	ARC (25-shot)	HellaSwag (10-shot)	MMLU (5-shot)	TruthfulQA (0-shot)
ArabianGPT-0.1B-FT	Arabic	31.2	25.2	25.0	**24.9**	**49.6**
ArabianGPT-0.3B-FT	Arabic	**31.4**	24.8	27.6	**24.9**	48.2
ArabianGPT-0.8B-FT	Arabic	**31.4**	**27.2**	**29.46**	23.41	45.54

particularly for TruthfulQA, might indicate that smaller models can sometimes be more resistant to generating false or misleading information, possibly due to their more constrained knowledge representation.

The 0.3B model shows balanced performance across tasks, often scoring between its smaller and larger counterparts. Its strong performance on HellaSwag (27.6) compared to the 0.1B model (25.0) suggests that the increased capacity allows for better modeling of sequential events and situational logic.

Table 22 displays a sample output of the three ArabianGPT models after fine tuning on the Question/Answering dataset. While ArabianGPT-0.1B commonly yields irrelevant, incomplete, or unclear responses, some of ArabianGPT-0.3B and ArabianGPT-0.8B responses are relevant and accurate, such as for the first and third example. However, their responses to more complex questions, such as photosynthesis (second example), lack clarity or are off-topic. Overall, the models demonstrate varying

Table 22 Sample output of the fine-tuned ArabianGPT models on the Question/Answering dataset

Question	ArabianGPT-0.1B-FT Answer	Comment	ArabianGPT-0.3B-FT Answer	Comment	ArabianGPT-0.8B-FT Answer	Comment
ما هي عاصمة مصر؟	6,449,42720 | - 202118-202150 | - 2021 | -20215 | - 2021 | -2021 | -2021 |	Irrelevant and incorrect.	.عاصمة مصر هي القاهرة	Accurate.	القاهرة او هو اسم العاصمة المصرية وأكبر مدنها	Correct and accurate.
ما هي عملية البناء الضوئي؟	يوجد 6 في عدة أقوال ، يمكنك الاستفادة من المواد الكيميائية ، التي تحتوي على نظام ، ومن عليها عملية الزرع. يوجد	Unclear, incomplete, and off-topic.	إذا لم تستطع أيضا أن تتكاثر، هل تستطيع أن تفعل ذلك؟ يجب أن تعرف الخلايا، وخاصة تلك الخلايا التي يوجد بها بعض الجينات التي يمكن أن تكتسب صفات وراثية نتيجة حدوث طفرة ما هي	Unclear and incomplete.	عملية كيميائية معقدة تستخدم الضوء والأكسجين لإنتاج NADPH و ATP	Inaccurate and incomplete.
ما هو أطول نهر في العالم؟	من 0 حروف ؟ يعتبر من 4 '..- : حروف الاجابة هي	Irrelevant and incomplete.	.نهر النيل	Correct and concise.	أطول نهر في العالم هو نهر ويعتبر النيل \n\n.النيل هو أطول نهر في العالم	Correct but redundant.
ما هي اللغة الرسمية للسعودية؟	لا يوجد كلمات متقاطعة المجموعة الثانية من الناحية البشرية، من الناحية الإسلامية.. ولا يوجد فيها أي شيء قدير. في حالة عدم انتشار أي كلمة كلمة أو حساب أو كلمة أو كلمة أو كلمة	Irrelevant and unclear.	يوجد باللغة العربية أكثر من مليون ونصف مليون نسمة، واللغة الإنجليزية هي الثانية في عدد السكان بعد اللغة العبرية، حيث يقطنها نحو 250 مليون نسمة أو أكثر، وهي تحتل المركز	Incorrect and irrelevant.	هي الغة الإنجليزية، و الغة العربية، كما أنها الغة الثانية في المملكة	Relevant but incorrect.

levels of effectiveness, with fine-tuning generally improving performance but still facing challenges in delivering precise and contextually appropriate responses.

10 Conclusion

The ArabianGPT-Base models represent a significant leap forward in NLP, introducing robust capabilities for both understanding and generating Arabic text. The architecture's strength, rooted in its training on a comprehensive corpus and the utilization of the customized tokenizer, AraNizer, collectively contribute to its effectiveness across various NLP tasks specific to the Arabic language context. Notably, the versatility demonstrated by the GPT-2 architecture is further emphasized by the successful adaptation to the complexities of the Arabic language through the ArabianGPT models.

The development of ArabianGPT is driven by the acknowledgment of a substantial gap in the availability of native LLMs, primarily due to the prevalence of English and Latin-based models. Existing models often suffer from the inclusion of English tokens, which compromises their efficacy in processing the intricate morphology and syntax inherent to native Arabic. This paper introduced ArabianGPT as a tailored solution within the ArabianLLM suite, explicitly designed to address this deficit. The suite includes models of varying sizes, from 0.1B to 0.8B, aligning with the nuanced linguistic characteristics of Arabic.

A key component of the ArabianGPT models is the AraNizer tokenizer, specifically crafted to address the unique morphological aspects of Arabic script. This ensures more accurate text processing, allowing the models to navigate the intricacies of the language with precision. The empirical results from fine-tuning these models on tasks like sentiment analysis and summarization showcase significant improvements. For instance, the fine-tuned ArabianGPT-0.1B model achieved an impressive accuracy of 95%, a substantial increase from the base model's 56%, in sentiment analysis. Similarly, in summarization tasks, fine-tuned models exhibited enhanced F1 scores, indicating improved precision and recall in generating concise summaries.

Comparative analyses of the fine-tuned ArabianGPT models against their base versions across various benchmarks reveal nuanced differences in performance. Fine-tuning is observed to have a positive impact on specific tasks, such as question answering, and summarization. These findings underscore the efficacy of fine-tuning in aligning ArabianGPT models more closely with specific NLP tasks, highlighting the potential of tailored transformer architectures in advancing Arabic NLP. As such, the ArabianGPT-Base models not only contribute to bridging the existing gap in Arabic LLMs but also pave the way for further advancements in the field of Arabic NLP.

Funding This research was funded by Prince Sultan University, grant number SEED-CCIS-2023-145.

Institutional Review Board Statement Not applicable.

Informed Consent Statement Not applicable.

Data Availability Statement The datasets used in this work are publicly available on https://huggingface.co/datasets.

Acknowledgements The authors thank Prince Sultan University for supporting this work.

References

1. A. Radford, K. Narasimhan, T. Salimans, I. Sutskever, Improving language understanding by generative pre-training. Adv. Neural Inf. Process. Syst. **30** (2018)
2. A. Radford, J. Wu, R. Child, D. Luan, D. Amodei, I. Sutskever, Language models are unsupervised multitask learners. OpenAI Blog **1**(8), 9 (2019)
3. O. ElJundi, W. Antoun, N. El Droubi, H. Hajj, W. El-Hajj, K. Shaban, hulmona: the universal language model in Arabic, in *Proceedings of the Fourth Arabic Natural Language Processing Workshop*, pp. 68–77, Aug 2019
4. W. Antoun, F. Baly, H. Hajj, Arabert: transformer-based model for Arabic language understanding. arXiv preprint arXiv:2003.00104 (2020)
5. V. Liu, J.R. Curran, Web text corpus for natural language processing, in *11th Conference of the European Chapter of the Association for Computational Linguistics* (2006), pp. 233–240
6. T. Brown, B. Mann, N. Ryder, M. Subbiah, J.D. Kaplan, P. Dhariwal, D. Amodei, Language models are few-shot learners. Adv. Neural Inf. Process. Syst. **33**, 1877–1901 (2020)
7. J. Achiam, S. Adler, S. Agarwal, L. Ahmad, I. Akkaya, F.L. Aleman, D. Almeida, J. Altenschmidt, S. Altman, S. Anadkat, et al., Gpt-4 technical report. arXiv preprint arXiv:2303.08774 (2023)
8. N. Sengupta, S.K. Sahu, B. Jia, S. Katipomu, H. Li, F. Koto, E. Xing, Jais and jais-chat: Arabic-centric foundation and instruction-tuned open generative large language models. arXiv preprint arXiv:2308.16149 (2023)
9. H. Huang, F. Yu, J. Zhu, X. Sun, H. Cheng, D. Song, Z. Chen, A. Alharthi, B. An, Z. Liu, et al., Acegpt, localizing large language models in Arabic. arXiv preprint arXiv:2309.12053 (2023)
10. W. Lan, Y. Chen, W. Xu, A. Ritter, An empirical study of pre-trained transformers for arabic information extraction. arXiv preprint arXiv:2004.14519 (2020)
11. M. Abdul-Mageed, A.R. Elmadany, E.M.B. Nagoudi, Arbert & marbert: deep bidirectional transformers for Arabic. arXiv preprint arXiv:2101.01785 (2020)
12. J. Devlin, M.-W. Chang, K. Lee, K. Toutanova, Bert: pre-training of deep bidirectional transformers for language understanding. arXiv preprint arXiv:1810.04805 (2018)
13. W. Antoun, F. Baly, H. Hajj, Araelectra: pre-training text discriminators for Arabic language understanding. arXiv preprint arXiv:2012.15516 (2020)
14. G. Inoue, B. Alhafni, N. Baimukan, H. Bouamor, N. Habash, The interplay of variant, size, and task type in Arabic pre-trained language models. arXiv preprint arXiv:2103.06678 (2021)
15. E.M.B. Nagoudi, A.R. Elmadany, M. Abdul-Mageed, AraT5: text-to-text transformers for Arabic language generation. arXiv preprint arXiv:2109.12068 (2021)
16. A. Abdelali, S. Hassan, H. Mubarak, K. Darwish, Y. Samih, Pre-training BERT on Arabic tweets: practical considerations. arXiv preprint arXiv:2102.10684 (2021)

17. A. Ghaddar, Y. Wu, S. Bagga, A. Rashid, K. Bibi, M. Rezagholizadeh, C. Xing, Y. Wang, X. Duan, Z. Wang, et al., Revisiting pre-trained language models and their evaluation for Arabic natural language processing, in *Proceedings of the 2022 Conference on Empirical Methods in Natural Language Processing* (2022), pp. 3135–3151
18. W. Antoun, F. Baly, H. Hajj, Aragpt2: pre-trained transformer for Arabic language generation. arXiv preprint arXiv:2012.15520 (2020)
19. A. Haouhat, H. Cherroun, S. Bellaouar, A. Nehar, Modos at araieval shared task: multimodal propagandistic memes classification using weighted sam, clip and arabiangpt, in *Proceedings of the Second Arabic Natural Language Processing Conference* (2024), pp. 483–488
20. M. Hasanain, M.A. Hasan, F. Ahmed, R. Suwaileh, M.R. Biswas, W. Zaghouani, F. Alam, Araieval shared task: propagandistic techniques detection in unimodal and multimodal Arabic content. arXiv preprint arXiv:2407.04247 (2024)
21. O. Najar, S. Sibaee, L. Ghouti, A. Koubaa, Aranizer 0.1.8. https://pypi.org/project/aranizer (2023)
22. T. Kudo, J. Richardson, Sentencepiece: a simple and language independent subword tokenizer and detokenizer for neural text processing. arXiv preprint arXiv:1808.06226 (2018)
23. Y. Shibata, T. Kida, S. Fukamachi, M. Takeda, A. Shinohara, T. Shinohara, S. Arikawa, Byte pair encoding: a text compression scheme that accelerates pattern matching. Technical Report DOI-TR-161, Department of Informatics, Kyushu University (1999)
24. A. Vaswani, N. Shazeer, N. Parmar, J. Uszkoreit, L. Jones, A.N. Gomez, I. Polosukhin, Attention is all you need. Adv. Neural Inf. Process. Syst. **30** (2017)
25. P. Clark, I. Cowhey, O. Etzioni, T. Khot, A. Sabharwal, C. Schoenick, O. Tafjord, Think you have solved question answering? try arc, the ai2 reasoning challenge. arXiv:1803.05457v1 (2018)
26. R. Zellers, A. Holtzman, Y. Bisk, A. Farhadi, Y. Choi, Hellaswag: can a machine really finish your sentence? arXiv preprint arXiv:1905.07830 (2019)
27. D. Hendrycks, C. Burns, S. Basart, A. Zou, M. Mazeika, D. Song, J. Steinhardt, Measuring massive multitask language understanding. arXiv preprint arXiv:2009.03300 (2020)
28. S. Lin, J. Hilton, O. Evans, Truthfulqa: measuring how models mimic human falsehoods. arXiv preprint arXiv:2109.07958 (2021)
29. CSEBUETNLP. XL-Sum Arabic Dataset. https://huggingface.co/datasets/csebuetnlp/xlsum/viewer/arabic (2024). Accessed on: 25/01/2024

Architectural Deep Dive into Large Language Models

Alaa Ghabi and Habib Hamam

Abstract This chapter presents an in-depth exploration of Large Language Models (LLMs), a cornerstone in the field of artificial intelligence and natural language processing. Beginning with an introduction to the fundamental concepts of language models, we trace the evolutionary trajectory from early statistical models to the sophisticated neural network-based models of today. The chapter explores the major designs that have influenced the evolution of LLMs, with a particular emphasis on the Transformer model, which has played a crucial role in more recent developments. We examine the mechanics of Transformer models, emphasizing their unique attention mechanisms and the role of encoder-decoder structures in language understanding and generation. The process of training these colossal models, including the challenges of scaling and data curation, is also scrutinized. Moving beyond architecture and training, the chapter explores advanced topics like fine-tuning, ethical considerations, and the emerging domain of multimodal models. Practical applications and case studies are discussed to demonstrate the real-world impact of LLMs. Finally, the chapter addresses the challenges and limitations facing LLMs, such as computational demands, environmental concerns, and the ongoing struggle with bias and reliability. This comprehensive overview aims not only to shed light on the technical intricacies of LLMs but also to provide insight into their profound implications for technology, society, and the future of AI-driven communication.

A. Ghabi (✉) · H. Hamam
Faculty of Engineering, Uni de Moncton, Moncton, Canada
e-mail: ghabi.alaa71@gmail.com

H. Hamam
International Institute of Technology and Management (IITG), Libreville, Gabon

Bridges for Academic Excellence, Tunis, Tunisia

Department of Electrical and Electronic Engineering Science, School of Electrical Engineering, University of Johannesburg, Johannesburg, South Africa

A. Koubaa et al. (eds.), *Generative AI and Large Language Models: Opportunities, Challenges, and Applications*, Studies in Computational Intelligence 1214,
https://doi.org/10.1007/978-3-031-90573-5_3

1 Introduction to Large Language Models

The rapid advancements in artificial intelligence (AI) have seen Large Language Models (LLMs) emerge as a groundbreaking and influential innovation [1]. Utilizing the prowess of machine learning combined with extensive data pools, these models have transformed our interaction with technology through natural language. Originating from simpler forms and evolving into their current complex and nuanced versions, LLMs have not only improved the machine's grasp of human language but have also forged new paths across various applications. These range from automated writing aids to advanced conversational agents. This section is dedicated to exploring the origins, evolution, and the sophisticated architecture of LLMs. It aims to provide an in-depth understanding of their functionalities and their significant role in the landscape of contemporary technology.

1.1 Overview of Language Models

At the heart of machine learning's approach to processing human language lies the technology of Language Models (LMs). These models, trained extensively on textual data, are adept at predicting word sequences, which allows them to carry out a range of tasks associated with Natural Language Processing (NLP). Such tasks include generating text, translating between languages, summarizing lengthy articles, and providing answers to questions posed in natural language [2].

Language Models function by grasping the statistical characteristics inherent in language. This grasp empowers them to anticipate subsequent words in a sentence or to supplement words that are missing. Consider the partial sentence "The cat sat on the…"; an LM that has been appropriately trained is likely to suggest "mat" as a fitting continuation.

The journey of LMs began with the adoption of simple statistical models, such as n-gram models. In this approach, the prediction of a word's occurrence was contingent solely on the preceding n-1 words. Despite their simplicity, these models faced significant challenges. They were notably deficient in their ability to apprehend dependencies over greater distances within text and to process the subtleties and complexities inherent in language structure.

1.2 Evolution of Language Models: From Simple to Complex

Language models have undergone a remarkable transformation, evolving through a series of significant technological advancements:

- The Inception of Statistical Methods: Initially, language models were grounded in statistical methodologies. N-gram models and hidden Markov models served as the progenitors in this domain. These foundational models, however, grappled with constraints imposed by the extent of their training data and struggled to account for dependencies that spanned across larger segments of text, which often led to less predictive power over complex language structures [3].

- The Advent of Neural Networks: Neural networks transformed language modeling. They began with Recurrent Neural Networks (RNNs). Later, Long Short-Term Memory (LSTM) networks emerged. These networks were advanced. They could grasp long text dependencies. This ability improved their language understanding. It also enhanced their predictive power [4].
- The Transformer Revolution: The advent of attention mechanisms brought about a paradigm shift, culminating in the creation of the Transformer model. These models, departing from the sequential processing characteristic of RNNs, embraced parallel data processing. This change facilitated a more nuanced handling of extended dependencies within text, thereby amplifying the power and efficiency of language models [5].
- The Phase of Pre-training and Fine-tuning: The introduction of extensive datasets for the initial training of language models, complemented by subsequent refinement for distinct tasks, has culminated in the development of sophisticated models like BERT and the GPT series [6]. These advanced models have demonstrated their prowess across various NLP tasks, establishing themselves as pillars in the field of modern natural language processing [7, 8].
- The Scaling Up Phenomenon: Language model development is currently characterized by an expansion to sizes that were once deemed unachievable, with certain models boasting parameter counts in the billions. Take GPT-3 as an instance; its staggering 175 billion parameters afford it the ability to produce text of remarkable coherence and pertinence on a multitude of subjects [9].
- Looking Forward: Research persists in its pursuit of training methodologies that are more efficient, in efforts to rectify biases inherent in language models, and in the pursuit of augmenting their capabilities to include multiple forms of data, like the amalgamation of textual and visual inputs. The goal of these advancements is to extend the range of what language models can understand and execute.

2 Foundational Concepts

2.1 Understanding Neural Networks

Neural networks are computational models inspired by the human brain's structure and function. They consist of artificial neurons or nodes that mimic the neurons in the brain. These nodes are organized into layers: an input layer that receives the data, hidden layers that process the data, and an output layer that delivers the final decision or prediction [10–12].

Key Concepts:

- Artificial Neurons: Serve as the basic processing units of a neural network, functioning similarly to biological neurons by receiving, processing, and transmitting information.
- Layered Architecture: Neural networks are structured in layers, with each layer responsible for extracting different levels of abstraction from the input data.
- Weighted Connections: Information flows through the network via connections between neurons, each associated with a weight that determines the strength of its influence on the next neuron.
- Activation Functions: Introduce non-linearity into the network, enabling it to learn complex patterns beyond simple linear relationships. Functions like ReLU, sigmoid, and tanh are commonly used.
- Training and Backpropagation: The process of adjusting the weights of the connections to minimize the difference between the actual output and the expected output. Backpropagation is a key algorithm for training neural networks.

2.2 The Role of Deep Learning

2.2.1 Deep Learning's Impact on Metadata Extraction from Online Job Postings

Deep learning, a subset of machine learning, has dramatically transformed how we approach complex tasks such as the multi-class classification of online job postings. Its ability to learn hierarchical representations from vast amounts of data makes it particularly suited for processing and understanding the intricacies of natural language, a critical aspect of job postings analysis [13, 14].

- Advancements in NLP for Job Posting Analysis: The advent of deep learning has enabled more nuanced analysis and classification of job postings, leveraging models such as BERT and GPT to understand context and semantics embedded in text. This has facilitated tasks like extracting metadata, identifying job categories, and even generating synthetic job postings to augment training datasets.
- Synthetic Data Generation for Training: A notable application of deep learning in job postings analysis is the generation of synthetic data to overcome the limitations of scarce or imbalanced real-world datasets. Models like text-davinci-003, Falcon 7b, Wizardlm 7b, and Vicuna 7b have been employed to create high-quality, annotated synthetic datasets. These datasets can either serve as standalone training material or augment existing data, enhancing the diversity and volume of training examples available for model training.
- Use-Case Scenarios for Synthetic Data:

  - Training Without Real Data: In situations where real data is scarce or unavailable, synthetic datasets generated by AI models can provide a viable alternative for training. This approach enables the development of models in domains where collecting annotated data is challenging.
  - Augmentation for Underrepresented Categories: Synthetic data can also be used to balance datasets where certain job categories are underrepresented. This augmentation helps in improving the model's accuracy and generalization capabilities by providing a more comprehensive view of the job market landscape.

- Evaluation of Deep Learning Models: The efficacy of deep learning models in job posting classification has been validated through comparative analysis against traditional datasets. Experiments have shown that models trained on synthetic data can match or even surpass the performance of those trained on real data. Furthermore, when synthetic data is used to augment underrepresented classes, there is a consistent improvement in classification accuracy, especially for simpler models like feedforward neural networks (FNN).
- BERT and Deep Learning Architectures: The BERT model, with its deep bidirectional architecture, has shown significant promise in enhancing the classification accuracy of job postings. Its ability to understand the context and generate embeddings that capture nuanced meanings makes it highly effective for this task. The use of BERT and similar models underscores the importance of selecting appropriate deep learning architectures and training strategies to optimize performance.

2.2.2 Challenges and Considerations

- Data Quality and Volume: The success of deep learning models in job posting analysis is closely tied to the quality and quantity of the training data. Ensuring data cleanliness and representativeness is crucial for achieving high accuracy.
- Model Complexity and Interpretability: Deep learning models, particularly those like BERT, are often complex, making them challenging to interpret. Balancing model complexity with the need for understandable and explainable outcomes is an ongoing area of research.
- Ethical Implications: The use of synthetic data and AI in job posting analysis raises ethical considerations, particularly concerning data privacy, bias, and fairness. Ensuring that AI models do not perpetuate existing biases or introduce new ones is paramount.

Stated in general terms, deep learning has revolutionized the analysis of online job postings, offering sophisticated tools for metadata extraction, classification, and even synthetic data generation. As we continue to harness the power of these technologies, it is imperative to address the challenges of data quality, model complexity, and ethical considerations, ensuring that the advancements in AI contribute positively to the understanding and functioning of the labor market.

2.3 Data Processing and Tokenization

Tokenization Methods: Modern LLMs use advanced tokenization methods such as Byte Pair Encoding (BPE) and WordPiece. BPE is a compression-based approach that combines frequent character sequences into subwords, optimizing vocabulary size and efficiency. WordPiece, used in models like BERT, iteratively refines subword units based on frequency and context. These methods enhance the model's ability to handle rare and compound words.

The process of preparing text data for LLMs involves critical steps of data preprocessing, tokenization, and feature extraction. These steps are essential for transforming raw text into a structured form that neural networks can efficiently process [15, 16].

2.3.1 Tokenization

Tokenization as pointed in Fig. 1, is a foundational step in text data processing, particularly for natural language processing (NLP) tasks. It involves breaking down text into smaller units, known as tokens, which can be words, characters, or subwords. This process is vital for understanding the context within the dataset and forms the basis for developing models capable of comprehending and generating human-like text.

Implementation: In the context of analyzing tweets or any textual content, tokenization helps in dissecting the text into manageable pieces, with a maximum observed length, for example, being 131 words. This granularity enables the model to analyze the semantic and syntactic structure of the language effectively.

2.3.2 Feature Extraction

Feature extraction is about identifying and extracting the most significant aspects of the text that contribute to its meaning. One of the most efficient ways to achieve this is through word embedding, a technique that represents words in a high-dimensional space where similar words are positioned closer to each other.

GloVe Model: For feature extraction, the GloVe (Global Vectors for Word Representation) model is extensively used. Developed by Stanford University, GloVe embeddings are particularly beneficial for capturing the context, semantics, and co-occurrence statistics of words within a document. The pre-trained GloVe Twitter word embeddings, derived from analyzing 2 billion tweets, provide a rich source for extracting meaningful features from text related to social media content.

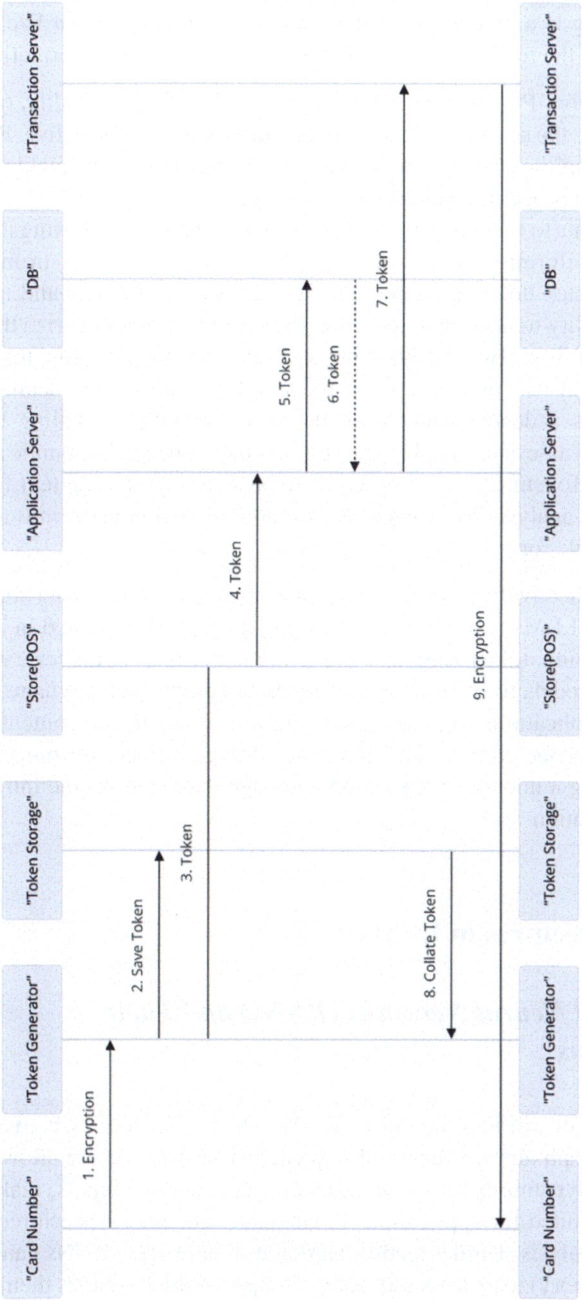

Fig. 1 Data processing and tokenization

2.3.3 Data Preprocessing

Data preprocessing is a crucial phase that aims to clean and standardize the input data, ensuring that the model is trained on high-quality, relevant information.

- Erase Punctuation: Punctuation, while crucial for human readability, often does not contribute to the sentiment or meaning from a model's perspective. Removing punctuation simplifies the text and eliminates unnecessary noise, making it easier for the model to process the underlying message.
- Convert Text Data to Lowercase: Standardizing text data by converting it to lowercase ensures uniformity, addressing the issue of case sensitivity in many NLP methods. This step helps in reducing the complexity of the data and improving the model's ability to accurately interpret the sentiment or categorize the text.
- Tokenization of Text: Beyond breaking text into individual words, tokenization can also segment text into phrases or meaningful chunks. This detailed breakdown facilitates a deeper analysis of the text's structure, enabling tasks like counting words, assessing word frequency, and more nuanced semantic and sentiment analysis. Tokenization serves as a critical step in preparing text for further processing and analysis by converting complex textual content into a simpler, more manageable form.

These steps collectively form the backbone of preparing text data for analysis and processing by LLMs and other deep learning models. Highlighted in Fig. 1, By meticulously cleaning, tokenizing, and extracting features from the text, we lay the groundwork for models to learn from and generate human-like language, enabling a wide array of applications from sentiment analysis to automated content creation. The use of advanced techniques like GloVe embeddings enriches the model's understanding, providing a nuanced view of the language that captures the intricacies of human communication.

3 Key Architectures in LLMs

3.1 Recurrent Neural Networks (RNNs) and Their Limitations

RNNs are a class of artificial neural networks where connections between nodes form a directed graph along a temporal sequence. This architecture allows them to exhibit temporal dynamic behavior and process sequences of inputs, making them particularly well-suited for applications in language modeling, speech recognition, and time series analysis. Unlike feedforward neural networks, RNNs can use their internal state (memory) to process sequences of inputs, which enables them to model dynamic temporal behavior.

In this comprehensive review [17] the authors delve into the practical challenges of training RNNs, including the difficulties of learning long-term dependencies due to vanishing and exploding gradients, and provide an overview of the various approaches developed to mitigate these issues. This paper is valuable for understanding both the capabilities and limitations of RNNs from a modern perspective, emphasizing the ongoing challenges and areas of research.

However, RNNs are not without limitations. Two primary challenges that they face are the vanishing gradient problem and the exploding gradient problem. The vanishing gradient problem occurs when the gradient of the loss function decreases exponentially with time, making it difficult for the model to learn correlations between distant events. Conversely, the exploding gradient problem occurs when the gradients increase exponentially, potentially leading to numerical instability and divergent model parameters during training. These issues are particularly pronounced in standard RNNs when dealing with long sequences, thereby limiting their ability to capture long-term dependencies [18]. Furthermore, RNNs inherently process data sequentially, which limits the parallelization capabilities and affects the training efficiency on modern computational platforms that excel at parallel processing. This sequential processing requirement becomes a significant bottleneck in scaling applications to longer sequences or larger datasets.

3.2 Introduction to Transformer Architecture

Comparison of Encoder, Decoder, and Encoder-Decoder Architectures:

Encoder-only models (e.g., BERT) are optimized for tasks like classification and feature extraction by processing inputs bidirectionally. Decoder-only models (e.g., GPT) excel in text generation by predicting sequential outputs. Encoder-decoder models (e.g., T5) combine these strengths for tasks requiring both comprehension and generation, such as translation and summarization.

The Transformer architecture, since its inception, has significantly influenced the domain of NLP, offering a departure from traditional models like RNNs and LSTMs. Recent scholarly works continue to explore and extend the capabilities of Transformer-based models, addressing both their potential and challenges across various applications.

One noteworthy paper, "End-to-End Transformer-Based Models in Textual-Based NLP," provides a comprehensive literature review on Transformer-based (TB) models, focusing on their application in textual-based tasks within NLP. This review discusses the fundamental concepts underpinning these models, their architecture, training modes, and the advantages and disadvantages of various techniques. It delves into how TB models have revolutionized NLP tasks by leveraging self-attention mechanisms for encoding long-range dependencies in input sequences, thus enhancing the model's understanding and generation of human language [5].

Another important contribution, "Transformers in the Real World: A Survey on NLP Applications," surveys the real-world applications of Transformers in NLP, particularly where text is the primary modality. This paper provides insights into the strengths and limitations of Transformers, highlighting their impact on research and the development of new NLP models. It emphasizes the role of Transformers in advancing NLP by enabling more sophisticated, human-like interactions between computers and humans across various applications, such as information retrieval, sentiment analysis, machine translation, and question answering. The paper also discusses challenges related to computational efficiency, interpretability, and ethical considerations [19].

Furthermore, the exploration of Transformers has extended beyond NLP to fields like computer vision, showcasing their versatility and general applicability. The paper "Exploring Recent Advancements of Transformer-Based Architectures in Computer Vision" illustrates how Transformers, traditionally associated with NLP, are now being applied to computer vision tasks, indicating their potential as a general computational approach across various modalities [20].

3.3 GPT Series: From GPT to GPT-4

The Generative Pre-Trained Transformer (GPT) series, developed by OpenAI, has been at the forefront of revolutionizing NLP technologies, significantly advancing the capabilities of artificial intelligence systems in understanding and generating human language. Since the introduction of the original GPT model in 2018, each subsequent iteration has not only expanded in size and complexity but has also demonstrated substantial improvements in linguistic task performance, setting new benchmarks in the field.

The inaugural GPT model leveraged a transformer-based architecture, pre-trained on an extensive corpus of text, which was then fine-tuned for specific NLP tasks. With its 117 million parameters, GPT showcased unparalleled versatility across various language tasks, including translation, question-answering, and text summarization, illustrating the potential of transformer technology in NLP [17].

The successor, GPT-2(2019), marked a significant expansion to 1.5 billion parameters, enhancing text generation capabilities and comprehension. Its proficiency in producing coherent and contextually rich text passages ignited discussions on AI ethics and the potential misuse of such technology in generating disinformation, highlighting the societal implications of advanced AI models [8].

GPT-3 (2020), with its 175 billion parameters, further raised the bar for language models, delivering robust performance across a diverse array of tasks with little to no task-specific data. Its achievements in areas ranging from content creation to code generation sparked widespread interest in the potential applications of AI and its impact on various industries, underscoring the model's versatility and power [9].

As of the latest updates, the specific details and capabilities of GPT-4 remain largely speculative within the AI community. However, there is a strong anticipation for further enhancements in the model's efficiency, comprehension, and text generation that closely mirrors human-like text. Expected improvements include a deeper

Table 1 GPT series

Model	Year	Key features	Parameters	Training data size	Note
GPT	2018	Transformer-based model, unidirectional context	117M	40GB	Introduced the use of transformers for language modeling
GPT-2	2019	Zero-shot learning, larger and more diverse dataset	1.5B	40TB	Significantly improved language understanding and generation
GPT-3	2020	Few-shot learning, massive scale model	175B	570GB	Advanced natural language understanding, generation, and task flexibility
GPT-4	2023	Improved few-shot learning, better handling of nuanced language	175B +	Larger than GPT-3	Enhanced performance across a wide range of tasks, reduced biases

nuanced understanding of complex subjects, reduced biases within the AI systems, and the introduction of innovative applications across multiple sectors, promising to push the boundaries of what AI can achieve in NLP and beyond.

This progression from GPT as we shown in Table 1, to its anticipated fourth iteration illustrates not just the rapid advancements in AI and NLP technologies but also reflects the growing expectations and ethical considerations surrounding the deployment of such powerful models. As the GPT series continues to evolve, it remains a pivotal area of research and application in the quest to bridge the gap between human and machine understanding of language.

3.4 BERT and Its Variants

Bidirectional Encoder Representations from Transformers (BERT), introduced by Google AI researchers in 2018, marked a paradigm shift in NLP. BERT's innovative approach, characterized by its deep bidirectionality, allows it to understand the context of a word by considering its entire surrounding context (both left and right), a stark contrast to prior models that processed text in a unidirectional manner and thus, had a constrained understanding of context.

The development of BERT encompassed two primary stages: pre-training and fine-tuning. In the pre-training phase, BERT was trained on a vast corpus, including Books Corpus with 800 million words and English Wikipedia with 2,500 million words, through two unsupervised tasks: masked language modeling (MLM) and next sentence prediction (NSP). This foundational training enabled BERT to acquire a deep understanding of language patterns and structures. Subsequently, the model was fine-tuned for specific NLP tasks using smaller, task-specific datasets, demonstrating remarkable adaptability across a wide array of NLP challenges.

A significant feature of BERT is its bidirectional context understanding, which drastically enhanced its performance on various tasks, such as question answering, language inference, and named entity recognition, thereby establishing new benchmarks in the field.

The success of BERT present in Table 2, spurred the development of several variants and improvements aimed at enhancing performance, reducing resource consumption, or tailoring the model to specific languages or domains:

RoBERTa (Robustly optimized BERT approach): Facebook AI's RoBERTa refined the BERT pre-training process by extending the training duration, utilizing more extensive datasets, and optimizing batch sizes, all while removing the NSP objective. These modifications significantly boosted performance across numerous benchmarks.

Distil BERT: Introduced by Hugging Face, DistilBERT presents a streamlined version of BERT that maintains 97% of the original model's performance on the GLUE benchmark but is 40% smaller and 60% faster. This efficiency is achieved through knowledge distillation, a method where the smaller model is trained to emulate the larger, original BERT model's behavior.

Table 2 BERT and its variants

Model	Year	Key features	Parameters	Training data size	Improvements
BERT	2018	Bidirectional Transformer, masked language model	110 M (Base), 340 M (Large)	Wikipedia + BookCorpus (2.5B words)	Revolutionized NLP benchmarks, introducing context to word meanings
RoBERTa	2019	Optimized BERT training process, larger datasets	Same as BERT, with variations	160GB of text	Improved on BERT by training on more data and removing the next-sentence prediction task
ALBERT	2019	Factorized embedding parameterization, cross-layer parameter sharing	12M (Base) to 18M (Large)	Same as BERT	Significantly reduced memory footprint while maintaining or improving performance
DistilBERT	2019	Distillation of BERT model, retaining 97% of performance with 40% of parameters	66M	Same as BERT	Made BERT more accessible for environments with limited resources

ALBERT (A Lite BERT): ALBERT implements parameter-reduction strategies to decrease memory demands and accelerate training. These strategies include factorized embedding parameterization and cross-layer parameter sharing. Furthermore, ALBERT revises the NSP task into a sentence-order prediction task, further refining its performance.

BERT for Specific Languages and Domains: Various BERT adaptations have been pre-trained for languages (such as CamemBERT for French) or specialized fields (like BioBERT for biomedical texts). These models leverage the core BERT architecture but are pre-trained on domain- or language-specific corpora, enhanc (feniofw, 2018)ing their efficacy in dealing with the intricacies and vocabularies of their focused areas.

4 Mechanics of the Transformer Model

4.1 Understanding Attention Mechanisms

The Transformer model's self-attention mechanism represents in Fig. 2, describe a pivotal innovation, enabling it to dynamically prioritize different segments of input data during processing. This mechanism functions by mapping a query and a set of key-value pairs to an output, with all elements represented as vectors. The output is calculated as a weighted sum of the values, determined by the query's compatibility with each key. Recent developments in Transformer technology have focused on extending its capabilities and applicability. The introduction of the Transformer-XL introduced a groundbreaking recurrence mechanism that allows the model to comprehend dependencies beyond predetermined lengths, significantly improving its handling of extended texts and intricate reasoning tasks without compromising on training efficiency. Concurrently, innovated by merging a time-adaptive dynamic convolutional layer with the Transformer, markedly enhancing its efficacy in assignments that require detailed understanding and temporal sensitivity. These advancements underscore the Transformer's adaptability and its potential for integration with other neural network frameworks to push the boundaries of its performance [21, 22].

4.2 Encoder-Decoder Structure

The Transformer model consists of two main parts: the encoder and the decoder. Both parts are composed of multiple layers that have similar but distinct structures. The architecture is designed to handle sequences of tokens by processing all tokens in parallel, which significantly improves efficiency over models that process sequences one token at a time.

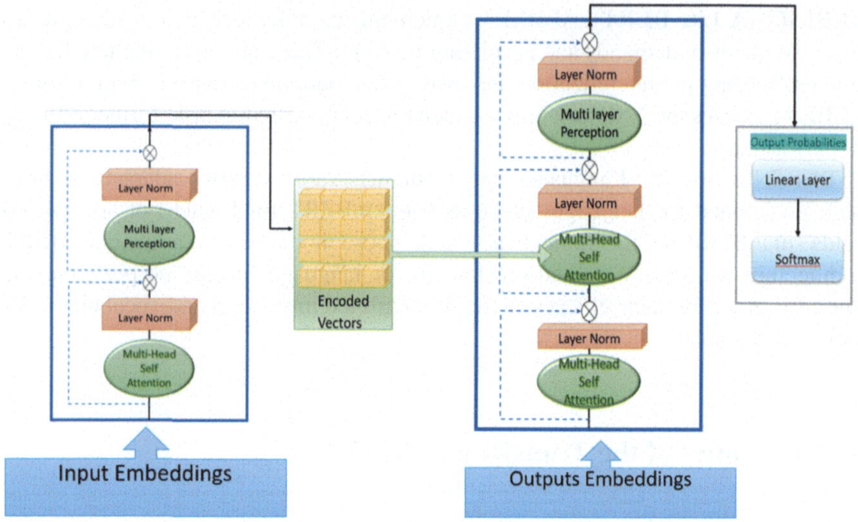

Fig. 2 Transformer architecture model

The encoder maps an input sequence of symbol representations (tokens) to a sequence of continuous representations, which then serves as input for the decoder. Each encoder layer consists of two sub-layers.

In the encoder, each layer is structured around two critical components. Figure 3, The first, a Multi-Head Self-Attention Mechanism, enables the model to consider each token's context by attending to all positions within the input sequence simultaneously. This multi-head approach allows the model to capture diverse contextual nuances by processing different aspects of the information in parallel. The second component, Position-wise Fully Connected Feed-Forward Networks, applies identical linear transformations to each sequence position independently, enabling complex feature integration across the sequence. Both components are enhanced with normalization and dropout, with residual connections that facilitate gradient flow and mitigate potential vanishing gradient issues [23].

Besides, the decoder pointed in Fig. 3, tasked with generating output sequences token by token, mirrors the encoder's complexity with three main sub-layers. It introduces a Masked Multi-Head Self-Attention Mechanism to ensure output predictions for a given position are dependent only on preceding elements, thereby preserving sequential integrity. Additionally, a layer of Multi-Head Attention over the encoder's output directs the decoder's focus to pertinent parts of the input sequence, enhancing relevance and accuracy in the generated text. The inclusion of Position-wise Fully Connected Feed-Forward Networks, akin to those in the encoder, allows for independent and parallel processing across positions. Similar to the encoder, these layers are supplemented with normalization, dropout, and residual connections, ensuring robust and efficient processing [23].

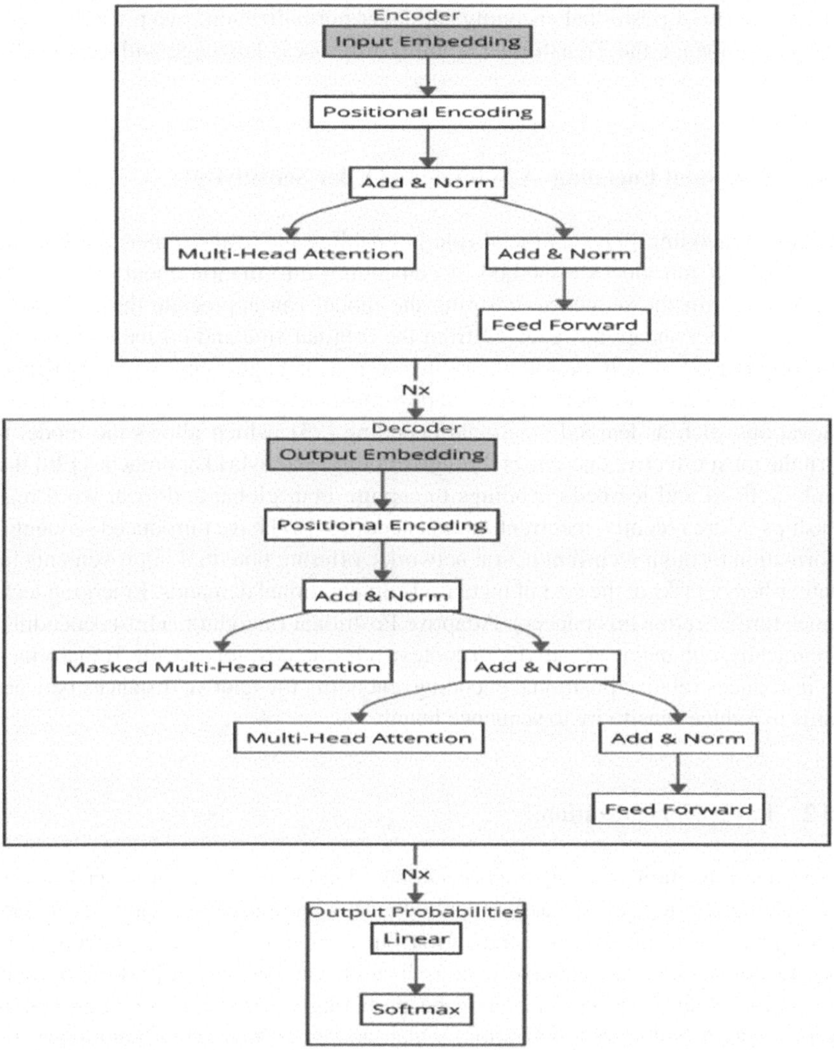

Fig. 3 Encoder and decoder

4.3 Positional Encoding and Layer Normalization

The Transformer model has revolutionized the field of natural language processing with its unique self-attention mechanism. Despite its prowess, the model initially lacks the capability to inherently understand the order of words, which is essential for grasping the full meaning of text. This challenge is ingeniously addressed through the integration of positional encoding and layer normalization, two pivotal components that enhance the Transformer's ability to process language with remarkable efficiency.

4.3.1 Positional Encoding: A Solution to Order Sensitivity

Positional encoding plays a crucial role in enabling the Transformer to recognize word order. It introduces a method of embedding information about the position of words within the sequence, ensuring the model can appreciate the sequence's structure. This concept has evolved from the original sine and cosine embeddings proposed in [24], which, despite their effectiveness in capturing long-range dependencies, sometimes fall short in nuanced positional nuances. The field has since seen innovations such as learned positional encoding [25], which allows the model to learn the most effective encodings through training, and hybrid approaches [26] that combine fixed and learned encodings to capture both global and local word relationships. More recently, recurrent-based encodings [27] have introduced sequential information through recurrent neural networks, offering potential improvements for context-heavy tasks at the cost of increased computational demands. Emerging techniques further refine this concept. Adaptive Positional Encoding, adjusts encodings dynamically, enhancing sensitivity to context-relevant positions, while Transformer-XL introduces relative positional encoding, encoding the relative distances between words to reduce sensitivity to sequence length.

4.3.2 Layer Normalization

Layer normalization is another cornerstone, stabilizing the training process by normalizing the activations across features. This technique not only accelerates convergence and enhances generalization but also smoothens the gradient flow, making the model less sensitive to initialization choices. In [28] which normalizes across channel groups and may be particularly beneficial for large-feature models, and Adaptive Normalization, which adapts normalization parameters for each specific context. Intriguingly, some studies, like that by Shen et al. [28], have explored the effects of omitting layer normalization, with findings suggesting potential performance improvements in certain architectures, indicating areas ripe for further research.

The optimal application of these techniques is highly dependent on various factors, including task requirements, dataset characteristics, model architecture, and computational resources. Through experimentation and thorough evaluation, researchers and practitioners can identify the most effective configurations for their specific needs.

4.4 *Ethical and Privacy Issues in AI-Driven Legal Systems*

In this section, we delve into the ethical and privacy implications inherent in deploying AI-driven legal systems, with a specific focus on the mechanics of the Transformer model. The Transformer model, a cornerstone of modern natural language processing (NLP), has been lauded for its ability to understand and generate human-like text by leveraging self-attention mechanisms. Its sophisticated architecture, which includes components such as positional encoding and layer normalization, enables it to process legal documents, case law, and legislation with unprecedented efficiency and accuracy.

However, the integration of such advanced AI technologies into legal systems raises significant ethical and privacy concerns. First, the use of positional encoding al., [4, 29] and adaptive techniques like those proposed by for dynamic positional adjustment introduce complex challenges related to data bias and fairness. The model's reliance on vast amounts of legal texts to learn patterns may inadvertently perpetuate historical biases present in the training data, leading to ethical dilemmas in judgment or legal advice provided by AI. Privacy issues are equally critical, especially considering the sensitive nature of legal documents and personal information they may contain. Layer normalization techniques and advanced adaptations such as No Layer Normalization [30] ensure the model's stable and efficient performance. Nonetheless, they also underscore the model's capacity to extract and process detailed personal data, potentially exposing individuals to privacy violations if proper safeguards are not implemented.

Moreover, the application of Transformer models in legal systems demands careful consideration of consent mechanisms, data anonymization practices, and the transparency of AI decision-making processes. The adaptability of the Transformer model, highlighted by innovations in positional encoding and normalization techniques, must be matched with equally robust ethical frameworks and privacy-preserving mechanisms to mitigate risks and uphold the integrity of legal adjudications.

5 Training Large Language Models

5.1 *Architectural Foundations and Pre-training*

5.1.1 Evolution of Pre-training Architectures: From RNNs to Transformers

The architecture of language models has undergone a substantial evolution, beginning with Recurrent Neural Networks (RNNs) which laid the groundwork for understanding sequential data. The introduction of Long Short-Term Memory (LSTM) networks, a special kind of RNN, marked a leap forward in the ability to capture long-range dependencies in text. However, the real breakthrough came with the development of the Transformer architecture, which abandoned recurrence in favor of self-attention mechanisms, enabling parallel processing of sequences and significantly improving performance on NLP tasks [5, 17].

5.1.2 Pre-training Objectives: Masked Language Modeling, Autoregressive Prediction

Pre-training objectives form the core of how language models learn to understand and generate text. Two primary objectives dominate: masked language modeling (MLM) and autoregressive prediction. MLM, introduced with BERT, involves hiding parts of the input text and training the model to predict them. This is effective in learning bidirectional context. In contrast, autoregressive models like GPT predict each next token in a sequence given all the previous tokens, fostering strong generative capabilities [9, 18].

5.1.3 Advancements in Pre-training: Attention Mechanisms, Sparsity, and Parallelism

Advancements in pre-training have largely been driven by the introduction of sophisticated attention mechanisms that allow models to weigh the importance of different parts of the input data. The Transformer model employs multi-head self-attention to process various parts of the input sequence in parallel, enhancing learning efficiency and model depth [5]. Innovations such as sparse attention mechanisms have allowed models like GPT-3 to scale up to hundreds of billions of parameters by efficiently handling longer sequences without attending to every single token [17]. Moreover, parallelism in both the attention and feed-forward layers has been leveraged to speed up training while maintaining or even improving model performance [9, 30].

5.2 Strategies for Fine-Tuning LLMs

5.2.1 Fine-Tuning Versus Full Model Training: Efficiency and Efficacy

The debate between fine-tuning pre-trained models versus training models from scratch hinges on efficiency and efficacy. Presented in Fig. 4, Fine-tuning involves adjusting a pre-existing model's weights slightly to adapt to specific tasks, conserving resources and time compared to full model training. This strategy has proven effective, as shown in models like T5, which adapt to various NLP tasks through fine-tuning, demonstrating the versatility of transformer-based architectures [18]. Full model training, however, can be advantageous for highly specialized tasks where pre-trained models may not provide a strong foundational knowledge base.

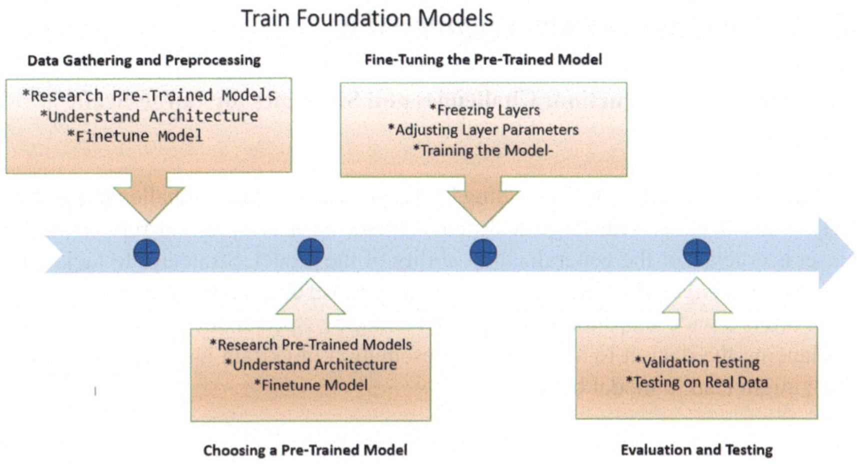

Fig. 4 Foundations models

5.2.2 Instruction Tuning: Enhancing Model Responsiveness to User Intent

Instruction tuning is a fine-tuning approach where models are specifically trained to follow user-given instructions, thus enhancing their responsiveness to user intent. This approach has been particularly successful with models like GPT-3, where fine-tuning with instruction-following prompts improves the model's ability to perform tasks as specified by the user, without the need for extensive task-specific data [9, 31, 32]. Instruction tuning aims to create models that are not just powerful in language understanding but also versatile in task execution.

5.2.3 Safe and Ethical Responses: Controlling Toxicity and Bias

As LLMs become more pervasive, ensuring their outputs are safe and ethical is paramount. Fine-tuning approaches have been developed to minimize toxicity and bias in model responses. Strategies such as targeted data augmentation, adversarial training, and controlled generation techniques are employed to steer models away from generating harmful content [33, 34]. For instance, models are fine-tuned on datasets specifically designed to promote ethical output or to recognize and avoid toxic language. This fine-tuning process is crucial in deploying LLMs in real-world applications where the quality and safety of the output are critical.

The strategies for fine-tuning LLMs are critical for their application in real-world scenarios. They ensure that the models are not only powerful and efficient but also behave in a way that aligns with user intent and societal norms. Fine-tuning also represents a balance between computational efficiency and model performance, making it a cornerstone of modern LLM deployment.

5.3 Dataset Curation and Preprocessing

5.3.1 Dataset Construction: Challenges and Strategies for Large-Scale Data

The construction of datasets for training LLMs presents numerous challenges, particularly when dealing with large-scale data. Ensuring a diverse and representative dataset is crucial for the generalization ability of the model. Strategies to tackle the scale and diversity of data include sourcing from broad corpora like Common Crawl, which offers a wide snapshot of the internet text [35]. Special attention is also given to balancing the dataset to avoid overrepresentation of certain languages or topics, which might lead to model bias.

5.3.2 Preprocessing Techniques: Tokenization, Sampling, and Vocabulary Selection

Preprocessing is a critical step in preparing data for LLM training. It involves tokenization, where text is broken down into manageable pieces; sampling, which ensures the model is exposed to a wide variety of text; and vocabulary selection, which impacts the model's ability to represent different words. Techniques like Sentence Piece have been utilized for tokenization, providing a flexible and language-agnostic approach [36]. The choice of vocabulary size is a trade-off between model complexity and coverage, with larger vocabularies offering finer granularity at the cost of increased model size.

5.3.3 Quality Assurance: Data Filtration and Deduplication

Quality assurance in dataset curation involves filtering out low-quality or irrelevant data and deduplicating content to prevent the model from overfitting to repetitive patterns. Sophisticated data filtration systems have been developed, such as the one used for the Chinese language model Yuan 1.0, which employs coarse and fine filtering techniques to process internet-sourced text [37]. Deduplication ensures that the model is trained on unique examples, which is essential for learning a wide range of linguistic phenomena without wasting computational resources on redundant data.

Curating and preprocessing datasets for LLMs is a complex task that requires careful consideration of the data sources, the methods used to process the text, and the techniques employed to ensure the quality of the training data. These steps are crucial to ensure that LLMs are trained efficiently and can generalize well across different tasks and domains.

5.4 Challenges in Scaling up Models

5.4.1 Computational Demands: Managing Energy and Hardware Constraints

Scaling up language models to improve performance also scales up their computational demands. As models like GPT-3 with 175 billion parameters emerge [9], the energy and hardware requirements for training and inference skyrocket. This not only requires more powerful and specialized hardware but also poses significant challenges in terms of energy consumption and heat dissipation. Techniques such as model parallelism, where the model is split across different GPUs or TPUs, and mixed-precision training, which uses lower-precision arithmetic for certain calculations, are employed to manage these constraints.

5.4.2 Environmental Considerations: Reducing Carbon Footprint

The environmental impact of training large-scale LLMs is a growing concern. The carbon footprint associated with the energy consumption of these models can be substantial. Researchers and organizations are exploring ways to mitigate this, such as using renewable energy sources for data centers and optimizing the efficiency of algorithms to reduce the total compute required [38]. Additionally, strategies like pruning, quantization, and knowledge distillation allow for smaller, more efficient models that retain much of the performance of their larger counterparts.

5.4.3 Model Bias and Catastrophic Forgetting: Mitigation Strategies

As LLMs scale, they become more prone to embedding biases present in their training data and forgetting previously learned information—a phenomenon known as catastrophic forgetting. To mitigate bias, techniques such as careful curation of training data, adversarial training, and bias evaluation benchmarks are used [33]. Catastrophic forgetting is particularly challenging when models are continuously updated; strategies like elastic weight consolidation, which protects previously learned knowledge, and rehearsal methods, which re-introduce old data during new learning phases, help in preserving earlier learning [39].

Scaling up models presents multifaceted challenges that extend beyond mere technicalities of machine learning. Addressing these challenges requires a concerted effort to balance the trade-offs between model performance and the broader impacts on society and the environment.

5.5 *Optimization and Efficiency in Training*

5.5.1 Training Efficiency: Sparse Activation and Memory Optimization

Enhancing training efficiency is crucial for scaling LLMs, with sparse activation and memory optimization techniques playing key roles. Sparse activation, as seen in models like GPT-3, allows the model to focus computation on a subset of neurons at any given time, significantly reducing the computational load without compromising performance [9, 30]. Memory optimization techniques, such as gradient checkpointing and offloading parts of the model to CPU during training, enable training larger models on hardware with limited memory, making efficient use of available resources.

5.5.2 Parallel Training and Distribution: Techniques for Acceleration

Parallel training and distribution strategies are essential for accelerating the training process of LLMs. Techniques such as data parallelism, where the dataset is split across multiple processors to train in parallel, and model parallelism, dividing the model's layers or parameters across different processors, are widely used. Distributed training frameworks also employ pipeline parallelism, where different stages of the model are trained on different processors sequentially. These approaches help in scaling training processes to accommodate very large models like GPT-3 and beyond [9, 30].

5.5.3 Hyperparameter Tuning at Scale: Balancing Performance and Costs

Hyperparameter tuning at scale is a delicate balance between achieving optimal model performance and managing computational costs. Automated machine learning (AutoML) techniques and Bayesian optimization are increasingly utilized to efficiently explore the hyperparameter space. For massive models like GPT-3, selecting the right set of hyperparameters—learning rates, batch sizes, and regularization terms—is critical and can significantly impact training time and model quality. Techniques to estimate the optimal hyperparameters, such as using the gradient noise scale, help in making informed decisions that balance performance and resource use [9, 38].

Optimizing the training of LLMs involves a suite of techniques aimed at making the process as efficient and effective as possible. These optimizations ensure that the growing computational demands of larger models do not become prohibitive, enabling continued innovation and application of these powerful tools in various domains.

5.6 *Model Evaluation and Benchmarking*

5.6.1 Performance Metrics for Pre-training and Fine-Tuning

Evaluating the performance of Large Language Models (LLMs) during both pre-training and fine-tuning phases is crucial for understanding their capabilities and limitations. Performance metrics vary depending on the task but commonly include accuracy, perplexity (for language models), F1 score, and BLEU score for translation tasks. For pre-trained models like BERT and GPT-3, metrics such as GLUE and SuperGLUE benchmarks provide a comprehensive assessment by aggregating results across multiple NLP tasks [9, 18]. These metrics help in gauging how well models understand language and generate text, guiding further improvements.

5.6.2 Benchmarking Across Domains: From Code Generation to Scientific Reasoning

Benchmarking LLMs across various domains, from code generation like Codex [40] to scientific reasoning as demonstrated by models trained on specialized datasets like Galactica [41], is essential for assessing their versatility and domain adaptability. Domain-specific benchmarks, such as HumanEval for code generation and various scientific knowledge tests, offer insights into a model's ability to generalize knowledge and apply it in contextually appropriate ways. These benchmarks reveal the strengths and weaknesses of models in specialized applications, informing future training and development efforts.

5.6.3 Comparative Analysis: Insights from Large-Scale Model Evaluations

Comparative analysis of LLMs through large-scale evaluations sheds light on the impact of model architecture, size, and training data on performance. Insights from comparing models like GPT-3, T5, and BERT across standard benchmarks highlight the importance of model design choices and training strategies. For example, the increase in parameters tends to improve performance, but gains vary across tasks [9, 18, 37]. Such analyses also explore the efficiency of different models, considering both their computational requirements and environmental impact, guiding the development towards more sustainable AI practices.

Model evaluation and benchmarking are integral to the iterative process of improving LLMs. By systematically assessing models across a range of metrics and domains, researchers and practitioners can identify the most effective strategies for model development, ensuring that LLMs continue to advance in capability and efficiency.

5.7 Future Directions in Training LLMs

5.7.1 Ethical and Responsible AI: Incorporating Values in Training

As LLMs become more integrated into society, the imperative to align their operations with ethical and responsible AI principles grows. Incorporating values into training involves not just mitigating biases or preventing harmful outputs but also ensuring that models can understand and reflect ethical considerations in their responses. This entails developing training regimes that prioritize fairness, transparency, and accountability. For example, efforts to incorporate ethical reasoning capabilities into models, as well as frameworks for ethical decision-making, are underway, aiming to make LLMs safer and more aligned with human values [33, 34].

5.7.2 Lifelong Learning Models: Towards Continuous Adaptation

Lifelong learning models represent a frontier in LLM development, aiming to enable models to continuously learn and adapt to new information without forgetting previously acquired knowledge—a challenge known as catastrophic forgetting. This involves creating models that can dynamically update their knowledge base, akin to human learning, allowing them to stay current with emerging trends, languages, and information. Techniques such as dynamic data sampling, model regularization, and memory replay are being explored to facilitate continuous learning, moving towards AI that can grow and evolve over time [30, 39].

5.7.3 Beyond Text: Preparing for Multimodal LLM Training

The future of LLM training is not confined to text but extends to understanding and generating content across multiple modalities, including images, audio, and video. Multimodal LLMs are being developed to process and integrate information from various sources, enabling richer and more context-aware AI applications. This requires not only novel model architectures that can handle diverse data types but also large-scale multimodal datasets for training. Efforts in this direction aim to create more versatile and capable AI systems that can understand and interact with the world in ways that mirror human cognitive abilities [9, 37].

The trajectory of LLM training is steering towards more ethical, adaptable, and comprehensive models. These future directions not only promise to enhance the capabilities and applications of LLMs but also address the ethical and societal challenges posed by AI, paving the way for more responsible and beneficial AI technologies.

6 Advanced Topics in LLMs

6.1 Fine-Tuning and Transfer Learning

Exploring the advanced realm of LLMs necessitates a deep dive into fine-tuning and transfer learning—two pivotal strategies that harness the power of pre-trained models for enhanced task performance [42–44]. This exploration deviates from traditional elaborations, incorporating a structured approach, illustrative diagrams, and Python code examples to enrich understanding.

6.1.1 Difference Between Fine-Tuning and Transfer Learning

Redefining the Approach

Transfer learning and fine-tuning serve as the cornerstone techniques for adapting pre-trained models to new, often more specialized, tasks. Their application extends beyond mere parameter adjustment, embodying the essence of modern AI's adaptability.

Conceptual Framework

- Transfer Learning: Embarks from a pre-trained model, repurposing its learned representations for a related but distinct task. It's akin to a seasoned chef applying their culinary skills to a new cuisine.
- Fine-Tuning: Advances this notion by meticulously adjusting the model's weights, refining its capability to excel in the specific nuances of the new task. It resembles sharpening a knife to ensure it meets the unique requirements of slicing different ingredients.

Illustrative Diagram

Consider a diagram that visually segregates the two processes:

1. Transfer Learning: Showcases a model with its core layers intact, highlighting the addition of new, task-specific layers atop the original architecture.
2. Fine-Tuning: Illustrates the subsequent phase where select layers of the model are unfrozen and adjusted to better align with the target task's data.

Practical Implementation: Python Code Snippets

- Transfer Learning Example:

```
from tensorflow.keras.applications import ResNet50
from tensorflow.keras.layers import GlobalAveragePooling2D, Dense
from tensorflow.keras.models import Model

# Load ResNet50 pre-trained on ImageNet, without the top classification
layer
base_model = ResNet50(weights='imagenet', include_top=False)

# Freeze the convolutional base
base_model.trainable = False

# Add custom layers
x = base_model.output
x = GlobalAveragePooling2D()(x)
x = Dense(1024, activation='relu')(x)
predictions = Dense(num_classes, activation='softmax')(x)

# Compile the model
model = Model(inputs=base_model.input, outputs=predictions)
```

– Fine-Tuning Example:

```
# Unfreeze the base model
base_model.trainable = True

# Fine-tune from this layer onwards
fine_tune_at = 100

# Freeze all layers before the 'fine_tune_at' layer
for layer in base_model.layers[:fine_tune_at]:
    layer.trainable = False

# Compile the model
model.compile(optimizer='adam', loss='categorical_crossentropy', met-
rics=['accuracy'])

# Continue training
model.fit(train_images, train_labels, epochs=5, valida-
tion_data=(val_images, val_labels))
```

Synthesizing Insights

Fine-tuning and transfer learning are not merely procedural; they encapsulate the strategic reapplication and refinement of knowledge. These methodologies facilitate the efficient deployment of LLMs across varied domains, from natural language understanding to complex image recognition tasks, underscoring the transformative potential of adaptable AI.

6.1.2 Advantages and Disadvantages of Transfer Learning and Fine-Tuning

Advantages

1. Reduced Training Time and Resource Requirements: Transfer learning significantly cuts down the need for extensive training from scratch, conserving both time and computational resources. It leverages pre-learned patterns, making the development of models for new tasks more efficient [45].
2. Improved Performance with Limited Data: Leveraging pretrained models allows for enhanced model performance on new tasks, even when faced with limited labeled data. The broad learning from massive datasets enables these models to adapt effectively to specific tasks [46].
3. Generalization and Transferability: The essence of transfer learning lies in its ability to generalize knowledge from one domain to another, facilitating the application of learned patterns to diverse tasks and thereby enhancing model versatility [47].
4. Accessible State-of-the-Art Performance: Pretrained models democratize access to cutting-edge AI, enabling practitioners without extensive computational resources or expertise to achieve state-of-the-art results in their respective domains [48].

Disadvantages

1. Domain Mismatch: Challenges arise when the domain of the pretrained model significantly diverges from the target task, potentially degrading performance due to the mismatch in domain-specific characteristics [49].
2. Overfitting Risk: Fine-tuning models on small, task-specific datasets heightens the risk of overfitting, where models perform well on training data but poorly generalize to unseen data [45].
3. Limited Interpretability: The complexity and depth of pretrained models often obscure their decision-making processes, presenting hurdles in model understanding and debugging [50].

Practical Examples

- Image Classification: Models such as VGG, ResNet, and Inception, fine-tuned on specific image classification tasks, showcase remarkable accuracy improvements with minimal additional training [51].

- NLP: Transformer models like BERT and GPT have been fine-tuned for a range of NLP tasks, setting new benchmarks in performance across sentiment analysis, entity recognition, and more [13].
- Autonomous Vehicles: Transfer learning is instrumental in the development of autonomous driving technologies, where models pretrained on extensive driving datasets are fine-tuned for specific environmental conditions [52].

Latest Work and Advancements

The landscape of transfer learning and fine-tuning is rapidly evolving, marked by innovations such as:

- Self-supervised Pretraining: Techniques like SimCLR and BYOL that leverage unlabeled data for pretraining enrich models with a broader understanding of data patterns, reducing reliance on large labeled datasets [53].
- Domain Adaptation: This strategy aims to minimize the gap between the source and target domains, enhancing the applicability of pretrained models to new tasks with limited labeled data [53].
- Multimodal Transfer Learning: Extending the principles of transfer learning to tasks involving multiple data types (e.g., text, images, audio) broadens the applicability and effectiveness of models across various multimedia tasks [54].

These advancements underscore the dynamic nature of transfer learning and fine-tuning, highlighting their critical role in the ongoing development and application of more adaptable, efficient, and capable machine learning models.

6.2 Addressing Bias and Ethical Considerations

The development and deployment of LLMs have ushered in transformative capabilities in artificial intelligence, coupled with significant ethical considerations. Addressing these considerations is paramount to ensuring that LLMs align with human values, rights, and societal well-being [54, 55].

6.2.1 Ethical Concerns in the Development and Application of LLMs

Control and Impact Assessment: The open-ended capabilities of LLMs pose challenges in control and assessment. The distinction between LLMs and Human-Level AI blurs, raising concerns over their governance and the prevention of misuse in applications like spam generation, fake news dissemination, and more.

Economic Exploitation in Model Training: The reliance on crowdworkers for dataset labeling, critical for LLM training, has raised issues of wage inequality and exploitation. This situation underscores the need for ethical labor practices within the "technoeconomy."

Language Bias and Lack of Diversity: LLMs face challenges in controlling generated text and ensuring diversity in training data. Predominantly English datasets risk propagating monocultural perspectives, sidelining other languages and cultures.

6.2.2 Mitigating Bias and Ensuring Ethical Deployment

Addressing the ethical implications of LLMs requires a multifaceted approach:

- Diverse and Representative Training Data: To mitigate bias and enhance fairness, it's crucial that LLM training data cover a broad spectrum of perspectives without inheriting societal biases.
- Transparency in Model Development: Implementing methods that enhance the interpretability of LLMs fosters trust and facilitates accountability in how decisions are generated.
- Ongoing Bias Detection and Correction: Continuous monitoring of LLM outputs for bias, coupled with the implementation of corrective measures, is vital for ethical deployment.
- Ethical Guidelines and Standards: The development and adherence to comprehensive ethical guidelines ensure that LLMs contribute positively to human discourse and decision-making, aligning with societal values.

Illustrative Example: Bias in LLMs

Consider an LLM trained on historical texts with skewed representations of certain demographics. Without appropriate bias mitigation strategies, the model might reproduce these biases, generating discriminatory outputs. To address this, developers can utilize debiasing techniques during training and actively monitor outputs for bias indications, making necessary adjustments to the model.

6.2.3 Strategies for Ethical AI Development

The path to ethically aligned LLMs involves proactive measures across all stages of AI development:

Purposeful AGI with Safeguards: Clearly defining the goals of AGI and embedding safeguards are essential to ensure that LLMs do not pose unforeseen societal risks.

Improving Labor Practices in Dataset Creation: Legislative frameworks at national and international levels are needed to protect the rights of workers involved in dataset creation, ensuring fair treatment and ethical labor conditions.

Promoting Linguistic and Cultural Diversity: Moving towards the development of multiple, context-specific models can help address biases towards a single culture, fostering a more inclusive representation of global values and perspectives.

6.2.4 Collaborative Ethical Framework

Solving the ethical challenges posed by LLMs requires collaboration across disciplines. Engineers, computer scientists, philosophers, and social scientists must unite to navigate these concerns, ensuring that AI development is guided by ethical principles and contributes to a future where technology amplifies human potential without compromising ethical standards.

By embedding ethical considerations into every phase of LLM development, we can create technologies that are not only technologically advanced but also socially responsible, reflective of the diverse spectrum of human values and cultures.

6.3 Exploring Multimodal Models

Multimodal LLMs represent an evolutionary leap in artificial intelligence, combining various types of data to enhance model understanding and interaction. Unlike traditional models that process a single data type, multimodal LLMs can interpret and relate information across text, images, videos, and even sensor data, offering a more holistic approach to machine learning tasks [56].

6.3.1 Understanding Multimodal Inputs

At the core of multimodal LLMs is the transformer model, renowned for its capacity to process sequential data through embeddings—numerical vectors representing input data. These embeddings encapsulate semantic and grammatical features, allowing the model to predict subsequent tokens in a sequence. Multimodal LLMs extend this mechanism to encode not just text but also other data types into a unified encoding space, enabling a single model to understand and process diverse inputs seamlessly.

6.3.2 Development Approaches and Innovations

Multimodal LLMs have been the subject of extensive research, leading to innovative models like Microsoft's Kosmos-1 and Google's PaLM-E. These models demonstrate the potential of multimodal LLMs to perform tasks beyond the reach of text-only models, such as image captioning, visual question answering, and even controlling robots.

- Kosmos-1: Microsoft's approach involves a transformer architecture that processes multiple data types through a single embedding module. Kosmos-1, with its 1.6 billion parameters, is trained on a variety of data, including text, images, and their combinations, showcasing remarkable capabilities in language understanding, image classification, and non-verbal reasoning tasks.

- PaLM-E: Developed by Google and TU Berlin, PaLM-E is designed as an "embodied multimodal language model," capable of integrating continuous inputs from an embodied agent's sensors. This model excels in tasks requiring sequential decision-making, such as robotic manipulation and task planning, demonstrating impressive generalization capabilities and data efficiency.

6.3.3 Challenges and Potential of Multimodal LLMs

While multimodal LLMs offer promising advancements, they also present unique challenges. The integration of multimodality in AI systems is complex, given the need to reconcile different data types into a coherent model architecture. Moreover, the comparison to human intelligence, where multimodality is intrinsically tied to bodily and sensory experiences, highlights potential limitations in AI's understanding and interaction with the world.

6.3.4 Future Directions and Conclusion

The exploration of multimodal LLMs is paving the way for AI systems that more closely mimic human cognitive processes, capable of understanding and interacting with the world in a more integrated and holistic manner. Future developments will likely focus on enhancing the compatibility of these models with human intelligence, improving their scalability, and addressing the challenges of data diversity and representation.

Multimodal LLMs stand at the forefront of AI research, offering a glimpse into a future where AI can understand and interact with the world across multiple dimensions. As these models evolve, they promise to unlock new possibilities in AI applications, from enhancing human–computer interaction to advancing robotics and beyond, marking a significant step towards more sophisticated and capable AI systems.

7 Applications and Case Studies

7.1 Natural Language Understanding and Generation

The integration of Natural Language Understanding (NLU) and Natural Language Generation (NLG) enables the development of advanced AI applications, including chatbots, virtual assistants, and automated content creators. NLU empowers computers to comprehend human language in its spoken or written form, deciphering intent, and context. On the other hand, (NLG) focuses on converting data into coherent, human-like text, crafting sentences and phrases that resonate with natural human communication [57, 58].

(NLU) systems are designed to interpret the intent and meaning behind the words, considering the context and nuances of language. This process involves various tasks such as sentiment analysis, entity recognition, and language identification, enabling machines to comprehend queries, instructions, or any textual content. The goal of NLU is to convert unstructured language data into a structured form that a machine can understand and act upon.

(NLG) systems can generate reports, write poetry, or compose responses in dialogue systems, mimicking human-like articulation of ideas and information. This requires understanding the rules of language syntax, semantics, and the context in which the communication occurs (Table 3).

Table 3 Applications and case

Application area	Use case	Case study	Key resources	Description
Customer service	Chatbots and virtual assistants	Domino's pizza	Chatbot platforms, NLU algorithms	Businesses employ chatbots powered by NLU to handle customer queries and orders, enhancing service availability and efficiency
Content generation	Automated journalism	The associated press	NLG tools, data analytics	Media outlets use NLG to automatically generate news stories from structured data, improving content production efficiency
Language translation	Machine translation	Google translate	Machine learning models, bilingual corpora	NLU and NLG enable accurate translations across languages, facilitating global communication and access to information
Personalized recommendations	E-commerce and streaming services	Netflix	Recommendation algorithms, user data analysis	Streaming services utilize NLU to tailor content recommendations based on user preferences and behaviors, enhancing user experience

(continued)

Table 3 (continued)

Application area	Use case	Case study	Key resources	Description
Educational tools	Language learning applications	Duolingo	Language processing algorithms, feedback systems	Language learning apps leverage NLU to provide real-time feedback and personalized learning experiences, making language education more interactive
Healthcare	Clinical documentation	IBM Watson	Medical NLU systems, clinical data analysis	NLG systems generate clinical documents from patient data, aiding in efficient and accurate medical record-keeping and decision-making
Financial services	Fraud detection	PayPal	Transaction analysis, pattern recognition algorithms	Financial platforms analyze communication and transactional data using NLU to detect and prevent fraud, enhancing security

7.2 Real-World Applications of LLMs

The advent of LLMs has marked a paradigm shift in artificial intelligence (AI) and natural language processing (NLP), facilitating a spectrum of real-world applications that leverage their advanced understanding, generation, and interaction capabilities with human language.

From creative writing and content generation to software development, customer service, and beyond, Lms have been transformative.

The realm of **Creative Writing and Content Generation**, LLMs such as OpenAI's GPT series catalyze the creative process by aiding in the generation of stories, poems, music lyrics, and more. They produce coherent and contextually relevant content, assisting writers and content creators in overcoming creative blocks and enhancing the richness of their work [59, 60]. Also, Research and Scientific Discovery benefit from LLMs' able to navigate through vast scientific literature, aiding in hypothesis generation, study summarization, and the drafting of research papers [61].

Software development, tools such as GitHub Copilot, powered by Codex, have shown remarkable proficiency in understanding and generating code, suggesting code snippets, automating documentation, and facilitating debugging to significantly enhance developer productivity. Similarly, in customer service, the deployment of LLMs in chatbots and virtual assistants has transformed the sector by providing round-the-clock, human-like support, reducing the need for constant human oversight while upholding high service standards [62].

The world of **Business Intelligence and Market Analysis**, LLMs analyze data to furnish businesses with actionable insights. They automate the generation of reports and can forecast trends, aiding strategic decision-making, and offering a competitive edge in the dynamic business landscape [59].

7.3 Future Directions and Potential

The rapid advancement of Large Language Models (LLMs) as pointed in Fig. 5, is setting the stage for their pivotal role in future scientific research and complex problem solving, with potential to significantly accelerate progress in fields ranging from climate change to healthcare. LLMs are expected to contribute to the discovery of new materials, drugs, and treatments, analyze large-scale environmental data, and offer solutions to critical global challenges [63].

Moreover, enhancing human-AI collaboration is envisioned as a key area of development, where LLMs could serve as co-creators in arts and sciences, research partners, and aid in decision-making across various sectors. The focus on improving intuitive interactions between humans and LLMs will be crucial in realizing this collaborative potential, opening new avenues for innovation and problem-solving.

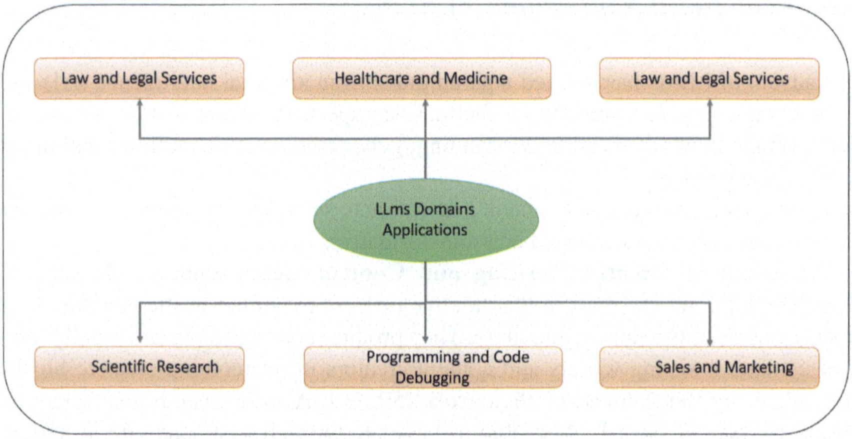

Fig. 5 LLM domains of application

8 Challenges and Limitations

The advent of LLMs has opened new frontiers in natural language processing, offering unprecedented capabilities in generating human-like text, understanding complex queries, and facilitating conversational AI systems. However, despite their impressive achievements, LLMs face significant challenges and limitations that may hinder their path toward achieving Artificial General Intelligence (AGI) and broader applicability [64]. Here, we delve into the core challenges and limitations confronting LLMs today.

8.1 Computational and Environmental Costs

LLMs demand substantial computational resources for training, often requiring millions of dollars in investment, thousands of compute hours, and significant energy consumption. This not only incurs high financial costs but also poses considerable environmental concerns. The training of models like GPT-3 has been associated with significant carbon emissions and water usage, raising questions about the sustainability of developing ever-larger models. Efforts to address these issues include optimizing training efficiency, adopting model parallelism, and exploring more energy-efficient algorithms. Yet, the balance between advancing model capabilities and minimizing environmental impact remains a pressing challenge [65].

8.2 Handling Ambiguity and Contextual Nuances

Despite their linguistic prowess, LLMs often struggle with ambiguity and the nuances of context-dependent language. Tokenization issues, such as handling non-familiar vocabulary or different spacing in multilingual environments, can lead to unexpected model responses. Additionally, LLMs' limited context length restricts their ability to interpret and analyze longer text segments comprehensively. Strategies like positional embedding schemes and efficient attention mechanisms offer potential solutions, yet the challenge of effectively capturing and processing nuanced contextual information persists [64].

8.3 Reliability and Misuse Concerns

The reliability of LLMs is compromised by several factors, including their susceptibility to adversarial attacks, the potential for biased outputs, and the occurrence of information hallucination. Adversarial tactics, such as prompt injection or data poisoning, can manipulate model outputs, while biases in training data can lead to discriminatory or unfair responses. Information hallucination, where models generate nonsensical or unfaithful text, further undermines reliability. Addressing these concerns requires robust mechanisms for detecting and mitigating bias, enhancing model security, and improving the interpretability of LLM outputs [66].

8.4 Additional Considerations

- Training Data Requirements: The vast datasets required for LLM pre-training introduce risks of data duplication, privacy breaches, and biased model behavior. Ensuring data quality and diversity is crucial for mitigating these risks [67–60].
- Fine-Tuning LLMs: Fine-tuning Large Language Models (LLMs) for specialized tasks by further training on task-specific datasets is a valuable strategy [13]. This process, however, demands substantial memory and computational power for the storage of model gradients, parameters, and activations, as well as the fine-tuned models themselves. Such requirements limit accessibility to a select number of organizations. To mitigate these challenges, Parameter-Efficient Fine-Tuning methods update only a fraction of the model's parameters. Techniques such as prefix fine-tuning [69], prompt tuning [70], and the use of adapters [71] represent efforts to reduce computational demands. Additionally, optimization strategies like Low-Rank Adaptation (LoRA) [72], LongLoRA [73], and QLoRA [74] aim to lower computational costs. Despite these advancements, the computational requirements of fine-tuning LLMs continue to pose significant obstacles.
- Inference Latency: The large size of LLMs leads to high inference latency, affecting real-time applications. Techniques like model quantization, pruning, and efficient attention can help reduce latency, but finding the right balance between performance and responsiveness remains a challenge [75–77].
- Knowledge Updating and Refinement: Keeping LLMs updated with the latest information is essential for maintaining their relevance and accuracy. Methods like model editing and accessing external knowledge sources are being explored, yet the challenge of continuously updating LLMs without extensive retraining is ongoing [78].
- Adversarial attacks: They present a significant security challenge to LLMs, manipulating their outputs through subtly modified inputs [79]. A prevalent form of such an attack is the text injection attack, where attackers embed meticulously crafted text snippets into the model's input. This manipulates the LLM into executing unintended actions. Instead of the "delete all files" command, consider an attacker injecting a command like "transfer funds" into a financial service's LLM, potentially causing unauthorized financial transactions [80]. Moreover, models that rely on visual prompts are not immune, suffering from corrupted inputs designed to derail their interpretation mechanisms [81].
- Ethical Considerations: LLMs raise significant ethical concerns related to privacy, data use, and the potential amplification of biases. Establishing ethical guidelines

and implementing responsible AI practices are imperative for mitigating these risks. These aspects were treated in a previous section.

As LLMs continue to evolve, addressing these challenges and limitations is paramount for their safe, responsible, and efficient deployment. Collaborative efforts among researchers, developers, ethicists, and policymakers are essential for navigating the complex landscape of LLM development and ensuring that these powerful tools contribute positively to society.

9 Conclusion

9.1 Summary of Key Insights

This exploration into LLMs has traversed the comprehensive landscape of their development, from foundational concepts and key architectural advancements to the challenges and ethical considerations that accompany their use. We've seen how neural networks and deep learning form the backbone of LLMs, enabling these models to understand and generate human-like text with an unprecedented level of sophistication. The evolution from simple statistical models to complex neural network-based models like the Transformer illustrates the rapid progress in the field of NLP.

We've delved into the mechanics of the Transformer model, highlighting its unique attention mechanisms and encoder-decoder structure, which have been instrumental in achieving significant breakthroughs in language understanding and generation. The training of these colossal models, including the challenges of scaling, data curation, and the fine-tuning strategies **that enhance their performance and applicability to specific tasks, has been scrutinized**.

Advanced topics such as addressing bias, ensuring ethical deployment, and exploring the potential of multimodal models have been discussed, showcasing the multifaceted nature of LLM development. Applications and case studies have demonstrated the transformative impact of LLMs across various domains, from creative writing and content generation to software development and customer service, highlighting their versatility and potential to revolutionize industries.

However, the journey of LLMs is not without its hurdles. Computational and environmental costs, the handling of ambiguity and contextual nuances, reliability and misuse concerns, and the ongoing struggle with bias and ethical deployment present significant challenges that need to be addressed as we continue to advance these models.

9.2 *The Road Ahead for LLMs*

Looking forward, the development of LLMs stands at a critical juncture. The path ahead involves not only technical advancements but also a deep consideration of the ethical implications and societal impacts of these models. The integration of ethical AI practices from the outset, the pursuit of environmentally sustainable training methodologies, and the continuous effort to mitigate bias and ensure fairness in LLM outputs are imperative.

The future of LLMs also points towards further exploration of multimodal capabilities, enabling AI to understand and interact with a broader range of human experiences beyond text. The potential for LLMs to enhance human-AI collaboration, serve as co-creators in various creative and scientific endeavors, and contribute to solving some of the world's most pressing challenges is immense.

Moreover, the development of models capable of lifelong learning, where LLMs can adapt and grow without the need for constant retraining, represents a promising direction. This adaptability would mark a significant step towards models that more closely mirror human learning processes, paving the way for more intuitive and effective human-AI interactions.

As we embark on this road ahead, the collaboration across disciplines—combining the expertise of technologists, ethicists, policymakers, and the broader community—will be crucial. Together, we can ensure that the advancements in LLMs not only push the boundaries of what's technically possible but also align with our collective values and contribute to a future where technology amplifies human potential without compromising our ethical standards. The journey of LLMs is far from complete, but the possibilities it heralds for the future of AI and society are boundless.

References

1. J. Clusmann et al., The future landscape of large language models in medicine. Commun. Med. **3**, 141 (2023). https://doi.org/10.1038/s43856-023-00370-1
2. Y.Y. Ghadi et al., Integration of federated learning with IoT for smart cities applications, challenges, and solutions. PeerJ Comput. Sci. **9**, e.1657. https://doi.org/10.7717/peerj-cs.1657
3. D. Jurafsky, J.H. Martin, Speech and Language Processing (2009)
4. S. Hochreiter, J. Schmidhuber, Long short-term memory. Neural Comput. (1997)
5. A. Vaswani et al., Attention is all you need. NIPS (2017)
6. M. Khadhraoui et al., Survey of BERT-base models for scientific text classification: COVID-19 case study. Appl. Sci. **12**(6), 2891 (2022). https://doi.org/10.3390/app12062891
7. J. Devlin et al., BERT: Pre-training of deep bidirectional transformers for language understanding. NAACL (2018)
8. A. Radford et al., Improving language understanding by generative pre-training (2018)
9. T.B. Brown et al., Language models are few-shot learners. Adv. Neural. Inf. Process. Syst. **33**, 1877–1901 (2020)
10. I. Goodfellow, Y. Bengio, A. Courville, *Deep learning* (MIT Press, 2016)
11. M.A. Nielsen, Neural networks and deep learning. Determination Press (2015)
12. F. Chollet, Deep learning with python. Manning Publications (2018)

13. J. Devlin et al., BERT: Pre-training of deep bidirectional transformers for language understanding (2018). Preprint at arXiv:1810.04805
14. S. Guefrachi, A. Echtioui, H. Hamam, Automated diabetic retinopathy screening using deep learning. Multimed. Tools Appl. (Q1) **81**, 18 (2024). https://doi.org/10.1007/s11042-024-18149-4
15. A. Iqbal et al., Sentiment analysis of consumer reviews using deep learning. Sustainability **14**(17), 10844 (2022). https://doi.org/10.3390/su141710844
16. L. Bendebane et al., A multi-class deep learning approach for early detection of depressive and anxiety disorders using twitter data. Algorithms **16**(12), 543 (2023). https://doi.org/10.3390/a16120543
17. R. Child, S. Gray, A. Radford, I. Sutskever, Generating long sequences with sparse transformers (2019). Preprint at arXiv:1904.10509
18. C. Raffel et al., Exploring the limits of transfer learning with a unified text-to-text transformer. J. Mach. Learn. Res. **21**(1), 5485–5551 (2020)
19. N. Patwardhan, S. Marrone, C. Sansone, Transformers in the real world: A survey on NLP applications. Information **14**, 242 (2023). https://doi.org/10.3390/info14040242
20. A. Rahali, M.A. Akhloufi, End-to-end transformer-based models in textual-based NLP. AI **4**, 54–110 (2023). https://doi.org/10.3390/ai4010004
21. Z. Dai et al., Transformer-xl: Attentive language models beyond a fixed-length context (2019). Preprint at arXiv:1901.02860
22. S. Jain, S. Wiegreffe, Y. Pinter, B.C. Wallace, Learning to faithfully rationalize by construction (2020). Preprint at arXiv:2005.00115
23. L. Liu, T. Li, C. Caramanis, High dimensional robust $ M $-estimation: Arbitrary corruption and heavy tails (2019). Preprint at arXiv:1901.08237
24. A. Haviv et al., Transformer language models without positional encodings still learn positional information (2022). Preprint at arXiv:2203.16634
25. X Chu et al., Conditional positional encodings for vision transformers (2021). Preprint at arXiv:2102.10882
26. J. Ma et al., Zero-shot neural passage retrieval via domain-targeted synthetic question generation (2020). Preprint at arXiv:2004.14503
27. F. Huang et al., Encoding recurrence into transformers. in *The Eleventh International Conference on Learning Representations* (2022)
28. Z. Liu, Q. Lv, Z. Yang, Y. Li, C.H. Lee, L. Shen, Recent progress in transformer-based medical image analysis. Comput. Biol. Med. 107268 (2023)
29. H. Surden, Artificial intelligence and law: An overview. Georg. State Univ. Law Rev. **35**, 19–22 (2019)
30. S. Black et al., Gpt-neox-20b: An open-source autoregressive language model (2022). Preprint at arXiv:2204.06745
31. H.W. Chung et al., Scaling instruction-finetune language models (2022). Preprint at arXiv:2210.11416
32. V. Sanh et al., Multitask prompted training enables zero-shot task generalization (2021). Preprint at arXiv:2110.08207
33. L. Ouyang et al., Training language models to follow instructions with human feedback. Adv. Neural Inf. Process. Syst. **35**(27), 730–744 (2022)
34. S. Iyer et al., Opt-IML: Scaling language model instruction meta learning through the lens of generalization (2022). Preprint at arXiv:2212.12017
35. L. Xue et al., mt5: A massively multilingual pre-trained text-to-text transformer (2020). Preprint at arXiv:2010.11934
36. Z. Zhang et al., CPM-2: Large-scale cost-effective pre-trained language models. AI Open **2**, 216–224 (2021)
37. S. Wu et al., Yuan 1.0: Large-scale pre-trained language model in zero-shot and few-shot learning (2021). Preprint at arXiv:2110.04725
38. S. McCandlish, J. Kaplan, D. Amodei, O.A.I.D. Team, An empirical model of large-batch training (2018). Preprint at arXiv:1812.06162

39. S. Yuan et al., Wudaocorpora: A super large-scale Chinese corpora for pre-training language models. AI Open **2**, 65–68 (2021)
40. M. Chen et al. Evaluating large language models trained on code (2021). Preprint at arXiv: 2107.03374
41. R. Taylor et al., Galactica: A large language model for science (2022). Preprint at arXiv:2211.09085
42. A. Kroon, K. Welbers, D. Trilling, W. van Atteveldt, Advancing automated content analysis for a new era of media effects research: The key role of transfer learning. Commun. Methods Meas. (2023). https://doi.org/10.1080/19312458.2023.2261372
43. Empirical Study of LLM Fine-Tuning for Text Classification in Legal Document Review. EDRM (2023). https://edrm.net/2023/12/empirical-study-of-llm-fine-tuning-for-text-classification-in-legal-document-review/
44. S. Minaee et al., Fine-tuning large language models: A survey (2023). https://arxiv.org/abs/2301.07521
45. R. Caruana, Learning transfer. in *Knowledge Transfer for Machine Learning*, (Springer, Boston, MA, 2009), pp. 1–46
46. S. Pan, Q. Yang, A survey on transfer learning. IEEE Trans. Knowl. Data Eng. **22**(10), 1345–1359 (2010)
47. D. Weiss et al., A survey of transfer learning. J. Artif. Intell. Res. **3**, 203–260 (2009)
48. J. Clark et al., Transformers: A unified model for sequence-to-sequence learning (2017). Preprint at arXiv:1706.03762
49. J. Blitzer et al., Domain adaptation with structural correspondence learning. in *Proceedings of the 27th International Conference on Machine Learning (ICML-10)*, (2010), pp. 120–127
50. Z.C. Lipton et al., An interpretable model for text classification with recurrent neural networks. in *Advances in Neural Information Processing Systems*, (2016), pp. 1615–1623
51. K. He et al., Deep residual learning for image recognition. in *Proceedings of the IEEE Conference on Computer Vision and Pattern Recognition*, (2016), pp. 770–778
52. M. Bojarski et al., End-to-end learning for self-driving cars. in *Proceedings of the IEEE Conference on Computer Vision and Pattern Recognition*, (2016), pp. 3166–3173
53. M. Brundage et al., The malicious use of artificial intelligence: Forecasting, prevention, and mitigation (2018). Preprint at arXiv:1802.07228
54. E.M. Bender, T. Gebru, (2021) On the dangers of stochastic parrots: Can language models be too big? (2021). Preprint at arXiv:2101.08237
55. W. Samek et al., Explainable artificial intelligence: Interpretability methods for machine learning (2019). Preprint at arXiv:1903.08759
56. J. Wu, W. Gan, Z. Chen, S. Wan, P. Yu, Multimodal large language models: A survey. in *2023 IEEE International Conference on Big Data (BigData)*, (Sorrento, Italy, 2023), pp. 2247–2256. https://doi.org/10.1109/BigData59044.2023.10386743
57. S.-Y. Su, C.-W. Huang, Y.-N. Chen, Dual supervised learning for natural language understanding and generation (2019). Preprint at arXiv:1905.06196
58. B.-H. Tseng et al., A generative model for joint natural language understanding and generation (2020). Preprint at arXiv:2006.07499
59. G. Franceschelli, M. Musolesi, On the creativity of large language models (2023). Preprint at arXiv:2304.00008
60. J. Kaddour, J. Harris, M. Mozes, H. Bradley, R. Raileanu, R. McHardy, Challenges and applications of large language models (2023). Preprint at arXiv:2307.10169
61. Q. He et al., Can large language models understand real-world complex instructions? (2023). Preprint at arXiv:2309.09150
62. C. Ziems et al., Can large language models transform computational social science (2023). Preprint at arXiv:2305.03514
63. S. Yao et al., Tree of thoughts: Deliberate problem solving with large language models. Adv. Neural Inf. Process. Syst. **36** (2024)
64. M.U. Hadi et al., Large language models: A comprehensive survey of its applications, challenges, limitations, and future prospects. TechRxiv (2023)

65. C.-J. Wu et al., Sustainable AI: Environmental implications, challenges and opportunitie. Proc. Mach. Learn. Syst. **4**, 795–813 (2022)
66. A. Lucchini, L. Sabatini, P. Spiga, A subexponential bound on the cardinality of abelian quotients in finite transitive groups. Bull. Lond. Math. Soc (2021)
67. J. Kaddour, The minipile challenge for data-efficient language models (2023). Preprint at arXiv: 2304.08442
68. J. Kaplan, A. Radford, J. Wu, D. Amodei et al., Scaling laws for neural language models (2020). Preprint at arXiv:2001.08361
69. X. L. Li, P. Liang, Prefix-tuning: Optimizing continuous prompts for generation (2021). Preprint at arXiv:2101.00190
70. H. Laurençon et al., The bigscience roots corpus: A 1.6 TB composite multilingual dataset. Adv. Neural. Inf. Process. Syst. **35**, 31809–31826 (2022)
71. N. Houlsby et al., Parameter-efficient transfer learning for NLP. in *International Conference on Machine Learning*, (PMLR, 2019), pp. 2790–2799
72. E.J. Hu et al., Lora: Low-rank adaptation of large language models (2021). Preprint at arXiv: 2106.09685
73. Y. Chen, S. Qian, H. Tang, X. Lai, Z. Liu, S. Han, J. Jia, Longlora: Efficient fine-tuning of long-context large language models (2023). Preprint at arXiv:2309.12307
74. T. Dettmers, A. Pagnoni, A. Holtzman, L. Zettlemoyer, Qlora: Efficient finetuning of quantized LLMS (2023). Preprint at arXiv:2305.14314
75. M. Pagliardini, D. Paliotta, M. Jaggi, F. Fleuret, Faster causal attention over large sequences through sparse flash attention (2023). Preprint at arXiv:2306.01160
76. S. Liu, Z. Wang, Ten lessons we have learned in the new "sparseland": A short handbook for sparse neural network researchers (2023). Preprint at arXiv:2302.02596
77. L. Chen, M. Zaharia, J. Zou, Frugalgpt: How to use large language models while reducing cost and improving performance (2023). Preprint at arXiv:2305.05176
78. A. Sedghi et al., Image registration: Maximum likelihood, minimum entropy and deep learning. Med. Image Anal. **69**, 101939 (2021)
79. F. Perez, I. Ribeiro, Ignore previous prompt: Attack techniques for language models (2022). Preprint at arXiv:2211.09527
80. Y. Liu et al., Prompt injection attack against LLM-integrated applications (2023). Preprint at arXiv:2306.05499
81. C. Zhang, C. Zhang, T. Kang, D. Kim, S.-H. Bae, I.S. Kweon, Attack-sam: Towards evaluating adversarial robustness of segment anything model (2023). Preprint at arXiv:2305.00866

Dual-Stage Global and Local Feature Framework for Image Dehazing

Anas M. Ali, Anis Koubaa, and Bilel Benjdira

Abstract Addressing the challenge of removing atmospheric fog or haze from digital images, known as image dehazing, has recently gained significant traction in the computer vision community. Although contemporary dehazing models have demonstrated promising performance, few have thoroughly investigated high-resolution imagery. In such scenarios, practitioners often resort to downsampling the input image or processing it in smaller patches, which leads to a notable performance degradation. This drop is primarily linked to the difficulty of effectively combining global contextual information with localized, fine-grained details as the spatial resolution grows. In this chapter, we propose a novel framework, termed the Streamlined Global and Local Features Combinator (SGLC), to bridge this gap and enable robust dehazing for high-resolution inputs. Our approach is composed of two principal components: the Global Features Generator (GFG) and the Local Features Enhancer (LFE). The GFG produces an initial dehazed output by focusing on broad contextual understanding of the scene. Subsequently, the LFE refines this preliminary output by enhancing localized details and pixel-level features, thereby capturing the interplay between global appearance and local structure. To evaluate the effectiveness of SGLC, we integrated it with the Uformer architecture, a state-of-the-art dehazing model. Experimental results on high-resolution datasets reveal a considerable

A. M. Ali (✉) · B. Benjdira
Robotics and Internet-of-Things Laboratory, Prince Sultan University, Riyadh, Saudi Arabia
e-mail: aaboessa@psu.edu.sa

B. Benjdira
e-mail: bbenjdira@psu.edu.sa

A. M. Ali
Department of Electronics and Electrical Communications Engineering, Faculty of Electronic Engineering, Menoufia University, Menouf, Egypt

A. Koubaa
College of Engineering, Alfaisal University, Riyadh, Saudi Arabia
e-mail: akoubaa@alfaisal.edu

B. Benjdira
SE & ICT Laboratory, LR18ES44, ENICarthage, University of Carthage, Tunis, Tunisia

A. Koubaa et al. (eds.), *Generative AI and Large Language Models: Opportunities, Challenges, and Applications*, Studies in Computational Intelligence 1214,
https://doi.org/10.1007/978-3-031-90573-5_4

improvement in peak signal-to-noise ratio (PSNR) when employing SGLC, indicating its potency in addressing haze in large-scale imagery. Moreover, the SGLC design is model-agnostic, allowing any dehazing network to be augmented with the proposed global-and-local feature fusion mechanism. Through this strategy, practitioners can harness both scene-level cues and granular details, significantly improving visual fidelity in high-resolution environments. Overall, this chapter provides an in-depth exposition of the SGLC framework, highlighting its fundamental concepts, implementation details, and empirical validation. Our findings underscore the importance of synergizing global and local features for effective high-resolution image dehazing, paving the way for future research directions and practical applications in domains where clarity and detail are paramount.

1 Introduction

Image dehazing, which aims to remove the atmospheric fog or haze present in digital imagery, has steadily attracted scholarly attention due to its direct impact on numerous computer vision applications. Tasks such as image classification, object detection, tracking, and semantic segmentation are particularly susceptible to performance degradation when haze is present in the scene. The fundamental challenge lies in handling the complex and non-homogeneous nature of haze particles, which causes uneven scattering and attenuation of light. Consequently, different regions of an image suffer from varying degrees of degradation and blur, rendering traditional computer vision pipelines less effective.

Early research endeavors in this field often concentrated on simpler scenarios of homogeneous haze [1–4]. However, realistic environments typically contain spatially varying haze distributions, thereby motivating the need for more sophisticated models. The widely accepted physical model for describing the propagation of light in hazy conditions can be formulated by:

$$I_h(x) = t(x)\, I_o(x) + \big(1 - t(x)\big) A(x), \tag{1}$$

where I_h denotes the observed hazy image, I_o is the latent (clean) scene, $t(x)$ is the medium transmission map capturing the haze-induced degradation, and $A(x)$ signifies the global atmospheric light [5–7]. The transmission map $t(x)$ is generally expressed as:

$$t(x) = e^{-\beta\, d(x)}, \tag{2}$$

where β is an atmospheric scattering parameter and $d(x)$ represents the scene depth as a function of pixel coordinates x.

Before the advent of deep learning, many studies relied on prior-based techniques [8–10] to approximate β, $A(x)$, and $t(x)$. Methods such as the dark channel prior and non-local prior were employed to derive estimations of the atmospheric scattering effects. However, these hand-crafted priors often fell short in handling the intricate dependencies between haze, depth, and environmental factors. The introduction of deep learning methods [11] brought about substantial improvements in haze removal performance. Subsequent models based on convolutional neural networks (CNNs) or transformer-based architectures demonstrated remarkable accuracy gains [12–18].

Despite these achievements, a significant limitation persists: state-of-the-art dehazing models are typically designed for small to medium input resolutions, making them computationally expensive or less effective when dealing with high-resolution images [19, 20]. Incorporating advanced network components (e.g., vision transformers) further escalates computational demands [21, 22], and attempts to balance performance and model efficiency pose ongoing challenges [23–25]. One of the central difficulties lies in effectively fusing global scene context with localized details in large-scale images. While low-resolution inputs can be processed in a single pass, high-resolution images often require downsampling or patch-wise processing. Such approaches risk losing crucial global features or fine-grained local details, thereby impairing the model's overall performance.

In this book chapter, we introduce the *Streamlined Global and Local Features Combinator* (SGLC) to enable more robust dehazing for high-resolution imagery. SGLC facilitates an end-to-end process that preserves both the global scene context and the local fine structure. Specifically, the method comprises two main components:

- **Global Features Generator (GFG):** Responsible for capturing broad, scene-level characteristics by operating on strategically subdivided patches from the high-resolution input.
- **Local Features Enhancer (LFE):** Dedicated to refining and augmenting localized details, thus ensuring the recovery of subtle edges, textures, and other critical high-frequency components.

Additionally, we introduce a Grid Patching process tailored for high-resolution images to capture their spatial variability more effectively. We also propose a customized loss function aimed at reinforcing the learning of high-frequency details, vital for visually pleasing and accurate dehazed outcomes. Finally, we confirm through comparative experiments that applying GFG prior to LFE aligns with the intuitive approach of first generating holistic context before focusing on fine detail enhancement.

Overall, this chapter addresses the persistent challenge of balancing global and local feature extraction in the dehazing of large-scale images. By systematically combining robust global scene understanding with fine-grained local refinements, the SGLC framework offers a scalable and computationally feasible pathway to high-resolution image dehazing, thereby broadening the applicability and impact of modern dehazing techniques.

2 Related Work

2.1 *Background on Image Dehazing*

Image dehazing, a critical pre-processing step for many computer vision tasks, has been extensively studied in both academic and industrial contexts. Early approaches generally used physical priors to estimate haze-relevant parameters. For instance, the dark channel prior [26] and the non-local prior [27] are classic techniques that derive statistics from hazy scenes to approximate the transmission map and atmospheric light. Other notable methods focused on modeling the interactions of light with haze particles via sophisticated atmospheric scattering equations [28, 29]. Although these hand-crafted strategies proved effective under certain assumptions, they often struggled to accommodate complex real-world conditions involving non-homogeneous and dense haze distributions.

With the advent of deep learning, data-driven dehazing models have rapidly gained traction and demonstrated superior performance. Early convolutional neural network (CNN) approaches [30, 31] replaced explicit priors with learned feature representations, allowing models to adapt to more diverse haze conditions. Further developments incorporated advanced architectures such as generative adversarial networks (GANs) [32, 33] and multi-scale designs [34–36] to handle varying levels of haze density and more complex scene structures. Moreover, several frameworks have pursued end-to-end training by integrating multiple loss terms (e.g., reconstruction loss, perceptual loss) to enhance image quality in both global appearance and fine details [37–40].

2.2 *High-Resolution Image Dehazing*

Although substantial progress has been made in single-image and multi-image dehazing for standard resolutions, the field has increasingly turned its attention to the challenges posed by large-scale, high-resolution imagery. When attempting to extend existing methods to high-resolution inputs, computational costs can become prohibitive, often necessitating either image downsampling or a patch-wise processing strategy. These approaches, however, risk discarding valuable global context and degrading local details.

Among the first attempts to specifically address the high-resolution dehazing problem, Sim et al. [33] proposed the Dehazing Generative Adversarial Network (DHGAN). Their method trains a generator on smaller hazy patches derived from downsampled input images, thereby reducing computational overhead and capturing essential global cues. To further strengthen the training process, they modified the cross-entropy loss to accommodate multiple outputs. Similarly, Ki et al. [32] introduced BEGAN (Boundary Equilibrium Generative Adversarial Network), which uses an enlarged receptive field and focuses on training the discriminator directly on high-resolution images, while conditioning the generation process on downscaled hazy counterparts.

In a related effort, Bianco et al. [41] presented HR-Dehazer (High-Resolution Dehazer), an encoder-decoder-based model combined with a specialized loss function that prioritizes semantic integrity and structural consistency. The network design is scale-invariant, enabling operation on large inputs without a considerable drop in accuracy. More recently, Zheng et al. [42] offered 4k-Dehazer, a model that employs three subnetworks working in tandem on a bilateral space. Each branch shares features bidirectionally, producing a robust global representation suitable for 4K images. To evaluate their system, the authors introduced a large 4K dataset, showing competitive results against contemporary methods.

Another strategy for high-resolution dehazing is illustrated by Chen et al. [37], who proposed H2RL-Net, a two-branch framework that separately addresses high-frequency and semantic information. By leveraging multi-resolution CNN streams along with a parallel cross-scale fusion (PCF) module, H2RL-Net dynamically integrates features across different scales, while the channel feature refinement (CFR) block recalibrates these features at the channel level. This division of labor aids in capturing both small-scale details and global context.

2.3 Limitations of Existing Approaches and Our Contribution

Although these high-resolution dehazing methods have reported notable gains, they generally rely on designing new architectures specifically tailored for large images. This tends to limit practitioners who may prefer to use well-established models designed for mid-resolution dehazing tasks. Moreover, most of the aforementioned techniques integrate the extraction of global and local features in a parallel manner, whether through bilateral latent spaces or concurrent multi-scale networks. While such parallel designs have proven effective, they can be computationally heavy and less modular for incremental improvements.

In this chapter, we address the gap between existing high-resolution dehazing methods and the extensive body of CNN- or transformer-based dehazing models. We propose a sequential, streamlined approach called *Streamlined Global and Local Features Combinator (SGLC)* to preserve both global scene semantics and fine-grained local details without demanding substantial architecture overhauls. Specifically, Our framework centers on two key modules. The first, termed the **Global Features Generator (GFG)**, processes large images via a Grid Patching mechanism to capture essential global statistics and contextual cues. The second, called the **Local Features Enhancer (LFE)**, refines the resulting representations by restoring high-frequency details and enhancing local structure, thereby boosting overall perceptual quality.

The sequential nature of SGLC simplifies the evaluation and potential refinement of each stage independently, offering a flexible and extensible architecture. By decoupling global feature generation from local feature enhancement, our method avoids common pitfalls associated with patch-wise downsampling and large memory footprints. Consequently, SGLC serves as a generic and versatile framework that can be integrated into existing state-of-the-art dehazing models, extending their applicability to high-resolution settings with minimal modifications and enhanced performance.

3 Proposed Methodology

In this section, we present the core components of our *Streamlined Global and Local Features Combinator (SGLC)* framework, whose primary objective is to enable high-resolution image dehazing while preserving both global scene context and fine local details. Figure 1 provides an overview of the proposed methodology, illustrating the two main blocks that compose the SGLC pipeline: the *Global Features Generator (GFG)* and the *Local Features Enhancer (LFE)*. In what follows, we describe each module and its associated algorithms in detail.

3.1 Global Features Generator (GFG)

The first stage of the SGLC framework, termed the Global Features Generator (GFG), aims to capture and reconstruct the essential global appearance of the hazy image. High-resolution images impose significant memory and computational

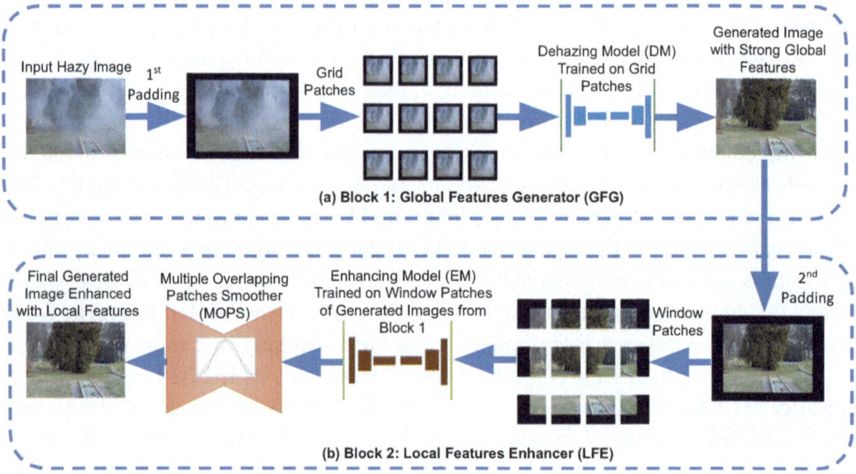

Fig. 1 SGLC diagram

burdens, making it impractical to process them holistically in a single forward pass. To address this issue, we introduce a *Grid Patching* mechanism that systematically samples the full-resolution image with minimal overlap, thereby maintaining a coarse-level representation of the entire scene.

3.1.1 Grid Patching for Global Features Learning

Consider an input image I of height H and width W. Our approach begins by introducing padding to ensure that H and W become divisible by a patch size $G \times G$. We denote the padded image by I' with height H' and width W'. If H (respectively W) is already divisible by G, then $H' = H$ (respectively $W' = W$); otherwise, H' and W' are computed as follows:

$$H' = \left(\lfloor H/G \rfloor + 1\right) \times G, \quad W' = \left(\lfloor W/G \rfloor + 1\right) \times G. \tag{3}$$

We define the number of vertical divisions as $n_h = H'/G$ and the number of horizontal divisions as $n_w = W'/G$. The total number of patches is then given by

$$N = n_h \times n_w. \tag{4}$$

Algorithm 1 outlines the Grid Patching procedure. Each generated patch, denoted P_k for $k = 0, \ldots, N-1$, is an array of size $G \times G \times 3$ (assuming RGB color channels). The central idea is to sample pixels from the padded image I' at intervals of n_h along the vertical axis and n_w along the horizontal axis. By doing so, the patch content collectively spans the entire image, capturing a broad overview of the scene and facilitating the learning of global features, as shown in Fig. 2.

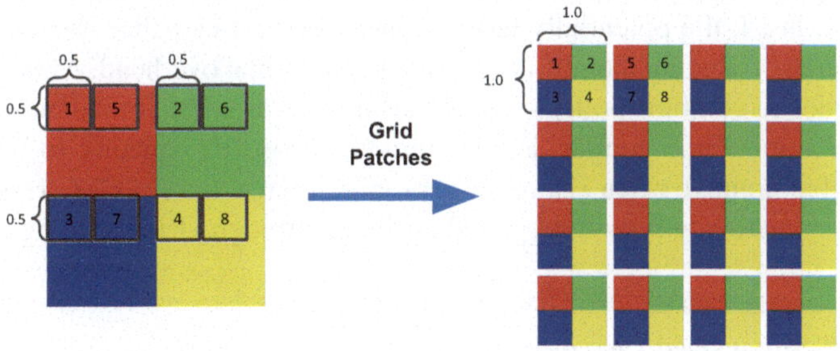

Fig. 2 Grid patches diagram

Algorithm 1 Grid Patching

Require: I': The padded image, n_h: The height number of divisions, n_w: The width number of divisions, N: The number of generated patches
Ensure: A set of grid patches $\{P_k\}_{k=0}^{N-1}$

for $k = 0$ to $N - 1$ **do**
    $P_k \leftarrow \text{zeros}(G, G, 3)$ ▷ Initialization to zeros
    for $i = 0$ to $G - 1$ **do**
        for $j = 0$ to $G - 1$ **do**
            $P_k[i, j] \leftarrow I'[i + n_w, j + n_h]$
        end for
    end for
end for

3.1.2 Dehazing Model (DM)

Once the grid patches are obtained, they are individually fed to a dehazing model (DM), which is designed to learn scene-level information from these patches. In this work, we adopt Uformer [43] as our baseline dehazing architecture, although SGLC is by no means limited to this particular choice. Uformer is built on a hierarchical U-shaped network with skip connections reminiscent of the classic U-Net structure. By incorporating LeWin Transformer blocks, Uformer captures both short-range and long-range dependencies using a locally enhanced window-based attention mechanism.

More concretely, let $i_h \in \mathbb{R}^{c \times h \times w}$ be a hazy patch. Uformer begins with a 3×3 convolution and LeakyReLU activation to derive initial feature maps, followed by several encoder stages each containing (1) a stack of LeWin Transformer blocks and (2) a down-sampling layer. The down-sampling layer halves the spatial dimensions while expanding the channel dimension. A decoder mirrors the encoder, employing up-sampling and additional LeWin Transformer blocks, as well as skip connections that fuse low- and high-level features across scales. The local attention windows, of size $m \times m$, break the potentially large feature maps into smaller non-overlapping regions, thereby reducing the self-attention computational overhead. Although originally designed to combine both global and local features within moderate-resolution imagery, Uformer (and similar architectures) can lose its capacity to holistically encode global context when scaled to very large image sizes. SGLC resolves this issue by systematically providing global coverage through Grid Patching.

3.1.3 Self-Supervised Learning

Inspired by works such as [1–4], we employ a preliminary self-supervised training step for both the Dehazing Model (DM) and the subsequent Enhancer Model (EM). Specifically, we generate an auxiliary dataset in which small squares of each clean image are randomly in-painted with white patches, simulating localized hazy regions. The model is trained to reconstruct the original, uncorrupted image from these artificially degraded samples. This procedure compels the network to capture the semantic context necessary for accurately filling in missing or obscured details, thereby enhancing its ability to handle realistic haze distributions.

3.1.4 Customized Loss Function

To train both the DM and the EM, we utilize a customized loss function that balances spatial fidelity, structural consistency, and robustness. In particular, we define

$$\mathcal{L} = \sqrt{\|I - \hat{I}\|^2 + \|\Pi(I) - \Pi(\hat{I})\|^2 + \varepsilon^2}, \tag{5}$$

where I denotes the ground-truth image, $\hat{I}$ is the predicted output, and $\Pi(\cdot)$ represents the Laplacian Pyramid operator [5]. The term $\|I - \hat{I}\|^2$ enforces pixel-level consistency between the prediction and the ground truth. Meanwhile, $\|\Pi(I) - \Pi(\hat{I})\|^2$ emphasizes the alignment of high-frequency details by comparing the Laplacian Pyramid representations of the images [6]. The small constant ε (set to 10^{-3} in our experiments) is included following the Charbonnier penalty function [7], making the training process more stable and less prone to vanishing gradients in the regime where $\|I - \hat{I}\|$ is small. Similar strategies have demonstrated enhanced robustness in various image restoration tasks [10].

3.1.5 Reverse Grid Patches Reconstruction

After the DM has processed each grid patch $\hat{P}_l$ (for $l \in \{0, \ldots, N-1\}$), we merge the predictions back into a single full-size image $\hat{I}_F$. This step, referred to as *Reverse Grid Patches Reconstruction*, is summarized in Algorithm 2. Initially, we initialize an empty array of the target size $G \times n_w \times G \times n_h$. We then assign each patch $\hat{P}_l$ to its corresponding spatial position in $\hat{I}_F$ based on horizontal and vertical offsets computed from l. Finally, the image is cropped to its original dimensions (removing any padding). The resulting image, $\hat{I}_F$, forms the first-stage dehazed output produced by GFG.

3.2 *Local Features Enhancer (LFE)*

While the Global Features Generator (GFG) described in the previous subsection provides a coherent large-scale representation of the scene, additional refinement is often required to restore subtle textures, edges, and other small-scale details. In the proposed *Streamlined Global and Local Features Combinator* (SGLC) framework, this role is fulfilled by the *Local Features Enhancer (LFE)*, which focuses on rectifying any deficiencies left unresolved by the GFG output, as shown in Fig. 3.

Algorithm 2 Reverse Grid Patches Reconstruction

Require: Predicted patches $\{\hat{P}_l\}_{l=0}^{N-1}$; grid size G; number of vertical divisions n_h; number of horizontal divisions n_w; total number of patches N.
Ensure: Reconstructed image $\hat{I}_F$.

$\hat{I}_F \leftarrow \text{zeros}(G \cdot n_w,\ G \cdot n_h,\ 3)$
for $l \in \{0, \ldots, N-1\}$ **do**
  $i_0 \leftarrow l \bmod n_w$ // horizontal offset
  $j_0 \leftarrow l \div n_w$ // vertical offset
  for $i_p \in \{0, \ldots, G-1\}$ **do**
    for $j_p \in \{0, \ldots, G-1\}$ **do**
      $x \leftarrow i_0 \times G + i_p$
      $y \leftarrow j_0 \times G + j_p$
      $\hat{I}_F[x, y] \leftarrow \hat{P}_l[i_p, j_p]$
    end for
  end for
end for

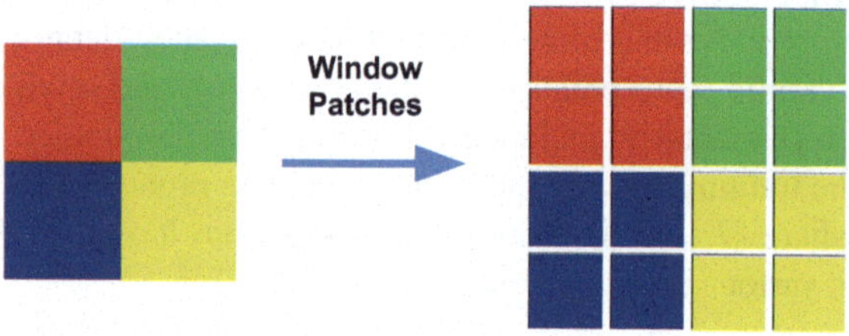

Fig. 3 Window patches diagram

3.2.1 Enhancer Model (EM)

Compared to the Dehazing Model (DM) used in the GFG stage, the Enhancer Model (EM) has two key differences. First, it operates on *window patches* (rather than grid patches) to capture spatially continuous regions. Second, it is trained on a new dataset composed of the GFG-generated dehazed images alongside their corresponding clean counterparts. Concretely, for each training sample, the GFG network is first used to generate an initial dehazed version of the hazy input. Both the resulting dehazed image and the corresponding ground-truth clean image are then padded to ensure divisibility by the chosen window size. Next, both padded images are subdivided into patches. These window patches form a new training set for the EM, which learns to focus on fine-grained improvements not fully resolved by the coarse-grained global predictions.

As in the GFG block, we employ the Uformer network [43] for the EM architecture, maintaining the same self-supervised initialization and the same customized loss function introduced earlier (Sect. 3). By following this multi-step training approach, the EM can better capture local feature nuances and ameliorate minor errors introduced by the GFG predictions. This strategy effectively transforms the dehazing process into two complementary stages, where coarse global restoration is tackled first, and fine local refinement is addressed second.

3.2.2 Multiple Overlapping Patches Smoother (MOPS)

A naive approach to reconstruct the final image from window patches-using a simple one-to-one arrangement-can introduce visible seams or discontinuities, commonly referred to as *vignetting artifacts* (see Fig. 4). These artifacts arise because convolutional neural networks (CNNs) often exhibit limited translational equivariance, a limitation exacerbated by zero-padding and strided operations [1]. Although such artifacts may be subtle, they can negatively impact both perceptual quality and quantitative dehazing metrics.

To mitigate these discontinuities, we adapt a *blending algorithm* [4] based on overlapping patch predictions and spatial window functions [2]. In the context of this study, we refer to this strategy as the *Multiple Overlapping Patches Smoother (MOPS)*. Specifically, the MOPS procedure aggregates multiple overlapping patches, each predicted by the EM under slight variations in the input (e.g., through rotations or reflections). By summing these predictions with a second-order spline window function, MOPS achieves a more seamless blending of local patch reconstructions and reduces boundary inconsistencies. Originally proposed for improving semantic segmentation results [2, 4], we adapt this technique here to enhance high-resolution dehazing outputs.

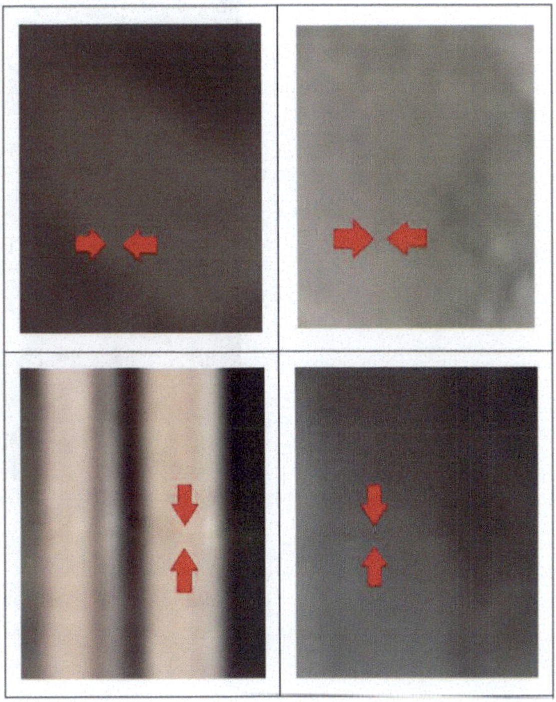

Fig. 4 Vignetting artifacts encountered during the application of Expectation Maximization (EM) in conjunction with the Reverse Window Patches Reconstruction algorithm

While MOPS significantly improves local consistency, its computational overhead is correspondingly increased. Therefore, Sect. 4 includes an ablation study evaluating the trade-off between the performance gains offered by MOPS and the additional computational cost incurred.

3.2.3 Visual Illustration and Final Output

After applying MOPS, the EM produces the final, refined dehazed image, which is expected to feature noticeably improved local detail and fewer perceptual artifacts compared to the GFG-only prediction. Figure 5 provides a qualitative comparison of a scene before and after local feature enhancement, underscoring the increased sharpness and clarity. Furthermore, the quantitative advantages of the LFE stage are evidenced in the metric evaluations described later, indicating that the two-stage SGLC pipeline delivers superior dehazing performance relative to single-stage approaches.

Fig. 5 The predicted image before and after applying the Local Feature Enhancement (LFE) block

(a) Before LFE

(b) After LFE

4 Experimental Evaluation

In this section, we examine the empirical effectiveness of the proposed *Streamlined Global and Local Features Combinator (SGLC)* on high-resolution dehazing tasks. We contrast its performance not only against the original Uformer [43] applied to downscaled images, but also with two additional methods representative of recent state-of-the-art research: *DW-GAN* [6], the winning approach in the 2021 NTIRE Non-Homogeneous Dehazing Challenge [5], and *4K-Dehazer* [20], an algorithm specifically tailored for large images. Notably, many existing methods are forced to perform an aggressive resize owing to GPU memory constraints, thereby sacrificing local detail. By contrast, our SGLC framework is designed to handle large image dimensions more efficiently, preserving both global and local information.

4.1 Experimental Configurations

4.1.1 Dataset Description

We evaluate our approach on the *HD-NH-HAZE* dataset, introduced as part of the *High-Resolution Non-Homogeneous Dehazing Challenge* at NTIRE 2023. This dataset consists of 50 images in extremely large dimensions (either 4000×6000 or 6000×4000), of which 40 are designated for training, 5 for validation, and 5 for testing. As the corresponding clean (ground-truth) images for the validation and test splits are withheld by the challenge organizers, we confined ourselves to the 40 images that include both hazy and clean pairs. From these 40 images, we separated out 36 for training and 4 for testing in our local experiments. No supplementary external data were utilized.

4.1.2 Implementation Details

All SGLC experiments were performed on a Lambda AI server equipped with 8 NVIDIA QUADRO 8000 GPUs, each providing 48 GB of GDDR6 memory. The server includes 512 GB of RAM and two Intel Xeon Silver 4216 CPUs (16 cores each). Additionally, we carried out some experiments on a single-workstation setup featuring a single NVIDIA QUADRO 8000 GPU. Our pipeline is implemented in `PyTorch`, with typical hyperparameter settings following established dehazing benchmarks (e.g., batch size between 2 and 8, depending on network depth, and initial learning rates in the range 1×10^{-4} to 2×10^{-4}).

4.2 Results and Analysis

We report quantitative results on the local test subset of 4 images using several standard metrics: Peak Signal-to-Noise Ratio (PSNR) and Structural Similarity Index Measure (SSIM). In addition, we measure the average inference time (in seconds per image) to capture computational overhead. All experiments employ a patch size of 1024×1024 for both *Grid Patching* in the Global Features Generator (GFG) and *Window Patching* in the Local Features Enhancer (LFE).

Table 1 compares our proposed SGLC approach (with different configurations) to Uformer [43], DW-GAN [5, 6], and 4K-Dehazer [20]. Uformer was tested on images downscaled to 1024×1024, as larger dimensions were infeasible due to memory constraints; similarly, 4K-Dehazer was restricted to a maximum resolution of 4000×4000.

The resizing approach severely hampers Uformer's ability to capture fine haze structures, leading to relatively poor performance (PSNR 14.90 dB, SSIM 0.6403). Although 4K-Dehazer and DW-GAN improve over this baseline, they do not significantly close the gap. By contrast, our proposed SGLC achieves markedly higher PSNR and SSIM, especially when both the Global Features Generator (GFG) and Local Features Enhancer (LFE) are deployed. We observe that the GFG alone already delivers a substantial boost over the other baselines (PSNR 24.49 dB, SSIM 0.8176). Further local refinement via the LFE block yields an additional improvement of about 0.89 dB in PSNR and 0.0335 in SSIM, The visual comparison in Fig. 6.

We also investigate the use of MOPS (Multiple Overlapping Patches Smoother), a blending algorithm to reduce patch-based vignetting artifacts. Although its contribution to the final PSNR and SSIM is modest (approximately 0.05 dB in PSNR and 0.0013 in SSIM), it can be critical for scenarios demanding the highest perceptual quality. The ablation study indicates that MOPS accounts for a significant fraction of

Table 1 Comparison of multiple dehazing methods on HD-NH-HAZE (our local test set). The inference time is computed per image of size 4000×6000

Model description	PSNR (dB)	SSIM	Inference time (sec/image)
Uformer on resized 1024 × 1024	14.90	0.6403	14.9
4K-Dehazer	17.46	0.7230	9.0
DW-GAN	17.48	0.7367	45.7
SGLC-GFG only	24.49	0.8176	46.0
SGLC w/o MOPS	25.38	0.8511	86.0
SGLC (GFG + LFE + MOPS)	**25.43**	**0.8524**	**553.5**
Inv-SGLC LFE only w/o MOPS	23.07	0.8307	40.0
Inv-SGLC LFE only + MOPS	23.25	0.8335	556.1
Inv-SGLC (LFE + GFG + MOPS)	24.39	0.8392	605.7

Fig. 6 Dehazing performance of SGLC compared to Uformer (1024*1024), DW-GAN, and the Ground Truth

the total computational cost (about 467.5 seconds out of the total 553.5 seconds per image), suggesting a trade-off between inference time and optimal visual fidelity.

Although the complete SGLC pipeline (including MOPS) requires more than 500 seconds to process a single 4000×6000 image, this overhead is still feasible for offline or batch-processing settings. If real-time or near-real-time performance is paramount, disabling MOPS reduces total inference to around 86 seconds per image with only a slight dip in performance (SGLC w/o MOPS achieves 25.38 dB PSNR versus 25.43 dB with MOPS).

We further investigate the impact of reversing the order of global and local feature modules. In the inverse setting (Inv-SGLC), the hazy images first pass through a local-focused module (LFE), followed by a global approach (GFG). As indicated in Table 1, every configuration of Inv-SGLC underperforms SGLC, highlighting the importance of first capturing global structure before refining local details.

4.3 Challenge Leaderboard

To verify the consistency of our local results, we also report the final standings from the *NTIRE 2023 Non-Homogeneous Dehazing Challenge* [44], shown in Table 2. SGLC placed 5th among 13 submitted solutions when considering multiple metrics (PSNR, SSIM, LPIPS, and MOS). When restricting attention to methods that did not employ extra external training data, SGLC attained the top PSNR and SSIM scores and the second-highest MOS.

Notably, SGLC tied for the best PSNR among entries that did not leverage additional training images, while also sharing the top SSIM score. Among all solutions, SGLC placed third in PSNR (0.69 dB behind first place) and fourth in SSIM (0.01 difference from the leader). The main shortfall lies in the LPIPS metric, hinting at potential avenues for future refinement (e.g., incorporating perceptual loss terms).

Table 2 Rankings in the NTIRE 2023 Non-Homogeneous Dehazing Challenge [44]. "No extra data" refers to solutions trained solely on the official challenge data

Model	PSNR	SSIM	LPIPS	MOS
SGLC Performance	22.27	0.70	0.439	7.40
Best "No Extra Data" Solution	22.27	0.70	0.384	7.65
SGLC Rank (No Extra Data)	1/12	1/12	7/12	2/12
Best Overall Performance	22.96	0.71	0.345	8.70
SGLC Rank (Overall)	3/17	4/17	10/17	5/17

Table 3 NTIRE 2023 Non-Homogeneous Dehazing Challenge [44] final leaderboard (with the *Username* column removed). Numbers in parentheses indicate each method's sub-ranking for that specific metric. **SGLC**, which corresponds to our SGLC framework, ranked 5th overall and 2nd among solutions that did not rely on external training data

Rank	Models	PSNR↑	SSIM↑	LPIPS↓	Params (M)	Device	Extra data
1	DWT-FFC GAN	22.87 (2)	0.71 (2)	0.346 (2)	373	RTX2080 Ti	Yes
2	ITB Dehaze	22.96 (1)	0.71 (1)	0.345 (1)	110	2×TitanXP	Yes
3	[Mask]	22.90 (4)	0.68 (9)	0.501 (13)	9.31	A100	No
4	NUSRIQC DEHAZING	21.97 (8)	0.68 (9)	0.380 (3)	25.58	4×RTX3090	Yes
5	**SGLC (our)**	**22.49 (6)**	**0.70 (3)**	**0.439 (10)**	**5.0**	**RTX8000**	**No**
6	NTU607-dehaze	21.86 (10)	0.73 (1)	0.442 (11)	101.5	V100	Yes
7	MIPCer	21.75 (12)	0.69 (6)	0.464 (8)	2.39	A100	No
8	iPAL-LightDehaze	22.09 (9)	0.67 (13)	0.556 (16)	n/a	TitanXP	No
9	Xsourse	22.09 (13)	0.65 (12)	0.556 (16)	n/a	A4000	No
10	Xiaofeng Cong	21.89 (14)	0.64 (14)	0.470 (9)	24.7	A100	No
11	MengFeiHome	20.96 (15)	0.62 (15)	0.515 (14)	n/a	RTX3060	No
12	CANT HAZE	20.95 (16)	0.69 (6)	0.415 (6)	80	T4/K80	No
13	IR-SDE	19.64 (17)	0.61 (16)	0.406 (7)	78	A100	No

Overall, these challenge results confirm the effectiveness and adaptability of SGLC in addressing high-resolution, non-homogeneous dehazing with minimal reliance on extra data.

As shown in Table 3, **SGLC** (our proposed SGLC framework) attained an overall rank of 5 across multiple metrics, including PSNR, SSIM, LPIPS, and MOS. Moreover, when considering only those solutions that did not rely on any external dataset for training, SGLC placed second in the competition. This performance underscores the ability of the SGLC methodology to accurately remove haze in large-scale, high-resolution images without necessitating additional data sources. Our experiments demonstrate that SGLC substantially improves dehazing performance for large-scale images compared to resizing-based baselines and specialized high-resolution methods. Despite the higher computational cost incurred by MOPS, the overall framework exhibits notable flexibility: one can omit MOPS if inference time is the priority, retaining most of the accuracy gains. Finally, the challenge leaderboard underscores SGLC's competitiveness, particularly in data-constrained settings where access to external training resources is limited.

5 Conclusion

In this chapter, we introduced the *Streamlined Global and Local Features Combinator* (SGLC) framework, designed to tackle the longstanding challenge of high-resolution image dehazing. By decomposing the dehazing process into two specialized stages-the *Global Features Generator* (GFG) and the *Local Features Enhancer* (LFE)-SGLC effectively addresses both large-scale atmospheric artifacts and fine-grained textual details. The GFG stage capitalizes on a *Grid Patching* strategy, enabling comprehensive scene-level restoration without exceeding computational resources, while the LFE stage adopts *Window Patching* and an overlapping-patch smoothing algorithm to refine localized structures and mitigate vignetting artifacts. Extensive experiments on high-resolution datasets indicate that SGLC consistently achieves superior quantitative and qualitative results compared to conventional single-stage approaches and alternative high-resolution dehazing methods. Our ablation studies reveal that (1) the GFG and LFE modules each independently contribute to performance gains, and (2) the sequential order of applying global and local refinement proves crucial. Additionally, although the Multiple Overlapping Patches Smoother (MOPS) algorithm incurs a computational cost, it can further enhance perceptual fidelity when ultimate image quality is required. Overall, this work highlights the importance of carefully orchestrating global and local feature extraction to handle complex, large-scale haze conditions. The modular nature of SGLC also encourages future extensions, such as integrating more efficient patch division schemes, incorporating advanced Transformer-based blocks, or exploring additional self-supervised pretraining paradigms. By providing a robust and flexible two-stage solution, SGLC serves as a promising framework that pushes the boundaries of high-resolution image dehazing in both research and real-world deployments.

Competing Interests The authors declare that they have no competing interests related to this work.

Ethics Approval This study does not involve any primary research with human or animal subjects. Therefore, ethical approval was not required.

Acknowledgements The authors thank Prince Sultan University for the support and funding in conducting this research.

References

1. C.O. Ancuti, C. Ancuti, Single image dehazing by multi-scale fusion. IEEE Trans. Image Process. **22**(8), 3271–3282 (2013)
2. B. Cai, X. Xu, K. Jia, C. Qing, D. Tao, Dehazenet: an end-to-end system for single image haze removal. IEEE Trans. Image Process. **25**(11), 5187–5198 (2016)
3. K. Tang, J. Yang, J. Wang, Investigating haze-relevant features in a learning framework for image dehazing, in *CVPR* (2014)
4. W. Ren, L. Ma, J. Zhang, J. Pan, X. Cao, W. Liu, M.H. Yang, Gated fusion network for single image dehazing, in *CVPR* (2018)
5. C.O. Ancuti, C. Ancuti, F.A. Vasluianu, R. Timofte, Ntire 2021 nonhomogeneous dehazing challenge report, in *CVPR* (2021)
6. M. Fu, H. Liu, Y. Yu., J. Chen, K. Wang, Dw-gan: a discrete wavelet transform gan for nonhomogeneous dehazing, in *CVPR* (2021)
7. Z. Chen, Q. Li, H. Feng, Z. Xu, Y. Chen, Nonuniformity dehaze network for visible remote sensing images, in *CVPR* (2022)
8. R. Fattal, Dehazing using color-lines. ACM Trans. Graph. (TOG) (2014)
9. J.S. Kaiming He, X. Tang, Single image haze removal using dark channel prior. IEEE Trans. Pattern Anal. Mach. Intell. **33**(12), 2341–2353 (2011)
10. Q. Zhu, J. Mai, L. Shao, A fast single image haze removal algorithm using color attenuation prior. IEEE Trans. Image Process. **24**(11), 3522–3533 (2015)
11. Y. LeCun, Y. Bengio, G. Hinton, Deep learning. Nature (2015)
12. B. Li, X. Peng, Z. Wang, J. Xu, D. Feng, Aod-net: all-in-one dehazing network, in *ICCV* (2017)
13. M. Hong, Y. Xie, C. Li, Y. Qu, Distilling image dehazing with heterogeneous task imitation, in *CVPR* (2020)
14. R. Li, J. Pan, Z. Li, J. Tang, Single image dehazing via conditional generative adversarial network, in *CVPR* (2018)
15. X. Qin, Z. Wang, Y. Bai, X. Xie, H. Jia, Ffa-net: feature fusion attention network for single image dehazing, in *AAAI* (2020)
16. Y. Qu, Y. Chen, J. Huang, Y. Xie, Enhanced pix2pix dehazing network, in *CVPR* (2019)
17. H. Zhang, V.M. Patel, Densely connected pyramid dehazing network, in *CVPR* (2018)
18. Z. Deng, L. Zhu, X. Hu, C.W. Fu, X. Xu, Q. Zhang, J. Qin, P.A. Heng, Deep multi-model fusion for single-image dehazing, in *ICCV* (2019)
19. J. Gui, X. Cong, Y. Cao, W. Ren, J. Zhang, J. Zhang, J. Cao, D. Tao, A comprehensive survey and taxonomy on single image dehazing based on deep learning. ACM Comput. Surv. (2022)
20. Z. Zheng, W. Ren, X. Cao, X. Hu, T. Wang, F. Song, X. Jia, Ultra-high-definition image dehazing via multi-guided bilateral learning, in *CVPR* (2021)
21. Y. Song, Y. Zhou, H. Qian, X. Du, Rethinking performance gains in image dehazing networks. arXiv preprint (2022)
22. Z. Chen, Z. He, Z.M. Lu, Dea-net: single image dehazing based on detail-enhanced convolution and content-guided attention. arXiv preprint (2023)
23. A. Vaswani, N. Shazeer, N. Parmar, J. Uszkoreit, L. Jones, A.N. Gomez, Ł. Kaiser, I. Polosukhin, Attention is all you need (2017)
24. C.O. Ancuti, C. Ancuti, F.A. Vasluianu, R. Timofte, Nitre 2023 challenge on nonhomogeneous dehazing, in *CVPR Workshops* (2023)
25. T. Ye, M. Jiang, Y. Zhang, L. Chen, E. Chen, P. Chen, Z. Lu, Perceiving and modeling density is all you need for image dehazing. arXiv preprint (2021)
26. K. He, J. Sun, X. Tang, Single image haze removal using dark channel prior. IEEE Trans. Pattern Anal. Mach. Intell. **33**(12), 2341–2353 (2011)
27. D. Berman, T. Treibitz, S. Avidan, Non-local image dehazing, in *Proceedings of the IEEE Conference on Computer Vision and Pattern Recognition (CVPR)* (2016), pp. 1674–1682
28. R. Fattal, Single image dehazing. ACM Trans. Graph. (TOG) **27**(3), 72:1–72:9 (2008)
29. R. Fattal, Dehazing using color-lines. ACM Trans. Graph. (TOG) **34**(1), 13:1–13:14 (2014)

30. B. Cai, X. Xu, K. Jia, C. Qing, D. Tao, Dehazenet: an end-to-end system for single image haze removal. IEEE Trans. Image Process. **25**(11), 5187–5198 (2016)
31. W. Ren, S. Liu, H. Zhang, J. Pan, X. Cao, M.H. Yang, Single image dehazing via multi-scale convolutional neural networks, in *European Conference on Computer Vision (ECCV)*. Springer (2016), pp. 154–169
32. S. Ki, H. Sim, J.S. Choi, S. Kim, M. Kim, Fully end-to-end learning based conditional boundary equilibrium gan with receptive field sizes enlarged for single ultra-high resolution image dehazing, in *Proceedings of the IEEE Conference on Computer Vision and Pattern Recognition Workshops*, pp. 817–824 (2018)
33. H. Sim, S. Ki, J.S. Choi, S. Seo, S. Kim, M. Kim, High-resolution image dehazing with respect to training losses and receptive field sizes, in *Proceedings of the IEEE Conference on Computer Vision and Pattern Recognition Workshops* (2018), pp 912–919
34. X. Qin, Z. Wang, Y. Bai, X. Xie, H. Jia, FFA-Net: feature fusion attention network for single image dehazing, in *Proceedings of the AAAI Conference on Artificial Intelligence*, vol. 34 (2020), pp. 11908–11915
35. Y. Qu, H. Cai, C. Xiao, D. Ren, W. Ma, Enhanced pix2pix dehazing network. arXiv preprint arXiv:1908.11035 (2019)
36. H. Zhang, V.A. Sindagi, V.M. Patel, Densely connected pyramid dehazing network, in *Proceedings of the IEEE Conference on Computer Vision and Pattern Recognition (CVPR)* (2018), pp. 3194–3203
37. X. Chen, Y. Li, L. Dai, C. Kong, Hybrid high-resolution learning for single remote sensing satellite image dehazing. IEEE Geosci. Remote Sens. Lett. **19**, 1–5 (2021)
38. B. Dong, G. Chen, X. Zong, T. Feng, Multi-scale boosted dehazing network with dense feature fusion. Pattern Recognit. **108**, 107563 (2020)
39. B. Li, X. Peng, Z. Wang, J. Xu, D. Feng, Aod-net: all-in-one dehazing network, in *Proceedings of the IEEE International Conference on Computer Vision (ICCV)* (2017), pp. 4770–4778
40. B. Li, X. Peng, Z. Wang, J. Xu, D. Feng, Y. Guo, Semi-supervised image dehazing. IEEE Trans. Image Process. **29**, 2766–2779 (2019)
41. S. Bianco, L. Celona, F. Piccoli, R. Schettini, High-resolution single image dehazing using encoder-decoder architecture, in *Proceedings of the IEEE/CVF Conference on Computer Vision and Pattern Recognition Workshops* (2019), pp. 0–0
42. Z. Zheng, W. Ren, X. Cao, X. Hu, T. Wang, F. Song, X. Jia, Ultra-high-definition image dehazing via multi-guided bilateral learning, in *2021 IEEE/CVF Conference on Computer Vision and Pattern Recognition (CVPR)*. IEEE (2021), pp. 16180–16189
43. Z. Wang, X. Cun, J. Bao, W. Zhou, J. Liu, H. Li, Uformer: a general u-shaped transformer for image restoration, in *Proceedings of the IEEE/CVF Conference on Computer Vision and Pattern Recognition* (2022), pp. 17683–17693
44. C.O. Ancuti, C. Ancuti, F.A. Vasluianu, R. Timofte, Nitre 2023 challenge on nonhomogeneous dehazing, in *CVPR Workshops* (2023)

Anas M. Ali is a Research Assistant within the Robotics and Internet of Things Laboratory at Prince Sultan University (PSU). He received his Master's Degree in Electrical Engineering from the Faculty of Electronic Engineering, Menoufia University, Egypt, in 2021. His primary research interests include computer vision, healthcare image analysis, deep learning, image restoration, medical image processing, and remote sensing.

Anis Koubaa is the Leader and Founder of the Robotics and Internet-of-Things Laboratory at Prince Sultan University. He is also the Executive Director of the Innovation Center at Prince Sultan University. He is a full professor of computer science and has been working on several research and development projects on robotics, unmanned systems, deep learning, and the Internet of Things. He is a Senior Fellow of the Higher Education Academy, U.K. He has presented numerous training programs on cutting-edge technologies. His research interests include developing artificial intelligence products and automated solutions for logistics, using drones and robots for delivery systems.

Bilel Benjdira is an Assistant Professor at Prince Sultan University (PSU) and a researcher within the Robotics and Internet of Things Laboratory at PSU. He received his Ph.D. degree in Electrical Engineering from ENICarthage, University of Carthage, Tunisia, in 2021. His primary research interests include computer vision, healthcare image analysis, deep learning, and remote sensing.

Diffusion Models in a Nutshell: A Tutorial

Anas M. Ali, Anis Koubaa, and Bilel Benjdira

Abstract This chapter presents a comprehensive examination of diffusion models, a significant innovation in deep generative modeling. Distinct from other generative approaches like Generative Adversarial Networks (GANs) and Variational Autoencoders (VAEs), diffusion models incrementally refine data representations through a series of noise addition and denoising steps, closely resembling natural diffusion processes. This methodical enhancement enables diffusion models to produce stable, high-quality outputs while avoiding the common pitfalls of instability in GANs and the complex posterior matching issues in VAEs. The theoretical underpinnings of diffusion models are elucidated, with an emphasis on demystifying advanced concepts and detailing their mathematical foundations. The chapter further delves into the latest advancements and research trends, providing a thorough analysis of state-of-the-art developments. Practical applications are explored across a variety of fields including computer vision, natural language processing, audio synthesis, and time series analysis, demonstrating the broad applicability and flexibility of diffusion models. Despite their computational intensity, recent breakthroughs in accelerated sampling and model distillation are discussed, which have markedly improved their efficiency. The chapter concludes by synthesizing key insights, discussing current challenges, and proposing directions for future research to continue advancing this dynamic field.

A. M. Ali · B. Benjdira
Robotics and Internet-of-Things Laboratory, Prince Sultan University, Riyadh, Saudi Arabia
e-mail: bbenjdira@psu.edu.sa

A. M. Ali (✉)
Department of Electronics and Electrical Communications Engineering, Faculty of Electronic Engineering, Menoufia University, Menouf, Egypt
e-mail: aaboessa@psu.edu.sa

A. Koubaa
College of Engineering, Alfaisal University, Riyadh, Saudi Arabia
e-mail: akoubaa@alfaisal.edu

B. Benjdira
SE and ICT Laboratory, LR18ES44, ENICarthage, University of Carthage, Tunis, Tunisia

A. Koubaa et al. (eds.), *Generative AI and Large Language Models: Opportunities, Challenges, and Applications*, Studies in Computational Intelligence 1214,
https://doi.org/10.1007/978-3-031-90573-5_5

1 Introduction

Diffusion models represent a pivotal milestone in deep generative modeling, offering a unique framework for high-quality sample generation across multiple domains. They operate by simulating data distributions through iterative noise addition and subsequent denoising. The term "diffusion" is derived from a physical process describing how particles spread through a medium [153], which serves as an analogy for the progressive refinement of data representations. Historically, diffusion-like processes were utilized in image processing tasks, including denoising and inpainting via differential equations. However, the advent of neural networks has allowed diffusion models to directly learn these transformations from data, enabling their evolution.

Unlike other generative modeling frameworks such as Generative Adversarial Networks (GANs) and Variational Autoencoders (VAEs), diffusion models emphasize gradual refinement over multiple steps instead of a single-pass generation. This approach promotes stability during training, consistently yielding high-quality outputs [37, 159].

Diffusion models form part of a broader family of deep generative models, alongside VAEs [82, 129], Energy-Based Models (EBMs) [87, 120], GANs [36, 49], and normalizing flows (NFs) [83, 140]. While these models are vital in enhancing machine learning's capacity to replicate human-like creativity, diffusion models have garnered significant traction by overcoming challenges inherent in other approaches, such as VAEs' posterior distribution mismatch, GANs' adversarial training instability, EBMs' reliance on computationally expensive Markov Chain Monte Carlo (MCMC), and NFs' architectural constraints.

Diffusion models utilize a forward process that reduces complex data distributions to simpler ones, often Gaussian, and a reverse process that relies on a neural network to iteratively revert this transformation. The forward process resembles Brownian motion with time-dependent coefficients [75], while the reverse process employs a denoising score-matching objective [154] to estimate the score function. This approach leads to stable training, avoiding many challenges seen in other generative architectures [37, 159]. Despite this stability, the iterative transformations involving Ordinary or Stochastic Differential Equations (ODE/SDE) or Markov processes demand numerous function evaluations, making sampling computationally intensive compared to GANs and VAEs.

Despite these computational demands, diffusion models excel across diverse domains, including computer vision [13, 61, 155], natural language processing [10, 94], time series [165, 192], audio processing [23, 135], graph generation [69, 124], and bioinformatics [115, 187]. Researchers have mitigated computational challenges through strategies such as sampling acceleration [110, 188, 201], model distillation [147, 155], innovative diffusion process design [40, 189, 190], and dimensionality reduction [142, 169]. Efforts to bridge arbitrary distributions have also been explored [2, 107].

This chapter is structured as follows: Sect. 2 delves into the fundamental principles and mathematical formulations of diffusion models. Section 3 provides a classification and comparative analysis, emphasizing the unique properties of each approach. Section 4 examines categorized applications based on the generation mechanism. Finally, Sect. 5 discusses key insights and future directions for diffusion models, summarizing the chapter's content while outlining current limitations and promising research opportunities.

2 Background and Basic Concepts

his section provides a comprehensive analysis of diffusion models within the field of deep learning, with a particular focus on their processes of progressively introducing and eliminating noise. It traces the evolution from traditional diffusion methods to those that leverage neural networks, elucidating their fundamental principles and diverse applications. Furthermore, it critically examines the strengths and limitations of these models, incorporating a practical case study on image generation to offer empirical insights.

2.1 Basic Concepts

In the domain of deep learning, diffusion models represent a class of generative models that simulate a data distribution through a sequential process of incremental noise addition and subsequent removal. The term "diffusion" is derived from the physical process in which particles disperse or "diffuse" within a medium, analogous to molecular movement in fluids or gases. These models have gained prominence as viable alternatives to established generative frameworks such as Generative Adversarial Networks (GANs) [49] and Variational Autoencoders (VAEs) [82]. Distinctively, diffusion models incrementally refine their outputs, gradually enhancing the quality through successive iterations.

Historically, the concept of diffusion has been integral to image processing applications including denoising and inpainting, where it was traditionally modeled using differential equations specifically designed for each task [18]. However, the advent of deep learning has revolutionized the application of diffusion processes, transitioning from reliance on bespoke differential equations to data-driven approaches facilitated by neural networks. This shift has enabled the diffusion process to be learned directly from data, significantly broadening the applicability and efficiency of diffusion models in various tasks within machine learning, particularly in generative modeling [75, 153, 154].

2.1.1 Principles Behind Diffusion Models

Diffusion models, a class of generative models in deep learning, can be intuitively understood by drawing parallels with natural phenomena such as random walks and Brownian motion. A random walk is a simple yet fascinating process where a particle moves in a random direction with equal likelihood of stepping left or right at consistent time intervals. This can be mathematically expressed as:

$$X_{n+1} = X_n + \epsilon_n, \quad \epsilon_n \sim \text{Uniform}(-1, 1)$$

where X_n represents the position of the particle at step n, and ϵ_n is the random step, which can be either -1 or 1 [86].

Expanding on this concept, Brownian motion-first documented by Robert Brown in 1827 when he observed the erratic movement of pollen grains in water-offers a deeper insight. As the step size in a random walk becomes infinitesimally small and the steps more frequent, the movement of the particle begins to exhibit fluid and chaotic dynamics. This phenomenon is modeled as a continuous stochastic process:

$$dX_t = \mu dt + \sigma dW_t$$

where μ is the drift coefficient, σ the volatility coefficient, and dW_t represents the differential of the Wiener process, capturing the essence of standard Brownian motion [125].

In the context of deep learning, diffusion models leverage these principles to generate new data. The process initiates by introducing noise to a simple initial data distribution, often Gaussian, akin to setting particles into random motion. This noisy state is gradually refined through a series of denoising steps, represented by the reverse stochastic differential equation:

$$dX_t = -f(X_t, t)dt + g(t)dW_t$$

Here, $f(X_t, t)$ denotes a neural network that predicts the reverse diffusion path at time t, and $g(t)$ controls the noise level at each step [153].

This iterative process of adding noise and then methodically removing it, using deep neural networks trained to predict and correct the data in each step, effectively reverses the noisy sequence back towards the original distribution. The approach differs significantly from traditional generative models that generate outputs in a single step. Instead, diffusion models construct their output progressively, allowing for refinement and error correction at each stage, often resulting in higher quality and more accurate samples [59].

Through the lens of random walks and Brownian motion, diffusion models exemplify how sophisticated algorithms can simulate and reverse natural dispersion processes. This step-by-step journey from chaos to order mirrors the physical phenomena of particles gradually settling after a series of erratic movements, making these models scientifically intriguing and remarkably effective for generating complex data distributions.

2.1.2 Generic Diffusion Models For Image Generation

Generic Diffusion Model for Training an Image Generation Model

Diffusion models are a powerful tool in deep learning, employed across a variety of domains including image synthesis and restoration, where they excel in producing high-quality images from random noise. These models apply a technique that iteratively refines this noise, effectively reconstructing the data to resemble the characteristics of training samples. Their capabilities are showcased in tasks such as enhancing image resolution, inpainting to fill gaps, and noise reduction [59]. Beyond imaging, diffusion models have proven versatile in generating coherent text, synthesizing and refining sounds, enhancing video content frame-by-frame [159], and even aiding in drug discovery by optimizing molecular structures in bioinformatics [156].

To illustrate how diffusion models can be applied in **training image generation**, consider the following simple algorithm:

Algorithm 1 Generic diffusion model for training an image generation model

---

Objective: Generate high-quality images from random noise using a diffusion model.
Initialization: Start with an image I_0 initialized as random noise from a normal distribution:

$$I_0 \sim \mathcal{N}(0, \mathbf{I})$$

Training: Train a neural network to act as the denoiser. The network learns to reverse the noise addition in the diffusion process.
Noise Addition: Incrementally add noise to the initial image over a series of steps t, adjusting the noise level with:

$$I_{t+1} = \sqrt{1-\beta_t} I_t + \sqrt{\beta_t}\epsilon, \quad \epsilon \sim \mathcal{N}(0, \mathbf{I})$$

where β_t represents the variance of the noise at step t.
Iterative Refinement: Use the trained network to progressively denoise the image across iterations, refining I_{t+1} back towards the cleaner image:

$$I_t = \text{Denoiser}(I_{t+1})$$

Completion: Continue the refinement until the image closely approximates the desired quality and detail.
Results: The final image should closely resemble the desired high-quality target, demonstrating the model's ability to reconstruct detailed images from noise.

For Training to Inference

Once the diffusion model has been trained to effectively denoise and reverse the noise addition process, it is ready to be used for generating new images through inference. The training phase involves teaching the model to progressively refine noisy images toward their original form by reversing the added noise. In contrast, the inference phase leverages this trained model to start with a completely noisy image and iteratively transform it into a detailed and high-quality image. This process showcases the practical application of diffusion models in generating images from noise, using the learned denoising capabilities to reconstruct or synthesize images that are visually close to real-world objects or scenes.

The inference algorithm, outlined below, describes the step-by-step application of the trained diffusion model to generate images from an initialized state of random noise. This phase emphasizes the model's ability to reverse the diffusion process, applying the learned transformations in a sequence that starts from the highest level of noise and progressively refines this noise into a coherent and detailed image.

Algorithm 2 Generic algorithm for inference with trained diffusion model for image generation

Objective: Generate high-quality images from random noise using a pre-trained diffusion model.
Initialization: Start with an image I_T initialized as random noise, where T is the total number of diffusion steps:

$$I_T \sim \mathcal{N}(0, \mathbf{I})$$

Iterative Refinement: Sequentially apply the trained denoising model to reduce noise and refine the image at each step from $t = T$ down to 1:

$$I_{t-1} = \text{Denoiser}(I_t, t), \quad \text{for } t = T, T-1, \ldots, 1$$

where Denoiser is the neural network trained to perform the reverse diffusion process, and t indicates the current diffusion step.
Completion: Continue the refinement process until $t = 0$, at which point the image I_0 should represent a high-quality approximation of the desired target image.
Results: The final image I_0 demonstrates the capability of the diffusion model to reconstruct or generate detailed images from noise, showcasing the efficacy of the trained model.

2.1.3 Advantages and Limitations

Diffusion models exhibit a number of significant strengths. The iterative refinement process allows for the generation of high-quality samples that closely resemble real data [59]. They demonstrate flexibility, being adaptable across domains like images, text, audio, and molecular structures. Their robustness derives from avoiding adversarial training, reducing vulnerabilities like mode collapse. Furthermore, the gradual construction of samples provides a clear understanding of the generation process, allowing for improved interpretability compared to other generative models [159].

However, these models are not without limitations. Their iterative nature can be computationally intensive, making sample generation slower compared to single-shot models [123]. Effective training requires careful hyperparameter tuning and sensitivity to the denoising process. The quality of generated samples depends significantly on the neural network-based denoisers employed, so suboptimal denoisers can result in poor performance. Lastly, their iterative approach may not be suitable for real-time generation tasks due to latency constraints [37].

2.2 *Understanding the Diffusion Equation*

This section provides a fundamental understanding of the diffusion equation, as illustrated in Fig. 1, which depicts a diffusion model. The sequence begins on the left with a glass of clear water and an ink droplet introduced at the center. Moving towards the right, the ink gradually disperses throughout the water, demonstrating the diffusion process in action.

2.2.1 Intuition Behind the Diffusion Equation

Overlaying the dispersion of ink in water is a mathematical equation that predicts both the rate and pattern of the ink's spread. This equation, known as Fick's second law [52], mathematically describes the phenomenon observed when substances naturally diffuse over time. It is formally expressed as:

$$\frac{\partial \phi}{\partial t} = D\nabla^2 \phi \tag{1}$$

Fig. 1 The figure illustrates simple sequence for Diffusion process

where:

- ϕ: Represents the concentration of the substance, similar to the density of ink in a specific region of water.
- t: Denotes time.
- $\frac{\partial \phi}{\partial t}$: Illustrates the rate at which the concentration changes over time, indicating the speed at which the ink spreads through the water.
- D: The diffusion coefficient, a parameter characterizing the intrinsic spread rate of the substance, explaining why some substances diffuse more quickly than others.
- $\nabla^2 \phi$: The Laplacian operator, representing the second spatial derivative, provides insight into the direction and extent of the substance's spread, helping determine where and how much the ink disperses.

For practical computational applications, the continuous diffusion equation is often discretized to integrate it into iterative frameworks, such as those used in deep learning. This process involves dividing both time and space into discrete intervals and units [80]. By discretizing the diffusion equation, we transform our continuous understanding of time and space into small, distinct segments, akin to converting a continuous gradient into separate color blocks. This discrete representation is essential in computational models, particularly in deep learning scenarios [59, 156]. Upon discretizing Eq. 1, we use the following notations for incremental changes:

- Δt: Represents the discrete time increment.
- Δx: Denotes the discrete spatial step in the x-direction (extendable to Δy, Δz for 2D or 3D settings).

The discretized diffusion equation can be expressed as follows:

$$\phi(x, t + \Delta t) - \phi(x, t) = D \left[\frac{\phi(x + \Delta x, t) - 2\phi(x, t) + \phi(x - \Delta x, t)}{\Delta x^2} \right] \tag{2}$$

Equation 2 suggests that the change in concentration at a point over a short interval Δt depends on the concentration differences at neighboring points, as determined by Δx, at the current time. Diffusion equations have proven relevant in generative modeling within deep learning. The data distribution is viewed as a "substance" that diffuses over time, starting from a simple distribution and gradually evolving to resemble a complex target distribution [19]. This process involves adding noise to simplify the data, followed by iterative refinement through neural networks to reverse this effect, effectively "denoising" the data step by step. Song and Ermon [156] present a framework for using this iterative process probabilistically: at each step, given the current state, a neural network estimates the transition probability to approximate the next state, thus enabling the generative process.

3 Theoretical Foundations

This probabilistic understanding of the diffusion process in generative modeling parallels the Markov process, where the future state relies only on the current state, reflecting how each step in generative diffusion depends solely on the current distribution.

3.1 *Markov Processes and Random Walks*

A Markov process, also known as a Markov chain, is a stochastic model used to predict sequences of possible events [71]. The model is characterized by the property of being "memoryless," meaning that the probability of each subsequent state depends solely on the current state, without consideration of the previous states. This concept is fundamental in understanding the behavior of Markov chains, as each transition is influenced only by the state immediately preceding it, rather than a series of prior states [50].

A random walk is a variant of a Markov process that involves a sequence of steps taken in random directions [71]. In a one-dimensional scenario, it typically refers to a particle that moves either left or right at each step with equal probability. The random walk exemplifies the probabilistic nature of Markov processes, as each step is determined independently based on the current position, without regard for previous movements.

3.2 *Brownian Motion In-Depth*

The mathematical description of Brownian motion [178], represented as $B(t)$, is a continuous-time stochastic process characterized by:

- $B(0) = 0$: It starts from zero.
- The increments are independent and follow a normal distribution: $B(t) - B(s) \sim N(0, t - s)$ for $0 \leq s < t$.

Properties

- Possesses stationary and independent increments.
- Continuous everywhere but non-differentiable anywhere.
- Represents the limit of scaled random walks as step size approaches zero and the number of steps increases indefinitely [126].

The connection between Brownian motion and diffusion is evident through the diffusion equation [178], which models the evolution of a particle's probability distribution as it undergoes Brownian motion. In summary, the diffusion equation governs the distribution of substances over space and time, showing how the rate of change in concentration at a point is linked to the curvature around that point [149]. It is solved through techniques like separation of variables or Fourier transforms to reveal the behavior of different initial distributions. The solutions significantly depend on boundary conditions (which govern the behavior of the substance at the edges of the domain) and initial conditions (which define the initial distribution). These factors shape how the substance evolves over time.

3.3 Forward Diffusion Process: From Image to Noise

Figure 2 illustrates the forward diffusion process, where a pristine image, denoted as x_0, progressively deteriorates through the application of Gaussian noise at each step from x_0 to x_T [59, 159]. The repeated addition of noise causes the image to gradually degrade, and by the time it reaches x_t and eventually x_T, the cumulative effect of the noise results in significant distortion or pixelation. The transition function, $q(x_t \mid x_{t-1})$, defines the rule by which Gaussian noise is applied at each step. This model serves to examine the impact of noise on data, simulate real-world scenarios, or evaluate the robustness of image processing algorithms against noise.

Mathematically, this diffusion process can be formulated as follows:

$$q(x_t \mid x_{t-1}) = \mathcal{N}(x_t; \mu_t = (1 - \beta_t)x_{t-1}, \Sigma_t = \beta_t I) \tag{3}$$

This is centered on the Gaussian (or Normal) distribution, which is essential in statistics and machine learning. The conditional probability distribution characterizes how $x(t)$ (the image at time t) is influenced by $x(t - 1)$ (the image from the preceding step) [153].

3.3.1 The Nature of the Noise: Gaussian Distribution

The noise in the diffusion process is modeled using a Gaussian distribution [91]. The mean (μ), central to this distribution, indicates the center of the noise. For each time step t, the mean μ_t is determined as follows:

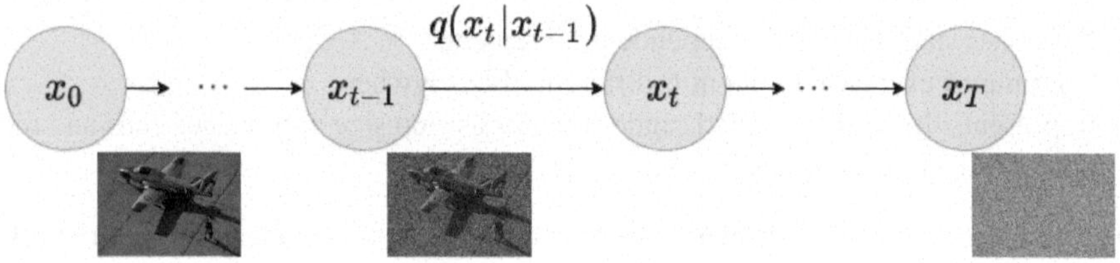

Fig. 2 The Forward Diffusion Process.

$$\mu_t = (1 - \beta_t)x_{t-1} \tag{4}$$

Thus, the principal component of our noise (its 'center') is essentially a scaled version of the preceding image, where the scaling factor is $(1 - \beta_t)$. This scaling implies that the image is subtly shifted by a factor dependent on β_t, which dictates the extent of deviation from the original. The variance, symbolized by Σ_t, represents the spread or dispersion of the noise. It is defined as follows:

$$\Sigma_t = \beta_t I \tag{5}$$

Here, I represents the identity matrix, which ensures that the noise is independently applied to each pixel or feature of the image. The coefficient β_t scales the noise; a larger β_t introduces more noise, whereas a smaller β_t results in less noise [59, 159].

The overarching concept of iterative refinement is that at each step t, the image undergoes a slight shift as dictated by the mean, and subsequently, noise is added around this shifted basis according to the variance. This procedure, repeated over several steps, results in an image that progressively becomes more noisy. Importantly, this noise is not random but structured, rooted in the original image itself. The efficacy of the diffusion model stems from its ability to invert this process. Through training on the forward noise application, the model acquires the capability to reverse this sequence: beginning with a noisy image, it iteratively refines and 'denoises' it, ultimately yielding a clear, coherent output [153].

The scaling factor $(1 - \beta_t)$ plays a crucial role in shifting the mean. This mean deviation in the diffusion process is vital to enabling the model's generative capacity. A blend of intuitive and technical insights can help elucidate this concept further.

$$\mu_t = (1 - \beta_t)x_{t-1} \tag{6}$$

The coefficient $(1 - \beta_t)$ acts as a stabilizer, ensuring that the image does not deviate excessively from its initial position. As β_t increases, this stabilizing factor becomes less restrictive, allowing for greater deviation. This deviation process is similar to adjusting a clay sculpture. If β_t is set too high initially, the modifications to the base model of the vehicle would be too severe, potentially losing any resemblance to the original structure. By managing the drift with β_t, we ensure that while the image of the vehicle evolves, it maintains its core characteristics and remains identifiable as a vehicle. As time progresses, this managed deviation enables the model to explore and create a varied set of images, analogous to how nuanced adjustments to a clay model can result in different facial representations starting from the same base.

3.3.2 Setting the Stage: Multi Dimensional Scenario

In a multi-dimensional setting, I denotes the identity matrix, implying that each dimension maintains a consistent standard deviation β_t. The conditional normal distribution $q(x_t \mid x_{t-1})$ is characterized by the mean μ_t and the variance Σ_t. The concept of tractability in progression becomes evident when starting with the initial data x_0 and transitioning toward x_T, as the transformation can be described across the entire sequence. This progression can be represented as follows:

$$q(x_{1:T} \mid x_0) = \prod_{t=1}^{T} q(x_t \mid x_{t-1}) \tag{7}$$

The colon symbol (:) in $x_{1:T}$ signifies that the function q is applied consecutively across all time steps from 1 to T. A significant challenge presents itself in this approach: to sample x_t at a given time, say $t = 500$, without proceeding to T, one would need to apply the function repeatedly 500 times. This method is markedly inefficient!

The *reparametrization trick*, described by [59, 81], provides an optimized approach to address this issue. Instead of directly adjusting β_t, we introduce $\alpha_t = 1 - \beta_t$ and compute its cumulative product up to time $\bar{\alpha}_t$. This technique enables us to represent x_t in terms of the initial point x_0 and added noise ϵ. Using this trick, a sample x_t can be derived directly as:

$$x_t \sim q(x_t \mid x_0) = \mathcal{N}(\alpha_t x_0, (1 - \alpha_t)I) \tag{8}$$

With β_t predefined, we can calculate α_t and $\bar{\alpha}_t$ for all timesteps in advance. To obtain a sample x_t at any timestep, simply sample the noise ϵ and substitute it into the equation provided earlier. In simpler terms, using the car analogy, it's like being able to recreate any customized version of the car without needing to recall all the previous customizations. Only the basic model and the current set of customizations are required. Imagine you have a remote-controlled toy car, but instead of directly steering it, you input commands into a computer that processes these commands and controls the car. The reparametrization trick is somewhat analogous to this: instead of directly sampling from a distribution, we use an alternate method to achieve our desired outcome.

The rationale for employing the reparametrization method stems from its computational efficiency. Although the approach may seem indirect, Once α_t and $\bar{\alpha}_t$ have been computed, it is possible to quickly generate a sample at any timestep without the need to retrace each preceding step. This method of operation is akin to fast-forwarding to any scene in a movie without having to watch everything that leads up to that point.

3.4 Reverse Diffusion: From Noise to Image

Diffusion processes can convert a clear image into a noisy one, while the reverse diffusion process works by starting with noise and tracing back to the original, effectively denoising or reconstructing the image. This process is depicted in Fig. 3, which

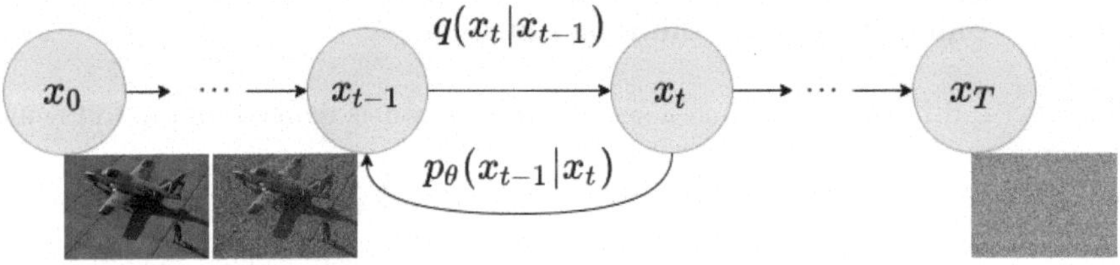

Fig. 3 The Reverse Diffusion Process.

illustrates the journey of denoising. The diagram starts with the heavily diffused image, labeled as x_T. The goal is to sequentially remove the noise from this image, retracing the path from x_T to x_0. This backward progression is characterized by $p(x_{t-1} \mid x_t)$, which defines the probabilistic rules guiding this reverse path [59].

The stages of denoising can be visually interpreted as follows:

- **Starting Point** (x_T): The starting image is highly distorted and noisy. This can be metaphorically likened to a glass filled with sand, if diffusion represents adding sand to clear water.
- **Intermediate States** ($x_t, x_{t-1},$ **etc.**): These transitional states between the distorted and original image gradually reduce the noise, akin to removing sand from water step-by-step to clarify it.
- **End Point** (x_0): The original, undistorted image, comparable to a glass of clear water after all the sand has been removed.

The following functions are central to understanding the forward and reverse diffusion processes:

- **Forward Diffusion** ($q(x_t \mid x_{t-1})$): Describes the forward diffusion process and defines the transition rule from one state to the next. It represents how noise is added, similar to adding sand to clear water.
- **Reverse Diffusion** ($p(x_{t-1} \mid x_t)$): Defines the reverse diffusion process, specifying how to retrace steps to remove noise. This is analogous to removing sand from water and clarifying the image.

In summary, the reverse diffusion process is essentially a reconstruction journey, demonstrating the transformative power of probabilistic models in handling complex transitions. The forward diffusion process resembles blurring a clear memory, while the reverse journey is akin to recollection and piecing scattered fragments back together to reconstruct the original image, reversing the entropy introduced in the forward process [153].

3.4.1 Decoding the Reverse Diffusion Equation

The reverse diffusion process at a specific time step t is defined by the following equation:

$$p_\theta(x_{t-1} \mid x_t) = \mathcal{N}(x_{t-1}; \mu_\theta(x_t, t), \Sigma_\theta(x_t, t)) \tag{9}$$

Components of the Equation

- $p_\theta(x_{t-1} \mid x_t)$: This conditional probability distribution represents the likelihood of the previous image x_{t-1} given the current image x_t. It determines the probability of transitioning from the current image to the previous one.
- $\mathcal{N}$: Stands for the Normal (Gaussian) distribution, chosen for its useful statistical properties and computational ease.
- $\mu_\theta(x_t, t)$: The mean of the Gaussian distribution, parameterized by θ, predicts the expected image at the previous step, serving as the starting point for reverse diffusion [59].
- $\Sigma_\theta(x_t, t)$: The variance of the Gaussian distribution, also parameterized by θ, indicates the range of possible images at x_{t-1}, given x_t. A larger variance implies a broader range, while a smaller one means a tighter cluster around the mean.

Think of this as watching a video in reverse: at any time step t, you see a noisy frame x_t. The reverse diffusion equation helps estimate the prior frame x_{t-1}, based on the current one. This probabilistic estimation, guided by parameter θ, facilitates accurate reconstruction and denoising of the original image.

4 Fundamentals and Mathematical Formulations of Diffusion Models

4.1 Notions and Definitions

4.1.1 Time and States

In diffusion models, the process evolves along a continuous or discrete timeline, where states represent distributions that outline the progression of the model. Noise is incrementally infused into the initial distribution, referred to as the starting state x_0, which is sampled from the data distribution p_0. This distribution gradually transitions to a known noise distribution, typically Gaussian, known as the prior state x_T. Intermediate states x_t lie between the starting and prior states, each having a marginal distribution p_t. This setup enables diffusion models to trace the evolution of the data distribution over time and generate samples that approximate the prior state x_T. The process proceeds through a sequence of intermediate states, each corresponding to a specific time point in the diffusion process.

4.1.2 Forward/Reverse Process and Transition Kernel

In diffusion models, the forward process F transforms the starting state into prior Gaussian noise, while the reverse process R denoises the prior state back to the starting state using transition kernels. As defined in DDPM [59], the discrete formulation of the diffusion model is generalized by establishing transition kernels for the diffusion and denoising processes:

$$F(x_0, \text{œ}) = F_T(x_{T-1}, \sigma_T) \cdots \circ F_t(x_{t-1}, \sigma_t) \cdots \circ F_1(x_0, \sigma_1) \tag{10}$$

$$R(x_T, \text{œ}) = R_1(x_1, \sigma_1) \cdots \circ R_t(x_t, \sigma_t) \cdots \circ R_T(x_T, \sigma_T) \tag{11}$$

where F_t and R_t represent the forward and reverse transition kernels at time t, with noise scale σ_t from the noise set **œ**. Unlike normalizing flow models, diffusion models leverage variable noise to gradually refine the distribution for a controlled transition towards the target distribution, providing a broader generation space and controllable generation. This discrete framework offers a practical implementation and computational efficiency by approximating the continuous diffusion process in discrete time.

4.1.3 From Discrete to Continuous

When the perturbation kernel is sufficiently small, the discrete processes ((10) and (11)) can be generalized into continuous processes. [159] demonstrated that diffusion models based on discrete Markov chains [59, 153] can be incorporated into a continuous Stochastic Differential Equation (SDE) framework, where the generative process reverses a fixed forward diffusion process. An equivalent reverse ODE marginally similar to the reverse SDE has also been derived [159]. The continuous process has better theoretical support and allows diffusion models to leverage established techniques in the ODE/SDE domain.

4.2 Background

In this subsection, we introduce three foundational formulations: Denoised Diffusion Probabilistic Models, SDE Formulation, and Conditional Diffusion Probabilistic Models.

4.2.1 Denoised Diffusion Probabilistic Models (DDPM)

DDPM Forward Process: Within the DDPM framework, a series of noise coefficients $\beta_1, \beta_2, \ldots, \beta_T$ are selected for the Markov transition kernels, often following constant, linear, or cosine schedules to enhance sample quality. As described by [59], the forward steps are defined as:

$$F_t(x_{t-1}, \beta_t) := q(x_t|x_{t-1}) := \mathcal{N}\left(x_t; \sqrt{1-\beta_t}x_{t-1}, \beta_t\mathbf{I}\right) \tag{12}$$

The composition of forward transition kernels from x_0 to x_T represents the Forward Diffusion Process, which injects Gaussian noise into the data through the Markov kernel $q(x_t|x_{t-1})$:

$$F(x_0, \{\beta_i\}_{i=1}^T) := q\left(x_{1:T} \mid x_0\right) := \prod_{t=1}^{T} q\left(x_t \mid x_{t-1}\right) \tag{13}$$

DDPM Reverse Process: The Reverse Process employs learnable Gaussian kernels parameterized by θ and is expressed as:

$$R_t(x_t, \Sigma_\theta) := p_\theta(x_{t-1}|x_t) := \mathcal{N}\left(x_{t-1}; \mu_\theta(x_t, t), \Sigma_\theta(x_t, t)\right) \tag{14}$$

where μ_θ and Σ_θ are the learnable mean and variance of the reverse Gaussian kernels defined by the reverse-step distribution p_θ. The reverse steps from x_T to x_0 are expressed as:

$$R(x_T, \Sigma_\theta) := p_\theta\left(x_{0:T}\right) := p\left(x_T\right)\prod_{t=1}^{T} p_\theta\left(x_{t-1} \mid x_t\right) \tag{15}$$

DDPM strives to estimate the data distribution p_0 through the joint probability distribution $p_\theta(x_0) = \int p_\theta(x_{0:T})dx_{1:T}$.

Diffusion Training Objective: The training objective is to minimize the variational bound on the negative log-likelihood via the KL-Divergence D_{KL}:

$$\begin{aligned} \mathbb{E}\left[-\log p_\theta\left(\mathrm{x}_0\right)\right] \leq \mathbb{E}_q[&\underbrace{D_{\mathrm{KL}}\left(q\left(x_T \mid x_0\right) \| p\left(x_T\right)\right)}_{L_T} \\ &+ \sum_{t>1} \underbrace{D_{\mathrm{KL}}\left(q\left(x_{t-1} \mid x_t, x_0\right) \| p_\theta\left(x_{t-1} \mid x_t\right)\right)}_{L_{t-1}} \\ &\underbrace{-\log p_\theta\left(x_0 \mid x_1\right)]}_{L_0} \end{aligned} \tag{16}$$

where L_T and L_0 represent the prior and reconstruction losses, respectively, while $L_{1:T-1}$ is the cumulative divergence between the forward and reverse-step posteriors. By simplifying L_{t-1}, the training objective simplifies to L_{simple} based on the posterior $q(x_{t-1}|x_t, x_0)$:

$$q\left(x_{t-1} \mid x_t, x_0\right) = \mathcal{N}\left(x_{t-1}; \tilde{\boldsymbol{\mu}}_t\left(x_t, x_0\right), \tilde{\beta}_t I\right) \tag{17}$$

where $\tilde{\beta}_t$ is dependent on β_t. Keeping this parameterization and reparameterizing x_t as $x_t(x_0, \sigma)$, L_{t-1} is given as an expectation of the ℓ_2 loss between two mean coefficients:

$$L_{t-1} = \mathbb{E}_q \left[\frac{1}{2\sigma_t^2} \left\| \tilde{\boldsymbol{\mu}}_t \left(\mathrm{x}_t, \mathrm{x}_0 \right) - \boldsymbol{\mu}_\theta \left(\mathrm{x}_t, t \right) \right\|^2 \right] + C \tag{18}$$

This is connected to the denoising score-matching discussed next. Simplifying L_{t-1} by reparameterizing μ_θ with ϵ_θ leads to the simplified training objective, named L_{simple}:

$$L_{simple} := \mathbb{E}_{x_0, \epsilon} \left[\frac{{\beta_t}^2}{2{\sigma_T}^2 \alpha_t (1 - \bar{\alpha}_t)} \right] \left\| \epsilon - \epsilon_\theta (\sqrt{\bar{\alpha}_t} x_0 + \sqrt{1 - \bar{\alpha}_t} \epsilon) \right\|^2 \tag{19}$$

Most diffusion models follow the DDPM training strategy. Improved DDPM combines L_{simple} with other objectives. Once trained, the prediction network ϵ_θ is used in the reverse process for ancestral sampling.

4.2.2 Score SDE Formulation

Score SDE [159] expands the DDPM discrete-time structure into a continuous-time framework using stochastic differential equations and proposes deterministic sampling frameworks based on ODE formulations.

Forward SDE: Score SDE [159] relates the continuous diffusion process to stochastic differential equations. The reverse process is connected to the solution of an Itô SDE [9], consisting of a drift term (for mean shift) and a Brownian motion (for noising):

$$dx = f(x, t)dt + g(t)dw, t \in [0, T] \tag{20}$$

where w_t represents the standard Wiener process, $f(\cdot, t)$ is the drift coefficient, and $g(\cdot)$ is a simplified diffusion coefficient independent of x. The marginal and prior distributions are $p_t(x)$ and p_T, respectively. If the coefficients are piecewise continuous, the forward SDE has a unique solution [126]. Two types of forward processes are introduced: Variation Preserving (VP) and Variation Explosion (VE) SDEs. The VP process is a continuous extension of the DDPM framework:

$$\text{VP:} \quad \mathrm{d}x = -\frac{1}{2}\beta(t)x \,\mathrm{d}t + \sqrt{\beta(t)}\mathrm{d}w$$

$$\text{VE:} \quad \mathrm{d}x = \sqrt{\frac{\mathrm{d}\left[\sigma^2(t)\right]}{\mathrm{d}t}}\,\mathrm{d}w$$

Reversed SDE: Sampling of diffusion models is achieved through the corresponding reverse-time SDE for the forward process (20) [7]:

$$\mathrm{d}x = \left[f(x,t) - g(t)^2\nabla_x \log p_t(x)\right]\mathrm{d}\bar{t} + g(t)\mathrm{d}\overline{w}, t \in [0, T] \tag{21}$$

where w_t is the standard Wiener process, $f(\cdot, t)$ is the drift coefficient, and $g(\cdot)$ is a simplified diffusion coefficient. If the coefficients are piecewise continuous, the forward SDE equation has a unique solution [173]:

$$L := \mathbb{E}_t\{\lambda(t)\mathbb{E}_{x_0}\mathbb{E}_{q(x_t|x_0)}[\|s_\theta(x_t,t) - \nabla_{x_t}\log p(x_t|x_0)\|_2^2]\} \tag{22}$$

where x_0 is sampled from p_0 and $\lambda(t)$ serves as the positive weighting function to maintain the time-dependent loss magnitude [159]. The Gaussian transition kernel associated with the forward process (20) is given by $q(x_t|x_0) = \mathcal{N}\left(x_t; x_0, \sigma^2(t)\mathbf{I}\right)$. The optimal solution to the denoising score-matching objective (22) matches the true score function $\nabla_x \log p_t(x)$ for nearly all x and t. Additionally, [191] demonstrates that the score function in the forward process of diffusion models can be divided into three phases. As data transition from the near field to the far field, perturbations influence more modes of the data distribution.

Probability Flow ODE: The probability flow ODE [159] provides a deterministic process that shares the same marginal probability density as the SDE. Drawing inspiration from Maoutsa et al. [117] and Chen et al. [24], any type of diffusion process can be rewritten as a special form of ODE. The corresponding probability flow ODE for (21) is:

$$dx = \{f(x,t) - \frac{1}{2}g(t)^2\nabla_x \log p_t(x)\}dt \tag{23}$$

Unlike the SDE, the probability flow ODE can be solved using larger step sizes since it lacks randomness. Consequently, methods like PNDMs [102] and DPM-Solver [109] achieve faster sampling speeds using advanced ODE solvers.

4.3 Conditional Diffusion Probabilistic Models

Diffusion models are versatile, capable of generating data samples from both unconditional p_0 and conditional $p_0(x|c)$ distributions, with c as a given condition such as a class label or text linked to data x [142]. The score network $s_\theta(x,t,c)$ integrates this condition during training. Various sampling algorithms, including classifier-free guidance [60] and classifier guidance [37], are designed for conditional generation.

Labeled Conditions Sampling with labeled conditions guides each sampling step's gradient. It typically requires an additional classifier with a UNet Encoder architecture to generate condition gradients for specific labels, which can be text, categorical, binary, or extracted features [12, 25, 37, 61, 67, 110, 118, 121, 130, 179]. The method, first presented by [37], underpins current conditional sampling techniques.

Unlabeled Conditions Unlabeled condition sampling uses self-information for guidance, often applied in a self-supervised manner [20, 68]. It is commonly used in denoising [32], paint-to-image [29], and inpainting tasks [165].

5 Classification and Comparative Analysis of Diffusion Models

Diffusion models are a class of probabilistic generative models designed to learn the reversal of a process that incrementally corrupts the structure of training data. The training involves two primary phases: the forward diffusion and the backward denoising process.

In the forward phase, the process comprises multiple steps where incremental levels of low-level noise are introduced to each input image, with the noise intensity varying at each stage. This process systematically degrades the training data until it transforms into pure Gaussian noise.

The backward phase entails reversing the process used in the forward phase. Employing the same iterative method but in reverse, this phase methodically eliminates the noise, thereby reconstructing the original image. Thus, during inference, images are generated by progressively restoring them from random white noise. The noise reduction at each step is computed using a neural network, generally employing a U-Net architecture [143], which helps maintain the dimensions.

Subsequently, three different formulations of diffusion models are introduced: denoising diffusion probabilistic models, noise conditioned score networks, and a method utilizing stochastic differential equations that extends the first two approaches. Each formulation is detailed in terms of the noise addition process, the learning method to invert this process, and the generation of new samples during inference. All three formulations are depicted within a unified framework in Fig. 4.

The final subsection is devoted to exploring the relationships with other deep generative models.

5.1 Denoising Diffusion Probabilistic Models (DDPMs)

Forward process. DDPMs [59, 153] gradually degrade the training data by applying Gaussian noise. Let $p(x_0)$ represent the data density, with the index 0 indicating that the data is in its original, uncorrupted state. Starting from an uncorrupted sample $x_0 \sim p(x_0)$, the noisy variants $x_1, x_2, \ldots, x_T$ are generated via the following Markovian sequence:

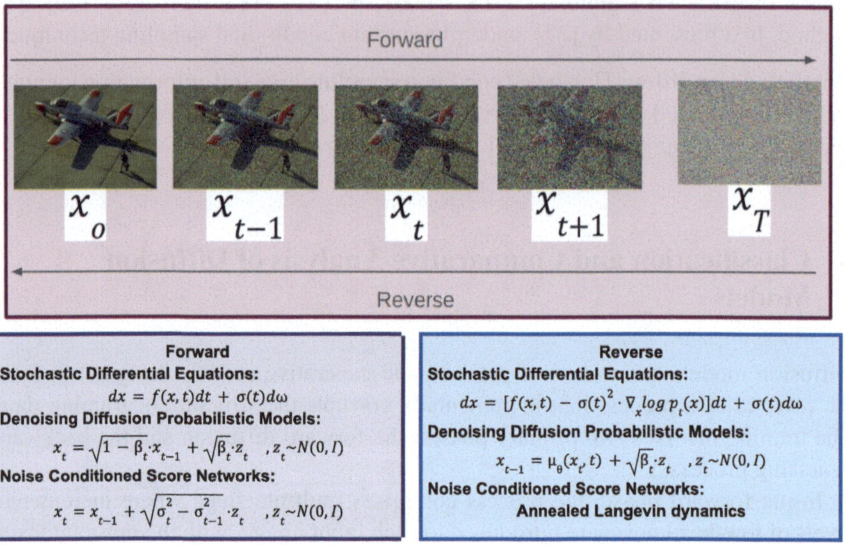

Fig. 4 The figure illustrates a framework for diffusion models, incorporating three approaches: stochastic differential equations (SDEs), denoising diffusion probabilistic models (DDPMs), and noise conditioned score networks (NCSNs). Each model features a forward process that transforms data into noise and a reverse process that reconstructs data from the noise. The SDE approach models data change over time as a function with a stochastic noise component, generalizing the other models. DDPMs and NCSNs introduce noise through iterative sampling and direct addition, respectively, with their reverse processes aimed at reconstructing the original data ffrom noise

$$p(x_t|x_{t-1}) = \mathcal{N}(x_t; \sqrt{1-\beta_t} \cdot x_{t-1}, \beta_t \cdot \mathbf{I}), \forall t \in \{1, \ldots, T\}, \tag{24}$$

where T denotes the number of diffusion steps, $\beta_1, \ldots, \beta_T \in [0, 1)$ are the hyper-parameters defining the variance at each step, $\mathbf{I}$ is the identity matrix sized to match the input image x_0, and $\mathcal{N}(x; \mu, \sigma)$ represents a normal distribution with mean μ and covariance σ that generates x. A key aspect of this recursive process is the ability to directly sample x_t from a uniformly distributed index t, i.e., $\forall t \sim \mathcal{U}(\{1, \ldots, T\})$:

$$p(x_t|x_0) = \mathcal{N}(x_t; \sqrt{\overline{\beta_t}} \cdot x_0, (1 - \overline{\beta_t}) \cdot \mathbf{I}), \tag{25}$$

where $\overline{\beta_t} = \prod_{i=1}^{t} \gamma_i$ and $\gamma_t = 1 - \beta_t$. This equation (25) implies that any noisy version x_t can be obtained directly from x_0 if the variance sequence β_t is known.

Sampling from $p(x_t|x_0)$ employs a reparametrization technique. Typically, to standardize a sample x from a normal distribution $x \sim \mathcal{N}(\mu, \sigma^2 \cdot \mathbf{I})$, one subtracts the mean μ and divides by the standard deviation σ, yielding a standardized sample $z = \frac{x-\mu}{\sigma}$ from the standard normal distribution $z \sim \mathcal{N}(0, \mathbf{I})$. The reparametrization reverses this process, generating x from z:

$$x_t = \sqrt{\overline{\beta_t}} \cdot x_0 + \sqrt{(1 - \overline{\beta_t})} \cdot z_t, \tag{26}$$

where $z_t \sim \mathcal{N}(0, \mathbf{I})$.

Properties of β_t. When β_t is structured such that $\overline{\beta}_T \to 0$, x_T is closely approximated by the standard Gaussian distribution $\pi(x_T) = \mathcal{N}(0, \mathbf{I})$, as per Eq. (25). Furthermore, when each $\beta_t \ll 1$, the reverse steps $p(x_{t-1}|x_t)$ mimic the form of the forward steps $p(x_t|x_{t-1})$ [153],?. This suggests that when x_t is generated through minimal changes, it's more probable that x_{t-1} will originate from a nearby region, which can be modeled with a Gaussian distribution. Ho *et al.*[59] selected a linearly increasing variance schedule $(\beta_t)_{t=1}^T$ from $\beta_1 = 10^{-4}$ to $\beta_T = 2 \cdot 10^{-2}$ with $T = 1000$.

Reverse process. Utilizing these characteristics, new samples from $p(x_0)$ can be synthesized starting from $x_T \sim \mathcal{N}(0, \mathbf{I})$ by retracing the reverse steps $p(x_{t-1}|x_t) = \mathcal{N}(x_{t-1}; \mu(x_t, t), \Sigma(x_t, t))$. To refine these steps, a neural network $p_\theta(x_{t-1}|x_t) = \mathcal{N}(x_{t-1}; \mu_\theta(x_t, t), \Sigma_\theta(x_t, t))$ is trained to estimate the mean $\mu_\theta(x_t, t)$ and the covariance $\Sigma_\theta(x_t, t)$ from the noisy image x_t and the time step embedding t.

Ideally, the neural network is trained using a maximum likelihood objective aiming to maximize the likelihood that the model $p_\theta(x_0)$ assigns to each training sample x_0. However, since $p_\theta(x_0)$ is computationally infeasible due to the need to consider all potential reverse paths, a variational lower-bound on the negative log-likelihood is minimized instead:

$$\begin{aligned} \mathcal{L}_{vlb} = & - \log p_\theta(x_0|x_1) + KL\left(p(x_T|x_0) \| \pi(x_T)\right) \\ & + \sum_{t>1} KL(p(x_{t-1}|x_t, x_0) \| p_\theta(x_{t-1}|x_t)), \end{aligned} \tag{27}$$

where KL represents the Kullback-Leibler divergence between two probability distributions. Through detailed analysis, it is evident that the second term can be omitted as it does not influence θ. The last term confirms that, at each time step t, the neural network training aligns $p_\theta(x_{t-1}|x_t)$ as close as possible to the true posterior of the forward process conditioned on x_0. It is also established that the posterior $p(x_{t-1}|x_t, x_0)$ is a Gaussian distribution, allowing closed-form solutions for the KL divergences.

Ho et al. [59] suggest fixing the covariance $\Sigma_\theta(x_t, t)$ at a constant value and modifying the mean $\mu_\theta(x_t, t)$ based on the noise profile as follows:

$$\mu_\theta = \frac{1}{\sqrt{\gamma_t}} \cdot (x_t - \frac{1 - \gamma_t}{\sqrt{1 - \overline{\beta}_t}} \cdot z_\theta(x_t, t)). \tag{28}$$

These simplifications enhance the clarity of the objective $\mathcal{L}_{vlb}$, which quantifies, for a randomly selected time step t of the forward process, the discrepancy between the actual noise z_t and the estimated noise $z_\theta(x_t, t)$ by the model:

$$\mathcal{L}_{simple} = \mathbb{E}_{t \sim [1,T]} \mathbb{E}_{x_0 \sim p(x_0)} \mathbb{E}_{z_t \sim \mathcal{N}(0,\mathbf{I})} \left\| z_t - z_\theta(x_t, t) \right\|^2, \tag{29}$$

where $\mathbb{E}$ denotes the expectation, and $z_\theta(x_t, t)$ is the predicted noise in x_t. It is highlighted that x_t is sampled using Eq. (26), utilizing a random image x_0 from the dataset.

The generative methodology remains defined by $p_\theta(x_{t-1}|x_t)$, yet the neural network does not compute the mean and covariance directly. Instead, it is optimized to predict the noise in the image, with the mean adjusted as per Eq. (28) and the covariance held constant.

5.2 Noise Conditioned Score Networks (NCSNs)

The score function associated with a data density $p(x)$ is the gradient of the log density with respect to the input, denoted as $\nabla_x \log p(x)$. This gradient directs the Langevin dynamics algorithm [156] to transition from an initial random sample (x_0) to samples (x_N) situated in areas of higher density. *Langevin dynamics*, originating from physics, is adapted here for sampling data. In physics, it predicts particle trajectories in molecular systems by accounting for interactions with surrounding molecules. The particle's movement is influenced by both a drag force and a stochastic force from rapid molecular interactions. Analogously, in our context, the gradient of the log density acts as a drag force pulling a random sample towards denser data regions, while the stochastic term ω_i helps to avoid local minima. The coefficient γ balances the influence of these forces, functioning as the friction coefficient in the simulated environment. The updates in Langevin dynamics are described by:

$$x_i = x_{i-1} + \frac{\gamma}{2}\nabla_x \log p(x) + \sqrt{\gamma} \cdot \omega_i, \tag{30}$$

where $i \in \{1, \ldots, N\}$, γ determines the update magnitude, x_0 originates from a prior distribution, $\omega_i \sim \mathcal{N}(0, \mathbf{I})$ combats local optima, and the process iterates for $N \to \infty$ steps. A generative model utilizes this method to sample from $p(x)$ after approximating the score with a neural network $s_\theta(x) \approx \nabla_x \log p(x)$. This network is trained via score matching, optimizing the objective:

$$\mathcal{L}_{sm} = \mathbb{E}_{x \sim p(x)} \|s_\theta(x) - \nabla_x \log p(x)\|_2^2 . \tag{31}$$

Directly minimizing this objective is infeasible as $\nabla_x \log p(x)$ is typically unknown; alternative methods like denoising score matching [173] and sliced score matching [157] are used instead.

Despite its utility in data generation, Song et al. [156] identify several challenges when implementing this method on real data, mainly due to the manifold hypothesis. Specifically, when data lies on a low-dimensional manifold, the score estimation $s_\theta(x)$ can be inconsistent, possibly causing Langevin dynamics to fail in converging to areas of high density. To address these issues, they propose introducing Gaussian noise at varying scales and training a noise conditioned score network (NCSN) for the resultant noisy distributions. The strategy in Eq. (30) is adapted for each noise scale during sampling.

Formally, with a sequence of increasing Gaussian noise levels $\sigma_1 < \sigma_2 < \cdots < \sigma_T$, we train an NCSN $s_\theta(x, \sigma_t)$ using denoising score matching to approximate $\nabla_x \log p_{\sigma_t}(x)$ for each noise level $t \in \{1, \ldots, T\}$. The gradient $\nabla_x \log p_{\sigma_t}(x)$ is derived as:

$$\nabla_{x_t} \log p_{\sigma_t}(x_t|x) = -\frac{x_t - x}{\sigma_t^2}, \tag{32}$$

based on:

$$p_{\sigma_t}(x_t|x) = \mathcal{N}(x_t; x, \sigma_t^2 \cdot \mathbf{I}) = \frac{1}{\sigma_t \cdot \sqrt{2\pi}} \cdot exp\left(-\frac{1}{2} \cdot \left(\frac{x_t - x}{\sigma_t}\right)^2\right), \tag{33}$$

where x_t is a noised version of x, and *exp* denotes the exponential function. Generalizing Eq. (31) for all $(\sigma_t)_{t=1}^T$ and utilizing the gradient form in Eq. (32), training of $s_\theta(x_t, \sigma_t)$ involves minimizing:

$$\mathcal{L}_{dsm} = \frac{1}{T} \sum_{t=1}^{T} \lambda(\sigma_t) \mathbb{E}_{p(x)} \mathbb{E}_{x_t \sim p_{\sigma_t}(x_t|x)} \left\| s_\theta(x_t, \sigma_t) + \frac{x_t - x}{\sigma_t^2} \right\|_2^2, \tag{34}$$

where $\lambda(\sigma_t)$ is a weighting function. The trained network $s_\theta(x_t, \sigma_t)$ subsequently estimates the scores $\nabla_{x_t} \log p_{\sigma_t}(x_t)$ based on the noised image x_t and the associated time step t.

At inference, Song *et al.*[156] employ annealed Langevin dynamics, initiating with white noise and iteratively applying Eq. (30) for a predefined number of steps. The necessary gradient (score) is supplied by the trained neural network, conditioned on the final time step T. The process sequentially propagates the output from one step to the next, with the ultimate sample being the output for $t = 0$.

5.3 *Stochastic Differential Equations (SDEs)*

Similarly to the methods previously discussed, the technique presented in [159] also transitions the data distribution $p(x_0)$ to noise, but it extends the earlier approaches by treating the diffusion process as continuous, formulated through a stochastic differential equation (SDE). As demonstrated by [7], the reversal of this diffusion is modeled via a reverse-time SDE, which necessitates computing the score function of the density at each time step. In this framework, Song et al. [159] use a neural network to approximate these score functions, generating samples from $p(x_0)$ by applying numerical SDE solvers. Similar to NCSNs, this network inputs perturbed data and time steps, outputting an estimate of the score function.

The forward diffusion SDE $(x_t)_{t=0}^{T}$ for $t \in [0, T]$ is given by:

$$\frac{\partial x}{\partial t} = f(x,t) + \sigma(t) \cdot \omega_t \Longleftrightarrow \partial x = f(x,t) \cdot \partial t + \sigma(t) \cdot \partial \omega, \tag{35}$$

where ω_t represents Gaussian noise, f defines the drift coefficient dependent on x and t, and σ is a function of time that determines the diffusion coefficient. The drift aims to incrementally negate the original data x_0, while the diffusion coefficient manages the addition of Gaussian noise. The associated reverse-time SDE, as defined in [7], is:

$$\partial x = \left[f(x,t) - \sigma(t)^2 \cdot \nabla_x \log p_t(x) \right] \cdot \partial t + \sigma(t) \cdot \partial \hat{\omega}, \tag{36}$$

where $\hat{\omega}$ signifies the reverse Brownian motion, spanning from T to 0. This reverse-time SDE illustrates that starting with pure noise, the data can be reconstructed by removing the drift that leads to data degradation, achieved by subtracting $\sigma(t)^2 \cdot \nabla_x \log p_t(x)$.

To train the neural network $s_\theta(x,t) \approx \nabla_x \log p_t(x)$, we optimize an objective similar to Eq. (34), modified for the continuous scenario:

$$\mathcal{L}^*_{dsm} = \mathbb{E}_t \left[\lambda(t) \mathbb{E}_{p(x_0)} \mathbb{E}_{p_t(x_t|x_0)} \left\| s_\theta(x_t, t) - \nabla_{x_t} \log p_t(x_t|x_0) \right\|_2^2 \right], \tag{37}$$

where $\lambda(t)$ is a weighting function and $t \sim \mathcal{U}([0, T])$. If the drift coefficient f is linear, $p_t(x_t|x_0)$ conforms to a Gaussian distribution. If not, denoising score matching may not be applicable, but sliced score matching [157] remains a viable alternative.

Sampling via this approach employs numerical methods on the SDE outlined in Eq. (36). Commonly, solvers adjust for the continuous nature by implementing the Euler-Maruyama method, which utilizes a small negative time step Δt and iterates from the initial time $t = T$ to $t = 0$. In this method, $\Delta\hat{\omega} = \sqrt{|\Delta t|} \cdot z$ with $z \sim \mathcal{N}(0, \mathbf{I})$. Song et al. [159] further refine sampling techniques, introducing the Predictor-Corrector sampler that enhances sample quality. This method initially employs a numerical solver for the reverse-time SDE, supplemented by a score-based corrector such as annealed Langevin dynamics. They also demonstrate that ordinary differential equations (ODEs) can model the reverse process, offering an alternative efficient sampling method based on numerical ODE solvers.

5.4 Relation to Other Generative Models

We explore the relationships between diffusion models and other types of generative models, beginning with likelihood-based methods and concluding with generative adversarial networks.

Diffusion models share several similarities with Variational Autoencoders (VAEs) [81]. In both methodologies, the data is encoded into a latent space, and the generative process learns to reconstruct data from these latent representations. Furthermore, the training objective in both models can be seen as a lower bound on the data likelihood. However, there are key differences. VAEs encode data into a compressed latent space containing essential information about the original data, whereas diffusion models progressively transform the data into pure noise. In diffusion models, the latent representations maintain the same dimensions as the original data, unlike VAEs which typically reduce the dimensionality for efficiency. Additionally, the mapping to latent space in VAEs is learnable, unlike the forward process in diffusion models, which is defined by the progressive addition of Gaussian noise. These similarities and differences have prompted research into integrating these models, such as works that incorporate diffusion models within the latent space of a VAE for enhanced efficiency [131, 169].

Autoregressive models [127, 148] interpret images as sequences of pixels, generating each pixel conditionally based on the previous ones. This sequential generation introduces a unidirectional bias, which is a known limitation of this model class. Esser et al. [42] view diffusion and autoregressive models as complementary, employing a technique that reverses a multinomial diffusion process through a Markov chain, where each transition is modeled as an autoregressive process. The global information for the autoregressive model is provided by the preceding step in the chain.

Normalizing flows [38, 39] transform a simple Gaussian distribution into a complex data distribution using a series of invertible functions that allow for an easily computed Jacobian determinant. These transformations result in architectural restrictions due to the need for invertibility and differentiability. While normalizing flows share with diffusion models the general concept of transforming data distributions into Gaussian noise, they differ significantly in that normalizing flows learn this transformation deterministically and involve a learnable, invertible mapping, imposing additional architectural constraints. DiffFlow [200] is an example of integrating diffusion models with normalizing flows, making both the forward and reverse processes trainable and stochastic.

Energy-based models (EBMs) [11, 87, 120, 161] estimate unnormalized density functions, known as energy functions, which provides them with a high degree of model flexibility. Despite this flexibility, EBMs are challenging to train. A common training approach is score matching [11, 161], which utilizes the score function for training and sampling, placing them within the broader framework of energy-based models.

Generative Adversarial Networks (GANs) [48], once considered state-of-the-art for the quality of generated samples, are known for their challenging training dynamics, including issues like mode collapse. In contrast, diffusion models offer a more stable training experience and greater sample diversity due to their likelihood-based nature. However, diffusion models are less efficient during inference, requiring multiple network evaluations [37]. Despite the lower-dimensional latent space of GANs, which contrasts with the dimension-preserving latent space of diffusion models, GANs have been shown to encode meaningful semantic subspaces [150]. This facilitates direct manipulation of visual attributes within the latent space [137, 150]. Conversely, diffusion models typically utilize guidance techniques for modifying sample attributes during generation [37, 60], which do not leverage semantic latent properties. However, Song et al. [159] have shown that diffusion model latent spaces do possess a structured organization, suggesting potential for future exploration in semantic manipulations similar to those used with GANs. This aspect remains under-explored compared to GANs but represents a promising direction for future research.

6 Applications of Diffusion Models

Leveraging their robust capability to produce lifelike samples, diffusion models have found extensive usage across diverse domains. The effectiveness of diffusion models in real-world scenarios hinges on their adaptation to the diffusion, denoising, and conditional sampling processes across a broad spectrum of data types. Motivated by this insight, the various applications of diffusion models can be categorized into several key areas, including **Image Generation**, **3D Generation**, **Video Generation**, **Medical Analysis**, **Text Generation**, **Time Series Generation**, **Audio Generation**, **Molecule Design**, and **Graph Generation**.

The Table 1 provide a comprehensive overview of various diffusion methods across multiple fields. In computer vision, techniques like CMDE [13], DDRM [76], and Palette [145] apply diffusion models to image inpainting, super-resolution, and colorization. Advanced architectures such as GLIDE [121] and VQ-Diffusion [51] explore text-to-image synthesis, while LSGM [169] supports both conditional and unconditional image generation. For video processing, diffusion models like VDM [61] and RVD [194] enable text-conditioned video generation and video compression.

Moving to multimedia, we see diffusion-based models like Score-MRI [33] for restoring MRI images, while Song's model [158] restores both MRI and CT images. In sequence modeling, Diffusion-LM [94] and Bit Diffusion [26] excel in text creation, while CSDI [165] and SSSD [3] address time series imputation and completion. Audio and speech processing leverage diffusion models like WaveGrad [23] and DiffWave [85] to synthesize waveforms, while GradTTS [135] specializes in waveform synthesis using SDEs. ProDiff [70] and BinauralGrad [89] extend these capabilities to text-to-speech and binaural audio synthesis.

In AI for science, diffusion methods like ConfGF [151] and GeoDiff [187] generate molecular conformations. Torsional Diffusion [73] and DiffDock [34] synthesize molecular structures, providing significant insights into drug discovery. ProteinSGM [88] and DiffFolding [180] tackle protein generation and folding via inverse folding. This collection of methods underscores the versatility and innovative potential of diffusion techniques in modern AI research. The performance of diffusion models is typically assessed using a variety of metrics that reflect the quality and fidelity of generated samples. In image-related tasks, the **Fréchet Inception Distance (FID)** [58] is commonly used to measure the distance between feature vectors calculated for real and generated images. For tasks involving time series and audio, metrics such as the **Signal-to-Noise Ratio (SNR)** [172] and **Log-Likelihood** [166] are employed to assess the clarity and fidelity of the generated outputs. In medical imaging, the **Struc-**

Table 1 Overview of diffusion methods in various applications

Technique	Year	Dataset	Algorithm type	Application	Source code
Computer vision					
CMDE [13]	2021	RGB-image	SDE	Inpainting, super-resolution, edge to image translation	[code]
DDRM [76]	2022	RGB-image	Diffusion	Super-resolution, deblurring, inpainting, colorization	[code]
Palette [145]	2022	RGB-image	Diffusion	Colorization, inpainting, uncropping, JPEG restoration	[code]
DiffC [167]	2022	RGB-image	SDE	Compression	–
SRDiff [92]	2021	RGB-image	Diffusion	Super-Resolution –	
RePaint [111]	2022	RGB-image	Diffusion	Inpainting, super-resolution, edge to image translation	[code]
FSDM [45]	2022	RGB-image	Diffusion	Few-shot generation	–
CARD [54]	2022	RGB-image	Diffusion	Conditional generation	[code]
GLIDE [121]	2022	RGB-image	Diffusion	Conditional generation	[code]
LSGM [169]	2022	RGB-image	SDE	UnConditional and conditional generation	[code]
SegDiff [5]	2022	RGB-image	Diffusion	Segmentation	–
VQ-Diffusion [51]	2022	VQ data	Diffusion	Text-to-image synthesis	[code]
DreamFusion [134]	2023	VQ data	Diffusion	Text-to-image synthesis	[code]
Text-to-Sign VQ [185]	2022	VQ data	Diffusion	Conditional pose generation	–
Improved VQ-Diff [164]	2022	VQ data	Diffusion	Text-to-image synthesis	–

(continued)

Table 1 (continued)

Technique	Year	Dataset	Algorithm type	Application	Source code
Luo's Model [112]	2021	Point cloud	Diffusion	Point cloud generation	[code]
PVD [202]	2022	Point cloud	Diffusion	Point cloud generation, Point-Voxel representation	[code]
PDR [116]	2022	Point cloud	Diffusion	Point cloud completion	[code]
Cheng's Model [28]	2022	Point cloud	Diffusion	Point cloud generation	[code]
Luo's Model[113]	2022	Point cloud	Score	Point cloud denoising	[code]
VDM [61]	2022	Video	Diffusion	Text-conditioned video generation	[code]
RVD [194]	2022	Video	Diffusion	Video forecasting, video compression	[code]
FDM [55]	2022	Video	Diffusion	Video forecasting, long-term video modeling	–
MCVD [174]	2022	Video	Diffusion	Video prediction, comprehensive video synthesis, video interpolation	[code]
RaMViD [64]	2022	Video	SDE	Directed video generation	–
Score-MRI [33]	2022	MRI	SDE	Restoration of MRI images	[code]
Song's Model [158]	2022	MRI, CT	SDE	Restoration of MRI and CT images	[code]
R2D2+ [30]	2022	MRI	SDE	MRI noise reduction	–
Sequence modeling					
Diffusion-LM [94]	2022	Text	Diffusion	Textual content creation	[code]
Bit diffusion [26]	2022	Text	Diffusion	Text creation triggered by image contexts	[code]
D3PM [10]	2021	Text	Diffusion	Text synthesis	–
Argmax [62]	2021	Text	Diffusion	Text segmentation and generation	[code]
CSDI [165]	2021	Time series	Diffusion	Time series completion	[code]
SSSD [3]	2022	Time series	Diffusion	Time series completion	[code]
CSDE [132]	2022	Time series	SDE	Time series imputation and prediction	–

(continued)

Table 1 (continued)

Technique	Year	Dataset	Algorithm type	Application	Source code
Audio and Speech					
WaveGrad [23]	2020	Audio	Diffusion	Generation of waveforms	[code]
DiffWave [85]	2021	Audio	Diffusion	Waveform synthesis for various conditions	[code]
GradTTS [135]	2021	Audio	SDE	Synthesis of waveforms	[code]
Diff-TTS [72]	2021	Audio	Diffusion	Mel-spectrogram synthesis, speech synthesis without auto-regression	–
DiffVC [136]	2022	Audio	SDE	Conversion of voices	[code]
DiffSVC [106]	2022	Audio	Diffusion	Voice conversion techniques	[code]
DiffSinger [101]	2022	Audio	Diffusion	Synthesis of singing voice	[code]
Diffsound [193]	2021	Audio	Diffusion	Synthesis from text to sound	[code]
EdiTTS [162]	2022	Audio	SDE	Pitch and content modification in audio	[code]
Guided-TTS [78]	2022	Audio	SDE	Targeted speech generation	–
Guided-TTS2 [79]	2022	Audio	SDE	Directed speech generation	–
Levkovitch's Model [90]	2022	Audio	SDE	Voice generation from spectrograms	[code]
SpecGrad [84]	2022	Audio	Diffusion	Voice synthesis from spectrograms	[code]
ItoTTS [182]	2022	Audio	SDE	Spectrogram-based voice synthesis	–
ProDiff [70]	2022	Audio	Diffusion	Synthesis of speech from text	[code]
BinauralGrad [89]	2022	Audio	Diffusion	Synthesis of binaural audio	–

(continued)

Table 1 (continued)

Technique	Year	Dataset	Algorithm type	Application	Source code
AI for science					
ConfGF [151]	2021	Molecular	Score	Generation of molecular conformations	[code]
DGSM [114]	2022	Molecular	Score	Generation of molecular conformations and sidechains	–
GeoDiff [187]	2022	Molecular	Diffusion	Molecular conformation synthesis	[code]
EDM [63]	2022	Molecular	SDE	Synthesis of molecular structures	[code]
Torsional Diff [73]	2022	Molecular	Diffusion	Molecular structure generation	[code]
DiffDock [34]	2022	Molecular	Diffusion	Molecular docking and conformation synthesis	[code]
CDVAE [186]	2022	Protein	Score	Generation of periodic Materials	[code]
Luo's Model [115]	2022	Protein	Diffusion	Complementarity-determining Region (CDR) Generation	–
Anand's Model [6]	2022	Protein	Diffusion	Generation of protein sequences and structures	–
ProteinSGM [88]	2022	Protein	SDE	de novo protein design	–
DiffFolding [180]	2022	Protein	Diffusion	Protein structure prediction via inverse folding	[code]

tural Similarity Index (SSIM) [175] and **Peak Signal-to-Noise Ratio (PSNR)** [65] are used to evaluate the similarity between the original and reconstructed images, reflecting the precision of the restoration. These metrics provide a comprehensive framework to objectively quantify the effectiveness of diffusion models across various domains, ensuring that their performance is consistently evaluated against rigorous standards.

6.1 Image Generation

Diffusion models have demonstrated exceptional capabilities in image generation tasks. This includes both traditional class-conditioned or unconditional generation approaches [37, 59, 146], as well as more complex conditions based on text or images [142, 199], and even their hybrids [17]. Moving forward, our discussion will focus on application settings that replicate real-world environments, organizing these applications by the types of conditional inputs they utilize.

6.1.1 Text Condition

Diffusion models excel in text-to-image generation, producing images that are not only photorealistic but also closely aligned with textual descriptions provided by users. Notable implementations include Imagen [146], Stable Diffusion [142], and DALL-E 2 [138]. These models enhance standard diffusion architectures by incorporating a cross-attention layer that integrates sequences of text embeddings directly into the diffusion process. Experimental evidence suggests that this text-conditioning approach successfully integrates textual data into the image generation process.

Moreover, the use of cross-attention facilitates numerous training-free image editing techniques. These methods manipulate the keys, values, or attention matrices within the cross-attention layers to modify image features. For instance, [57] alters concepts within source images by replacing or adding new feature maps to the outputs of cross-attention layers; [144] customizes new concepts by introducing novel text embeddings into the cross-attention mechanism. Additionally, [21] ensures that the cross-attention covers all relevant tokens from the text prompt, amplifying their influence to guarantee that the model accurately renders all subjects mentioned in the text.

6.1.2 Image Condition

Diffusion models can also incorporate image-based conditions, such as images targeted for editing, depth maps, or human skeletons, into their generative processes. This involves embedding encoded image features into the core diffusion architecture. For instance, [17] integrates encoded features from a source image directly into the first convolutional layer, facilitating image-to-image transformations guided by textual prompts. Likewise, [199] employs depth maps, Canny edges, or human skeletons to guide the spatial composition of the output images.

6.2 3D Generation

Diffusion models utilize primarily two approaches for 3D generation. The first involves directly training the models on 3D data, while the second leverages 2D diffusion priors to generate 3D content due to the scarcity of comprehensive 3D datasets.

6.2.1 3D Data Condition

Diffusion models have been effectively applied to a variety of 3D data representations, including NeRF, point clouds, voxels, and Gaussian splatting. Notable examples include works that generate point clouds for 3D objects [112, 113, 202]. For efficient shape processing, a hybrid point-voxel representation was used in PDR [116], which introduces a new approach to point cloud completion. Expanding upon this, Point-E [122] integrates image synthesis into point cloud diffusion models as an additional conditional input.

Conversely, Shape-E [74] applies diffusion models to NeRF representations of 3D objects. Zero-1-to-3 [105] explores training viewpoint-conditioned diffusion models to facilitate novel view synthesis and then utilizes these models to optimize a NeRF across various camera angles. Building on this, [104] incorporates a pose estimation phase into the Zero-1-to-3 framework.

6.2.2 2D Diffusion Prior

An innovative research direction involves extracting 3D structures from a 2D diffusion model. DreamFusion [134] employs score distillation sampling (SDS) to distill a NeRF from pretrained text-to-image models, optimizing a randomly initialized NeRF through gradient descent to minimize the loss in rendered images from multiple perspectives. [97] expands on DreamFusion by introducing a two-stage coarse-to-fine optimization framework, enhancing the speed of the generation process.

6.3 Video Generation

Video diffusion models enhance traditional 2D image generation diffusion models by incorporating an additional temporal dimension. This approach introduces a temporal layer to the existing 2D diffusion frameworks to effectively capture the dependencies between frames. Noteworthy contributions in this area include Video Diffusion Models [61], Make-A-Video [152], AnimatedDiff [53], RVD [194], FDM [55], and MCVD [174]. Additionally, RaMViD [64] enhances image-based diffusion models for video applications by utilizing 3D convolutional neural networks and introduces sophisticated conditioning techniques to support video prediction, infilling, and upscaling tasks.

6.4 Medical Analysis

Diffusion models offer solutions to the unique challenges of medical analysis, particularly in environments where acquiring large-scale, high-quality annotated datasets is difficult. These models excel in tasks that require in-distribution analysis and cross-distribution generation.

6.4.1 In-Distribution Analysis

In the medical imaging, diffusion models leverage their capacity to integrate strong prior knowledge, thereby enhancing performance across various tasks. They have shown notable success in super-resolution [31, 32], classification [195], and enhancing noise robustness [33, 66]. For instance, Score-MRI [33] utilizes pixel guidance SDE sampling to expedite MRI reconstruction. Similarly, Diff-MIC [195] uses Dual-granularity guidance and Maximum-Mean Discrepancy to achieve accurate classification across different imaging modalities. Additionally, MCG [31] introduces manifold correction during sampling for CT super-resolution, significantly reducing errors and improving the process efficiency.

6.4.2 Cross-Distribution Generation

The use of multimodal guidance has significantly enhanced the generative capabilities of diffusion models in medical settings. These models integrate class-specific [93, 184] and pixel-level guidance [46, 77, 93] to facilitate image translation across various types of scarce medical images, including those of high-quality, healthy, and unbiased categories. For example, FNDM [93] employs a non-Markovian framework with hybrid-condition guidance for accurate brain anomaly detection, while Diffuse-Morph [77] handles MR image registration through continuous diffusion sampling conditioned on pairs of moving and fixed images. Furthermore, diffusion models are being used to enrich training datasets by generating realistic medical images from a limited number of high-quality samples [30, 41, 133]. Notably, Latent Diffusion Models trained on 31,740 samples have synthesized a semantically rich dataset of 100,000 instances with an impressive FID score of 0.0076 [133].

6.5 Text Generation

Text generation is pivotal for bridging the gap between humans and advanced AI systems by producing natural, coherent language. While autoregressive language models generate text sequentially, ensuring high semantic coherence at the cost of slower generation speed [128], diffusion models facilitate parallel text generation, which speeds up the process but may compromise on semantic coherence [95, 204]. The challenges of generating discrete tokens are addressed primarily through two approaches: **Discrete Generation** and **Latent Generation**.

6.5.1 Discrete Generation

This approach involves models that take discrete words as inputs and employ advanced techniques, parameterizations, and pretrained models. Pioneering efforts to link diffusion models with discrete generation, such as D3PM [10] and Argmax [62], treat words as categorical vectors, establishing forward and backward processes via a discrete transition matrix. These models consider the generated data as part of a stationary distribution. DiffusionBERT [56] merges diffusion models with pretrained language models, improving text generation capabilities. It introduces an innovative noise schedule and explores the integration of time steps into BERT for reverse diffusion processes.

6.5.2 Latent Generation

The latent generation approach generates text in the latent space of tokens, capturing the continuous nature of the diffusion process. It includes enhanced loss functions [44, 183, 198], diverse generation types [47, 160], and advanced model architectures [94, 99]. For instance, LM-Diffusion [94] employs transformer-based graphical models for controlled generation, showing superior performance across various text generation tasks. GENIE [99] introduces a large-scale language model based on the diffusion framework, incorporating a novel Continuous Paragraph Denoise (CPD) loss for improved denoising and paragraph-level coherence. This model demonstrates the potential of diffusion-based decoders in text generation and lays a strong foundation for future research. Along with advancements in conditional sampling, token-level capturing, and post-refinement, diffusion models in NLP are poised to enhance embedding space modeling [95, 160], integrate with large pretrained language models, and support cross-modality generations [163, 196, 204].

6.6 *Time Series Generation*

Effective time series modeling is essential for accurate trend prediction, decision-making, and real-time analysis. Diffusion models enhance this process by incorporating modules specifically designed for time series data, enabling superior analysis and diverse generative capabilities [98]. These models can handle tasks categorized into inpainting and prediction, distinguished by their masking strategies. In inpainting tasks, observed states serve as prior conditions [96, 103, 108, 165], enhanced by context-based modules. The CSDI approach introduces a self-supervised training framework utilizing bidirectional CNN modules, which significantly improves the continuous generation of healthcare and environmental data [165]. In prediction tasks, prior states are converted into user-defined features and latent embeddings that act as self-conditions [139, 177, 192]. Integrating temporal-spatial modules like Graph UNet and RNN, DiffSTG and TimeGrad effectively achieve spatio-temporal probabilistic learning for time series [139, 177]. The success of these applications depends on precise modeling of time-dependent series and robust self-conditional guidance during sampling, signaling promising directions for future developments in this field [43, 98, 177].

6.7 Audio Generation

The synthesis of high-quality simulated speech supports various applications in music composition, virtual reality, game development, and voice-assisted technologies, enhancing personalized and immersive audio experiences and improving human-computer interaction. Diffusion models, adept at handling the unique characteristics of audio data, employ strong priors and manage high-dimensional, time-dependent information effectively. Hybrid conditions that combine text and control tags are used to achieve specific semantic or sound features. Techniques such as WaveGrad [23], DiffSinger [101], and DiffSVC [106] utilize Mel-Spectrograms for conditional guidance, while BinualGrad [89] processes audio based on mono input. These methods lay the groundwork for general waveform generation, with additional features like loudness, melody, and phonetic posteriorgrams enabling controllable style generation [101, 106, 136, 182]. Text-based and music-based generation, including text-to-speech and acoustic generation, depend on spectrogram features. Diffusion models incorporate text and rhythm as latent variables, utilizing spectrogram features and multi-view labels during sampling. Guided-TTS [78] and Diff-TTS [72] include a speaker text encoder, duration predictor, and phoneme classifier for generating content and guiding speech style. Guide-TTS2 [79] extends this approach to generating untranscribed speech using a classifier-free speaker encoder. Additional guidance factors include emotion, noise level, and music style, enhancing the richness and diversity of the generated audio [84, 101, 162, 193].

6.8 Molecule Design

Molecules, the fundamental building blocks of life, are crucial in myriad biological processes. The design of functional molecules has historically been a challenging endeavor [119]. Generative models have transformed molecular design by providing a more efficient alternative to traditional methods of enumeration and experimental validation. By capturing specific modal distributions and functional domains, these models can produce novel and potentially effective drug molecule structures, broadening the scope of drug design [16]. In drug discovery, diffusion models effectively explore vast compound spaces, speeding up the identification of potential drug candidates. This not only boosts the efficiency of the drug discovery process but also unveils complex compound relationships that deepen our understanding of drug mechanisms. The strategies in molecule design typically fall into two categories: unconditional generation and cross-modal generation.

6.8.1 Unconditional Generation

Unconditional molecule generation via diffusion models emphasizes speed and high-quality modeling. One method involves generating the positions of molecules in three-dimensional space, capturing molecular conformations [6, 63, 187]. However, this method may suffer from lower diversity and higher errors due to the non-uniform and irregular distribution of molecular structures. Alternatively, models that capture multiple features and structural feature distributions in high-dimensional space can yield more diverse distributions and enhance interpretability [73, 88, 180, 197]. To improve diversity, [35] introduces a repulsion force between samples.

6.8.2 Cross-Modal Generation

In molecular design, cross-modal generation integrates functionality as a conditioning factor. Diffusion-based methods excel at incorporating conditions and using denoising models across different modalities to boost modeling capabilities. Sequence-based cross-modal generation methods employ protein sequences and multiple sequence alignments (MSA) to train denoising models, integrating specific structural information and functional labels to guide the generation process [1, 100]. Structure-based cross-modal methods utilize prior knowledge from structure prediction models for precisely guided generation, combining protein sequences with functional information [176]. Molecular docking and antibody design methods leverage structural priors of target molecules to guide the docking process and identify favorable binding configurations [34, 115], using this prior knowledge to enhance generation and achieve promising conformations.

6.9 Graph Generation

The impetus for utilizing diffusion models in graph generation is driven by the desire to study and simulate diverse real-world networks and propagation processes. This methodology enhances understanding and problem-solving capabilities in complex issues by enabling researchers to probe the interactions and information propagation within intricate systems. It reveals hidden patterns and correlations and aids in predicting possible outcomes. Its applications range from social network analysis to the exploration of biological neural systems, and the generation and evaluation of graph datasets.

Traditional methods for graph generation typically involve creating an adjacency matrix or node features through discrete diffusion processes [69, 124, 171].

However, these unconditionally generated graphs often suffer from limited scalability and practical applicability. Consequently, the focus in graph generation has shifted towards creating graphs based on specific conditions and requirements.

Diffusion-based graph generation, directed by various conditions, supports the scaling up of graphs, enhancement of graph features, and addressing of dataset-specific challenges. PCFI [168] utilizes partial graph features and shortest path distances to predict pseudo confidence, which guides the generation process. Meanwhile, EDGE [27] and DiffFormer [181] employ node degree and energy constraints, respectively, to facilitate the discrete and continuous generation of adjacency matrices and latent embeddings, thus expanding the scope of generation capabilities. Additionally, D4Explainer [22] uses the distribution of graph data as a condition and combines distribution loss with counterfactual loss to investigate counterfactual instances, enhancing the depth and breadth of graph analysis.

7 Critical Insights and Future Directions

As diffusion models evolve, it is imperative to not only advance their capabilities but also address the impending challenges they face across various applications. One of the primary concerns is data scarcity which often limits the efficacy and scalability of these models. This issue manifests notably in contexts where high-quality, large-scale data sets are not readily available or are too costly to procure. The following sections delve into the complications arising from insufficient data and explore strategies to mitigate these challenges, ensuring that diffusion models can maintain their robustness and adaptability in diverse environments.

7.1 Challenges Due to Data Scarcity

Beyond the low inference speed, diffusion models often struggle to detect patterns and regularities in low-quality data, impairing their ability to generalize to new scenarios or datasets. Additionally, managing large-scale datasets leads to computational issues, including prolonged training periods, excessive memory consumption, or inability to converge as expected, ultimately constraining the model's scale and complexity. Moreover, biased or uneven sampling can hinder the model's ability to produce outputs that generalize effectively across diverse domains and demographic groups. Data scarcity significantly affects the performance and scalability of diffusion models, particularly in fields where acquiring large-scale, high-quality datasets is challenging or expensive. For instance, in medical imaging, the lack of extensive annotated datasets can severely restrict the training and effectiveness of diffusion models used for generating synthetic medical images for training purposes [4, 8, 15]. Similarly, in the domain of natural language processing, diffusion models trained on limited linguistic data often struggle to produce coherent and contextually accurate language outputs, particularly for under-represented languages [14]. Another example is in environmental modeling, where insufficient historical data can hinder the model's ability to accurately simulate complex climate systems, impacting the reliability of predictions [141]. Addressing these challenges requires innovative solutions such as synthetic data generation, transfer learning, and semi-supervised learning techniques, which help to augment the available data and improve model performance [170, 203].

7.2 Controlled Generation Based on Distributions

Enhancing the model's capability to interpret and generate samples within specified distributions is crucial for improved generalization when data is limited. By identifying patterns and correlations in the data, the model can generate samples that are representative of the training data while meeting specific criteria. Achieving this requires effective data sampling and usage strategies, alongside optimizing model structures and parameters. This refined understanding enables more accurate and controlled sample generation, resulting in improved generalization.

7.3 Advanced Multi-modal Generation Using LLMs

The future of diffusion models includes developing multi-modal generation through the incorporation of Large Language Models (LLMs). This integration allows the generation of outputs combining text, images, and other modalities. By integrating LLMs, the model's comprehension of inter-modal relationships is improved, leading to outputs that are more realistic and diverse. Furthermore, LLMs improve the efficiency of prompt-based generation by leveraging connections between text and other modalities. They also bolster diffusion models' generation capabilities, broadening the range of domains where different modalities can be produced.

7.3.1 Integration with Machine Learning Disciplines

Blending diffusion models with traditional machine learning frameworks offers novel possibilities for enhancing performance across various tasks. Semi-supervised learning is particularly beneficial for overcoming challenges like generalization, facilitating effective conditional generation even with limited data. Leveraging unlabeled data, it strengthens diffusion models' generalization capabilities to achieve optimal performance under specific conditions.

Reinforcement learning, on the other hand, is vital for employing fine-tuning algorithms to guide the sampling process, ensuring focused exploration and controlled generation. Additionally, incorporating feedback enriches reinforcement learning, improving conditional generation and control capabilities in diffusion models.

7.4 Mitigating Computational Challenges in Diffusion Models

As diffusion models gain prominence across various applications, addressing their computational demands becomes essential. This section discusses specific strategies that have recently been employed to mitigate these challenges, thereby providing a clearer roadmap for future research and development.

Efficient training techniques have been instrumental in reducing computational overhead. *Pruning* and *Quantization* are notable examples, where pruning helps in reducing the model size by eliminating unnecessary model weights, and quantization involves lowering the precision of the numbers used within the model, which speeds up computation and reduces memory use without a substantial loss in performance. Moreover, leveraging distributed and parallel computing has proven effective in alleviating training challenges. Techniques such as distributing computation across multiple GPUs or across different nodes in a cluster can drastically reduce training times. Frameworks like TensorFlow's *Distributed Strategy* and PyTorch's *DataParallel* allow for efficient scaling of training processes.

Optimized data handling also plays a crucial role, especially in managing large-scale datasets. Techniques like *Data Sharding* and *On-the-fly Data Loading* minimize memory consumption and enhance training speed by ensuring that only a portion of the dataset is loaded into memory at any given time. Adaptive sampling methods further address issues of data scarcity and improve model generalization. These methods, such as *Importance Sampling*, prioritize computational resources towards more significant data points, thereby improving both efficiency and accuracy of the model.

Additionally, the use of hardware accelerators such as TPUs (Tensor Processing Units) and FPGAs (Field-Programmable Gate Arrays) can significantly boost the computational efficiency of diffusion models. These devices are optimized for large-scale matrix operations and deep learning computations, offering faster processing times and enabling the training of more complex models. By implementing these strategies, researchers can effectively tackle the computational challenges posed by diffusion models, enhancing their scalability and practical applicability across diverse domains.

8 Conclusion

Diffusion models have increasingly cemented their significance in deep learning. This survey harnesses their potential by delivering a comprehensive, up-to-date review, offering detailed insights into fundamental theories, advanced algorithms, and practical applications. We discussed the progress made in enhancing diffusion models, from theoretical improvements to innovative architectures and optimization techniques. These advancements underscore their versatility in various domains, such as image generation, language modeling, and scientific computing. The review also explores emerging challenges and future directions, highlighting scalability, interpretability, and ethical considerations as crucial areas for further research. We hope this work serves as a valuable resource for scholars and practitioners alike, shedding light on the landscape of diffusion models and providing crucial perspectives on their real-world applications, encouraging deeper exploration and innovation.

Competing Interests The authors declare that they have no competing interests related to this work.

Ethics Approval This study does not involve any primary research with human or animal subjects. Therefore, ethical approval was not required.

Acknowledgements The authors thank Prince Sultan University for the support and funding in conducting this research.

References

1. S. Alamdari, N. Thakkar, R. van den Berg, A.X. Lu, N. Fusi, A.P. Amini, K.K. Yang, Protein generation with evolutionary diffusion: sequence is all you need. bioRxiv 2023-09 (2023)
2. M.S. Albergo, E. Vanden-Eijnden, Building normalizing flows with stochastic interpolants. ArXiv arXiv:2209.15571 (2022)
3. J.M.L. Alcaraz, N. Strodthoff, Diffusion-based time series imputation and forecasting with structured state space models. arXiv:2208.09399 (2022)
4. S. AlHalawani, B. Benjdira, A. Ammar, A. Koubaa, A.M. Ali, Diffplate: a diffusion model for super-resolution of license plate images. Electronics **13**(13) (2024). https://doi.org/10.3390/electronics13132670. https://www.mdpi.com/2079-9292/13/13/2670
5. Amit, T., Nachmani, E., Shaharbany, T., Wolf, L.: Segdiff: Image segmentation with diffusion probabilistic models. arXiv:2112.00390 (2021)
6. N. Anand, T. Achim, Protein structure and sequence generation with equivariant denoising diffusion probabilistic models. arXiv:2205.15019 (2022)
7. B.D. Anderson, Reverse-time diffusion equation models. Stochast. Process. Appl. **12**(3), 313–326 (1982)
8. A. Antoniou, A. Storkey, H. Edwards, Data augmentation generative adversarial networks. arXiv preprint arXiv:1711.04340 (2017)
9. L. Arnold, *Stochastic Differential Equations* (New York, 1974)
10. J. Austin, D.D. Johnson, J. Ho, D. Tarlow, R. Van Den Berg, Structured denoising diffusion models in discrete state-spaces. NeurIPS **34**, 17981–17993 (2021)

11. F. Bao, K. Xu, C. Li, L. Hong, J. Zhu, B. Zhang, Variational (gradient) estimate of the score function in energy-based latent variable models, in *International Conference on Machine Learning*(PMLR, 2021), pp. 651–661
12. D. Baranchuk, I. Rubachev, A. Voynov, V. Khrulkov, A. Babenko, Label-efficient semantic segmentation with diffusion models. arXiv:2112.03126 (2021)
13. G. Batzolis, J. Stanczuk, C.B. Schönlieb, C. Etmann, Conditional image generation with score-based diffusion models. arXiv:2111.13606 (2021)
14. E.M. Bender, T. Gebru, A. McMillan-Major, S. Shmitchell, On the dangers of stochastic parrots: can language models be too big? *Proceedings of FAccT* (2021)
15. B. Benjdira, M. Ali, A., Koubaa, A., Ammar, W. Boulila, Dm-ahr: a self-supervised conditional diffusion model for ai-generated hairless imaging for enhanced skin diagnosis applications. Cancers **16**(17) (2024). https://doi.org/10.3390/cancers16172947. https://www.mdpi.com/2072-6694/16/17/2947
16. C. Bilodeau, W. Jin, T. Jaakkola, R. Barzilay, K.F. Jensen, Generative models for molecular discovery: recent advances and challenges. Wiley Interdiscip. Rev.: Comput. Mol. Sci. **12**(5), e1608 (2022)
17. T. Brooks, A. Holynski, A.A. Efros, Instructpix2pix: learning to follow image editing instructions, in *CVPR* (2023), pp. 18392–18402
18. A. Buades, B. Coll, J.M. Morel, A review of image denoising algorithms, with a new one. Multiscale Model. Simul. **4**(2), 490–530 (2005)
19. Y. Chahibi, Molecular communication for drug delivery systems: a survey. Nano Commun. Netw. **11**, 90–102 (2017)
20. C.H. Chao, W.F. Sun, B.W. Cheng, C.Y. Lee, Quasi-conservative score-based generative models. arXiv:2209.12753 (2022)
21. H. Chefer, Y. Alaluf, Y. Vinker, L. Wolf, D. Cohen-Or, Attend-and-excite: attention-based semantic guidance for text-to-image diffusion models. ACM Trans. Graph. (TOG) **42**(4), 1–10 (2023)
22. J. Chen, S. Wu, A. Gupta, R. Ying, D4explainer: in-distribution GNN explanations via discrete denoising diffusion. arXiv preprint arXiv:2310.19321 (2023)
23. N. Chen, Y. Zhang, H. Zen, R.J. Weiss, M. Norouzi, W. Chan, Wavegrad: estimating gradients for waveform generation, in *ICLR* (2020)
24. R.T. Chen, Y. Rubanova, J. Bettencourt, D.K. Duvenaud, Neural ordinary differential equations. NeurIPS **31** (2018)
25. S. Chen, P. Sun, Y. Song, P. Luo, Diffusiondet: diffusion model for object detection. arXiv:2211.09788 (2022)
26. T. Chen, R. Zhang, G. Hinton, Analog bits: generating discrete data using diffusion models with self-conditioning. arXiv:2208.04202 (2022)
27. X. Chen, J. He, X. Han, L.P. Liu, Efficient and degree-guided graph generation via discrete diffusion modeling. arXiv preprint arXiv:2305.04111 (2023)
28. A.C. Cheng, X. Li, S. Liu, M. Sun, M.H. Yang, Autoregressive 3d shape generation via canonical mapping. arXiv:2204.01955 (2022)
29. J. Choi, S. Kim, Y. Jeong, Y. Gwon, S. Yoon, Ilvr: conditioning method for denoising diffusion probabilistic models, in *CVPR* (2021), pp. 14367–14376
30. H. Chung, E.S. Lee, J.C. Ye, Mr image denoising and super-resolution using regularized reverse diffusion. IEEE Trans. Med. Imaging **42**(4), 922–934 (2022)
31. H. Chung, B. Sim, D. Ryu, J.C. Ye, Improving diffusion models for inverse problems using manifold constraints. NeurIPS **35**, 25683–25696 (2022)
32. H. Chung, B. Sim, J.C. Ye, Come-closer-diffuse-faster: Accelerating conditional diffusion models for inverse problems through stochastic contraction, in *CVPR* (2022)
33. H. Chung, J.C. Ye, Score-based diffusion models for accelerated MRI. Med. Image Anal. **80**, 102479 (2022)
34. G. Corso, H. Stärk, B. Jing, R. Barzilay, T. Jaakkola, Diffdock: diffusion steps, twists, and turns for molecular docking. arXiv:2210.01776 (2022)

35. G. Corso, Y. Xu, V. De Bortoli, R. Barzilay, T. Jaakkola, Particle guidance: non-IID diverse sampling with diffusion models. arXiv preprint arXiv:2310.13102 (2023)
36. A. Creswell, T. White, V. Dumoulin, K. Arulkumaran, B. Sengupta, A.A. Bharath, Generative adversarial networks: an overview. IEEE Signal Process (2018)
37. P. Dhariwal, A. Nichol, Diffusion models beat Gans on image synthesis. NeurIPS **34**, 8780–8794 (2021)
38. L. Dinh, D. Krueger, Y. Bengio, Nice: non-linear independent components estimation. arXiv preprint arXiv:1410.8516 (2014)
39. L. Dinh, J. Sohl-Dickstein, S. Bengio, Density estimation using real NVP. arXiv:1605.08803 (2016)
40. T. Dockhorn, A. Vahdat, K. Kreis, Score-based generative modeling with critically-damped Langevin diffusion. arXiv:2112.07068 (2021)
41. Z. Dorjsembe, S. Odonchimed, F. Xiao, Three-dimensional medical image synthesis with denoising diffusion probabilistic models, in *Medical Imaging with Deep Learning* (2022)
42. P. Esser, R. Rombach, A. Blattmann, B. Ommer, Imagebart: bidirectional context with multinomial diffusion for autoregressive image synthesis. Adv. Neural Inform. Process. Syst. **34**, 3518–3532 (2021)
43. M. Fahim Sikder, R. Ramachandranpillai, F. Heintz, Transfusion: Generating long, high fidelity time series using diffusion models with transformers. arXiv e-prints arXiv-2307 (2023)
44. Z. Gao, J. Guo, X. Tan, Y. Zhu, F. Zhang, J. Bian, L. Xu, Difformer: empowering diffusion model on embedding space for text generation. arXiv preprint arXiv:2212.09412 (2022)
45. G. Giannone, D. Nielsen, O. Winther, OFew-shot diffusion models. arXiv:2205.15463 (2022)
46. S. Gong, C. Chen, Y. Gong, N.Y. Chan, W. Ma, C.H.K. Mak, J. Abrigo, Q. Dou, Diffusion model based semi-supervised learning on brain hemorrhage images for efficient midline shift quantification, in *International Conference on Information Processing in Medical Imaging* (Springer, 2023), pp. 69–81
47. S. Gong, M. Li, J. Feng, Z. Wu, L. Kong, Diffuseq: sequence to sequence text generation with diffusion models, in *ICLR* (2022)
48. I. Goodfellow, J. Pouget-Abadie, M. Mirza, B. Xu, D. Warde-Farley, S. Ozair, A. Courville, Y. Bengio, Generative adversarial nets. Adv. Neural Inform. Process. Syst. **27** (2014)
49. I. Goodfellow, J. Pouget-Abadie, M. Mirza, B. Xu, D. Warde-Farley, S. Ozair, A. Courville, Y. Bengio, Generative adversarial networks. Commun. ACM (2020)
50. G. Grimmett, D. Stirzaker, *Probability and Random Processes* (Oxford University Press, 2020)
51. S. Gu, D. Chen, J. Bao, F. Wen, B. Zhang, D. Chen, L. Yuan, B. Guo, Vector quantized diffusion model for text-to-image synthesis, in *CVPR* (2022), pp. 10696–10706
52. S. Guenneau, T. Puvirajesinghe, Fick's second law transformed: one path to cloaking in mass diffusion. J. Roy. Soc. Interface **10**(83), 20130106 (2013)
53. Y. Guo, C. Yang, A. Rao, Y. Wang, Y. Qiao, D. Lin, B. Dai, Animatediff: animate your personalized text-to-image diffusion models without specific tuning. arXiv preprint arXiv:2307.04725 (2023)
54. X. Han, H. Zheng, M. Zhou, Card: classification and regression diffusion models. arXiv:2206.07275 (2022)
55. W. Harvey, S. Naderiparizi, V. Masrani, C. Weilbach, F. Wood, Flexible diffusion modeling of long videos. arXiv:2205.11495 (2022)
56. Z. He, T. Sun, K. Wang, X. Huang, X. Qiu, Diffusionbert: improving generative masked language models with diffusion models. arXiv preprint arXiv:2211.15029 (2022)
57. A. Hertz, R. Mokady, J. Tenenbaum, K. Aberman, Y. Pritch, D. Cohen-Or, Prompt-to-prompt image editing with cross attention control. arXiv preprint arXiv:2208.01626 (2022)
58. M. Heusel, H. Ramsauer, T. Unterthiner, B. Nessler, S. Hochreiter, Gans trained by a two time-scale update rule converge to a local Nash equilibrium. Adv. Neural Inform. Process. Syst. **30** (2017)
59. J. Ho, A. Jain, P. Abbeel, Denoising diffusion probabilistic models. NeurIPS (2020)
60. J. Ho, T. Salimans, Classifier-free diffusion guidance. arXiv:2207.12598 (2022)

61. J. Ho, T. Salimans, A. Gritsenko, W. Chan, M. Norouzi, D.J. Fleet, Video diffusion models (2022)
62. E. Hoogeboom, D. Nielsen, P. Jaini, P. Forré, M. Welling, Argmax flows and multinomial diffusion: learning categorical distributions. NeurIPS **34**, 12454–12465 (2021)
63. E. Hoogeboom, V.G. Satorras, C. Vignac, M. Welling, Equivariant diffusion for molecule generation in 3d, in *ICML* (PMLR, 2022), pp. 8867–8887
64. T. Höppe, A. Mehrjou, S. Bauer, D. Nielsen, A. Dittadi, Diffusion models for video prediction and infilling. arXiv:2206.07696 (2022)
65. A. Hore, D. Ziou, Image quality assessment techniques. PNR **2010**, 1687–6962 (2010)
66. D. Hu, Y.K. Tao, I. Oguz, Unsupervised denoising of retinal oct with diffusion probabilistic model, in *Medical Imaging 2022: Image Processing*, vol. 12032 (SPIE, 2022), pp. 25–34
67. M. Hu, Y. Wang, T.J. Cham, J. Yang, P.N. Suganthan, Global context with discrete diffusion in vector quantised modelling for image generation, in *CVPR* (2022), pp. 11502–11511
68. V.T. Hu, D.W. Zhang, Y.M. Asano, G.J. Burghouts, C.G. Snoek, Self-guided diffusion models. arXiv:2210.06462 (2022)
69. H. Huang, L. Sun, B. Du, Y. Fu, W. Lv, Graphgdp: generative diffusion processes for permutation invariant graph generation. arXiv:2212.01842 (2022)
70. R. Huang, Z. Zhao, H. Liu, J. Liu, C. Cui, Y. Ren, Prodiff: progressive fast diffusion model for high-quality text-to-speech. arXiv:2207.06389 (2022)
71. O. Ibe, *Markov Processes for Stochastic Modeling* (Newnes, 2013)
72. M. Jeong, H. Kim, S.J. Cheon, B.J. Choi, N.S. Kim, Diff-TTS: a denoising diffusion model for text-to-speech, in *Proceedings of the Interspeech 2021* (2021), pp. 3605–3609. https://doi.org/10.21437/Interspeech.2021-469
73. B. Jing, G. Corso, R. Barzilay, T.S. Jaakkola, Torsional diffusion for molecular conformer generation, in *ICLR* (2022)
74. H. Jun, A. Nichol, Shap-e: generating conditional 3d implicit functions. arXiv preprint arXiv:2305.02463 (2023)
75. T. Karras, M. Aittala, T. Aila, S. Laine, Elucidating the design space of diffusion-based generative models. arXiv:2206.00364 (2022)
76. B. Kawar, M. Elad, S. Ermon, J. Song, Denoising diffusion restoration models, in *ICLR Workshop* (2022)
77. B. Kim, I. Han, J.C. Ye, Diffusemorph: unsupervised deformable image registration using diffusion model, in *ECCV* (Springer, 2022), pp. 347–364
78. H. Kim, S. Kim, S. Yoon, Guided-tts: a diffusion model for text-to-speech via classifier guidance, in *ICML* (PMLR, 2022), pp. 11119–11133
79. S. Kim, H. Kim, S. Yoon, Guided-tts 2: a diffusion model for high-quality adaptive text-to-speech with untranscribed data. arXiv:2205.15370 (2022)
80. D. Kingma, T. Salimans, B. Poole, J. Ho, Variational diffusion models. NeurIPS **34**, 21696–21707 (2021)
81. D.P. Kingma, M. Welling, Auto-encoding variational bayes. arXiv:1312.6114 (2013)
82. D.P. Kingma, M. Welling et al., An introduction to variational autoencoders. Found. Trends® Mach. Learn. (2019)
83. I. Kobyzev, S.J. Prince, M.A. Brubaker, Normalizing flows: an introduction and review of current methods. IEEE TPAMI (2020)
84. Y. Koizumi, H. Zen, K. Yatabe, N. Chen, M. Bacchiani, Specgrad: diffusion probabilistic model based neural vocoder with adaptive noise spectral shaping. arXiv:2203.16749 (2022)
85. Z. Kong, W. Ping, J. Huang, K. Zhao, B. Catanzaro, Diffwave: a versatile diffusion model for audio synthesis, in *ICLR* (2020)
86. T.M. Le, E. Kanso, P.K. Newton, Brownian motion of boomerang colloidal particles. Phys. Rev. Lett. **109**(16), 164501 (2012)
87. Y. LeCun, S. Chopra, R. Hadsell, M. Ranzato, F. Huang, A tutorial on energy-based learning. Predicting Struct. Data (2006)
88. J.S. Lee, P.M. Kim, Proteinsgm: score-based generative modeling for de novo protein design. bioRxiv (2022)

89. Y. Leng, Z. Chen, J. Guo, H. Liu, J. Chen, X. Tan, D. Mandic, L. He, X.Y. Li, T. Qin et al., Binauralgrad: a two-stage conditional diffusion probabilistic model for binaural audio synthesis. arXiv:2205.14807 (2022)
90. A. Levkovitch, E. Nachmani, L. Wolf, Zero-shot voice conditioning for denoising diffusion tts models. arXiv:2206.02246 (2022)
91. C. Li, P. Shen, Y. Liu, Z. Zhang, Diffusion information theoretic learning for distributed estimation over network. IEEE Trans. Signal Process. **61**(16), 4011–4024 (2013)
92. H. Li, Y. Yang, M. Chang, S. Chen, H. Feng, Z. Xu, Q. Li, Y. Chen, Srdiff: single image super-resolution with diffusion probabilistic models. Neurocomputing (2022)
93. J. Li, H. Cao, J. Wang, F. Liu, Q. Dou, G. Chen, P.A. Heng, Fast non-Markovian diffusion model for weakly supervised anomaly detection in brain MR images, in *MICCAI* (Springer, 2023), pp. 579–589
94. X. Li, J. Thickstun, I. Gulrajani, P.S. Liang, T.B. Hashimoto, Diffusion-lm improves controllable text generation. NeurIPS **35**, 4328–4343 (2022)
95. Y. Li, K. Zhou, W.X. Zhao, J.R. Wen, Diffusion models for non-autoregressive text generation: a survey. arXiv preprint arXiv:2303.06574 (2023)
96. H. Lim, M. Kim, S. Park, N. Park, Regular time-series generation using SGM. arXiv preprint arXiv:2301.08518 (2023)
97. C.H. Lin, J. Gao, L. Tang, T. Takikawa, X. Zeng, X. Huang, K. Kreis, S. Fidler, M.Y. Liu, T.Y. Lin, Magic3d: high-resolution text-to-3d content creation, in *CVPR*, pp. 300–309 (2023)
98. L. Lin, Z. Li, R. Li, X. Li, J. Gao, Diffusion models for time series applications: a survey. arXiv preprint arXiv:2305.00624 (2023)
99. Z. Lin, Y. Gong, Y. Shen, T. Wu, Z. Fan, C. Lin, N. Duan, W. Chen, Text generation with diffusion language models: A pre-training approach with continuous paragraph denoise, in *ICML* (PMLR, 2023), pp. 21051–21064
100. S.L. Lisanza, J.M. Gershon, S.W.K. Tipps, L. Arnoldt, S. Hendel, J.N. Sims, X. Li, D. Baker, Joint generation of protein sequence and structure with Rosettafold sequence space diffusion. bioRxiv 2023-05 (2023)
101. J. Liu, C. Li, Y. Ren, F. Chen, Z. Zhao, Diffsinger: singing voice synthesis via shallow diffusion mechanism, in *AAAI*, vol. 36 (2022), pp. 11020–11028
102. L. Liu, Y. Ren, Z. Lin, Z. Zhao, Pseudo numerical methods for diffusion models on manifolds. arXiv:2202.09778 (2022)
103. M. Liu, H. Huang, H. Feng, L. Sun, B. Du, Y. Fu, Pristi: a conditional diffusion framework for spatiotemporal imputation. arXiv preprint arXiv:2302.09746 (2023)
104. M. Liu, C. Xu, H. Jin, L. Chen, Z. Xu, H. Su et al., One-2-3-45: any single image to 3d mesh in 45 seconds without per-shape optimization. arXiv preprint arXiv:2306.16928 (2023)
105. R. Liu, R. Wu, B. Van Hoorick, P. Tokmakov, S. Zakharov, C. Vondrick, Zero-1-to-3: Zero-shot one image to 3d object, in *ICCV (ICCV)* (2023), pp. 9298–9309
106. S. Liu, Y. Cao, D. Su, H. Meng, Diffsvc: a diffusion probabilistic model for singing voice conversion, in *IEEE ASRU* (2021)
107. X. Liu, C. Gong, Q. Liu, Flow straight and fast: learning to generate and transfer data with rectified flow. ArXiv arXiv:2209.03003 (2022)
108. J.M. Lopez Alcaraz, N. Strodthoff, Diffusion-based time series imputation and forecasting with structured atate apace models. Trans. Mach. Learn. Res. 1–36 (2023)
109. C. Lu, Y. Zhou, F. Bao, J. Chen, C. Li, J. Zhu, Dpm-solver: a fast ode solver for diffusion probabilistic model sampling in around 10 steps. arXiv:2206.00927 (2022)
110. C. Lu, Y. Zhou, F. Bao, J. Chen, C. Li, J. Zhu, Dpm-solver++: fast solver for guided sampling of diffusion probabilistic models. arXiv:2211.01095 (2022)
111. A. Lugmayr, M. Danelljan, A. Romero, F. Yu, R. Timofte, L. Van Gool, Repaint: inpainting using denoising diffusion probabilistic models, in *CVPR* (2022), pp. 11461–11471
112. S. Luo, W. Hu, Diffusion probabilistic models for 3d point cloud generation, in *CVPR* (2021), pp. 2837–2845
113. S. Luo, W. Hu, Score-based point cloud denoising, in *ICCV* (2021), pp. 4583–4592

114. S. Luo, C. Shi, M. Xu, J. Tang, Predicting molecular conformation via dynamic graph score matching. NeurIPS **34**, 19784–19795 (2021)
115. S. Luo, Y. Su, X. Peng, S. Wang, J. Peng, J. Ma, Antigen-specific antibody design and optimization with diffusion-based generative models for protein structures, in *NeurIPS* (2022)
116. Z. Lyu, Z. Kong, X. Xu, L. Pan, D. Lin, A conditional point diffusion-refinement paradigm for 3d point cloud completion. arXiv:2112.03530 (2021)
117. D. Maoutsa, S. Reich, M. Opper, Interacting particle solutions of Fokker-Planck equations through gradient-log-density estimation. Entropy **22**(8), 802 (2020)
118. C. Meng, R. Gao, D.P. Kingma, S. Ermon, J. Ho, T. Salimans, On distillation of guided diffusion models. arXiv:2210.03142 (2022)
119. S. Min, B. Lee, S. Yoon, Deep learning in bioinformatics. Briefings Bioinformatics **18**(5), 851–869 (2017)
120. J. Ngiam, Z. Chen, P.W. Koh, A.Y. Ng, Learning deep energy models, in *ICML* (2011)
121. A. Nichol, P. Dhariwal, A. Ramesh, P. Shyam, P. Mishkin, B. McGrew, I. Sutskever, M. Chen, Glide: towards photorealistic image generation and editing with text-guided diffusion models. arXiv:2112.10741 (2021)
122. A. Nichol, H. Jun, H., Dhariwal, P., Mishkin, P., Chen, M.: Point-e: a system for generating 3d point clouds from complex prompts. arXiv preprint arXiv:2212.08751 (2022)
123. A.Q. Nichol, P. Dhariwal, Improved denoising diffusion probabilistic models, in *ICML* (2021)
124. C. Niu, Y. Song, J. Song, S. Zhao, A. Grover, S. Ermon, Permutation invariant graph generation via score-based generative modeling, in *AISTATS* (PMLR, 2020), pp. 4474–4484
125. J.R. Norris, *Markov Chains* (Cambridge University Press, 1998)
126. B. Oksendal, *Stochastic Differential Equations: An Introduction with Applications* (Springer Science & Business Media, 2013)
127. A. Van den Oord, N. Kalchbrenner, L. Espeholt, O. Vinyals, A. Graves, et al.: Conditional image generation with Pixelcnn decoders. Adv. Neural Inform. Process. Syst. **29** (2016)
128. D.W. Otter, J.R. Medina, J.K. Kalita, A survey of the usages of deep learning for natural language processing. IEEE Trans. Neural Netw. Learn. Syst. **32**(2), 604–624 (2020)
129. A. Oussidi, A. Elhassouny, Deep generative models: survey, in *ISCV* (IEEE, 2018)
130. K. Packhäuser, L. Folle, F. Thamm, A. Maier, Generation of anonymous chest radiographs using latent diffusion models for training thoracic abnormality classification systems. arXiv:2211.01323 (2022)
131. K. Pandey, A. Mukherjee, P. Rai, A. Kumar, Vaes meet diffusion models: efficient and high-fidelity generation, in *NeurIPS 2021 Workshop on Deep Generative Models and Downstream Applications* (2021)
132. S.W. Park, K. Lee, J. Kwon, Neural Markov controlled SDE: stochastic optimization for continuous-time data, in *ICLR* (2021)
133. W.H. Pinaya, P.D. Tudosiu, J. Dafflon, P.F. Da Costa, V. Fernandez, P. Nachev, S. Ourselin, M.J. Cardoso, Brain imaging generation with latent diffusion models, in *MICCAI Workshop on Deep Generative Models* (Springer, 2022), pp. 117–126
134. B. Poole, A. Jain, J.T. Barron, B. Mildenhall, Dreamfusion: text-to-3d using 2d diffusion. arXiv:2209.14988 (2022)
135. V. Popov, I. Vovk, V. Gogoryan, T. Sadekova, M. Kudinov, Grad-tts: a diffusion probabilistic model for text-to-speech, in *ICML* (PMLR, 2021), pp. 8599–8608
136. V. Popov, I. Vovk, V. Gogoryan, T. Sadekova, M.S. Kudinov, J. Wei, Diffusion-based voice conversion with fast maximum likelihood sampling scheme, in *ICLR* (2021)
137. A. Radford, L. Metz, S. Chintala, Unsupervised representation learning with deep convolutional generative adversarial networks. arXiv preprint arXiv:1511.06434 (2015)
138. A. Ramesh, P. Dhariwal, A. Nichol, C. Chu, M. Chen, Hierarchical text-conditional image generation with clip latents. arXiv:2204.06125 (2022)
139. K. Rasul, C. Seward, I. Schuster, R. Vollgraf, Autoregressive denoising diffusion models for multivariate probabilistic time series forecasting, in *ICML* (PMLR, 2021), pp. 8857–8868
140. D. Rezende, S. Mohamed, Variational inference with normalizing flows, in *ICML* (2015)

141. D. Rolnick, P.L. Donti, L.H. Kaack, K. Kochanski, A. Lacoste, K. Sankaran, A.S. Ross, N. Milojevic-Dupont, N. Jaques, A. Waldman-Brown et al., Tackling climate change with machine learning. AI Mag. **41**(4), 5–15 (2020)
142. R. Rombach, A. Blattmann, D. Lorenz, P. Esser, B. Ommer, High-resolution image synthesis with latent diffusion models, in *CVPR* (2022), pp. 10684–10695
143. O. Ronneberger, P. Fischer, T. Brox, U-net: convolutional networks for biomedical image segmentation, in *Medical Image Computing and Computer-Assisted Intervention–MICCAI 2015: 18th international conference, Munich, Germany, October 5-9, 2015, Proceedings, part III 18* (Springer, 2015), pp. 234–241
144. N. Ruiz, Y. Li, V. Jampani, Y. Pritch, M. Rubinstein, K. Aberman, Dreambooth: fine tuning text-to-image diffusion models for subject-driven generation, in *CVPR* (2023), pp. 22500–22510
145. C. Saharia, W. Chan, H. Chang, C. Lee, J. Ho, T. Salimans, D. Fleet, M. Norouzi, Palette: image-to-image diffusion models, in *ACM SIGGRAPH* (2022), pp. 1–10
146. C. Saharia, W. Chan, S. Saxena, L. Li, J. Whang, E. Denton, S.K.S. Ghasemipour, B.K. Ayan, S.S. Mahdavi, R.G. Lopes, et al., Photorealistic text-to-image diffusion models with deep language understanding. arXiv:2205.11487 (2022)
147. T. Salimans, J. Ho, Progressive distillation for fast sampling of diffusion models. arXiv (2022)
148. T. Salimans, A. Karpathy, X. Chen, D.P. Kingma, Pixelcnn++: improving the pixelcnn with discretized logistic mixture likelihood and other modifications. arXiv preprint arXiv:1701.05517 (2017)
149. P.N. Sen, Time-dependent diffusion coefficient as a probe of geometry. Concepts Magn. Reson. Part A: Educ. J. **23**(1), 1–21 (2004)
150. Y. Shen, J. Gu, X. Tang, B. Zhou, Interpreting the latent space of Gans for semantic face editing, in *Proceedings of the IEEE/CVF Conference on Computer Vision and Pattern Recognition* (2020), pp. 9243–9252
151. C. Shi, S. Luo, M. Xu, J. Tang, Learning gradient fields for molecular conformation generation, in *ICML* (PMLR, 2021), pp. 9558–9568
152. U. Singer, A. Polyak, T. Hayes, X. Yin, J. An, S. Zhang, Q. Hu, H. Yang, O. Ashual, O. Gafni, et al., Make-a-video: Text-to-video generation without text-video data. arXiv:2209.14792 (2022)
153. J. Sohl-Dickstein, E. Weiss, N. Maheswaranathan, S. Ganguli, Deep unsupervised learning using nonequilibrium thermodynamics, in *International Conference on Machine Learning* (PMLR, 2015), pp. 2256–2265
154. J. Song, C. Meng, S. Ermon, Denoising diffusion implicit models, in *ICLR* (2020)
155. Y. Song, P. Dhariwal, M. Chen, I. Sutskever, Consistency models. ArXiv **abs/2303.01469** (2023)
156. Y. Song, S. Ermon, Generative modeling by estimating gradients of the data distribution. Adv. Neural Inform. Process. Syst. **32** (2019)
157. Y. Song, S. Garg, J. Shi, S. Ermon, Sliced score matching: A scalable approach to density and score estimation, in *Uncertainty in Artificial Intelligence* (PMLR, 2020), pp. 574–584
158. Y. Song, L. Shen, L. Xing, S. Ermon, Solving inverse problems in medical imaging with score-based generative models, in *ICLR* (2021)
159. Y. Song, J. Sohl-Dickstein, D.P. Kingma, A. Kumar, S. Ermon, B. Poole, Score-based generative modeling through stochastic differential equations. arXiv preprint arXiv:2011.13456 (2020)
160. R. Strudel, C. Tallec, F. Altché, Y. Du, Y. Ganin, A. Mensch, W. Grathwohl, N. Savinov, S. Dieleman, L. Sifre et al., Self-conditioned embedding diffusion for text generation. arXiv preprint arXiv:2211.04236 (2022)
161. K. Swersky, M. Ranzato, D. Buchman, N.D. Freitas, B.M. Marlin, On autoencoders and score matching for energy based models, in *Proceedings of the 28th International Conference on Machine Learning (ICML-11)* (2011), pp. 1201–1208
162. J. Tae, H. Kim, T. Kim, Editts: score-based editing for controllable text-to-speech. arXiv:2110.02584 (2021)

163. T. Tang, Y. Chen, Y. Du, J. Li, W.X. Zhao, J.R. Wen, Learning to imagine: visually-augmented natural language generation. arXiv preprint arXiv:2305.16944 (2023)
164. Z. Tang, S. Gu, J. Bao, D. Chen, F. Wen, Improved vector quantized diffusion models. arXiv:2205.16007 (2022)
165. Y. Tashiro, J. Song, Y. Song, S. Ermon, Csdi: conditional score-based diffusion models for probabilistic time series imputation. NeurIPS **34**, 24804–24816 (2021)
166. L. Theis, A.V.D. Oord, M. Bethge, A note on the evaluation of generative models. arXiv preprint arXiv:1511.01844 (2015)
167. L. Theis, T. Salimans, M.D. Hoffman, F. Mentzer, Lossy compression with gaussian diffusion. arXiv:2206.08889 (2022)
168. D. Um, J. Park, S. Park, J. Young Choi, Confidence-based feature imputation for graphs with partially known features, in *ICLR* (2022)
169. A. Vahdat, K. Kreis, J. Kautz, Score-based generative modeling in latent space. NeurIPS **34**, 11287–11302 (2021)
170. J.E. Van Engelen, H.H. Hoos, A survey on semi-supervised learning. Mach. Learn. **109**(2), 373–440 (2020)
171. C. Vignac, I. Krawczuk, A. Siraudin, B. Wang, V. Cevher, P. Frossard, Digress: discrete denoising diffusion for graph generation. arXiv:2209.14734 (2022)
172. E. Vincent, A measure of signal: Noise ratio improvements for ICTD speech enhancement. IEICE Trans. Fundam. Electron. Commun. Comput. Sci. **89**(3), 1096–1103 (2006)
173. P. Vincent, A connection between score matching and denoising autoencoders. Neural Comput. **23**(7), 1661–1674 (2011)
174. V. Voleti, A. Jolicoeur-Martineau, C. Pal, Mcvd-masked conditional video diffusion for prediction, generation, and interpolation. NeurIPS **35**, 23371–23385 (2022)
175. Z. Wang, A.C. Bovik, H.R. Sheikh, E.P. Simoncelli, Image quality assessment: from error visibility to structural similarity. IEEE Trans. Image Process. **13**(4), 600–612 (2004)
176. J.L. Watson, D. Juergens, N.R. Bennett, B.L. Trippe, J. Yim, H.E. Eisenach, W. Ahern, A.J. Borst, R.J. Ragotte, L.F. Milles et al., De novo design of protein structure and function with rfdiffusion. Nature 1–3 (2023)
177. H. Wen, Y. Lin, Y. Xia, H. Wan, R. Zimmermann, Y. Liang, Diffstg: probabilistic Spatio-temporal graph forecasting with denoising diffusion models. arXiv preprint arXiv:2301.13629 (2023)
178. U.F. Wiersema, *Brownian Motion Calculus* (John Wiley & Sons, 2008)
179. J. Wolleb, F. Bieder, R. Sandkühler, P.C. Cattin, Diffusion models for medical anomaly detection. arXiv:2203.04306 (2022)
180. K.E. Wu, K.K. Yang, R.V.D. Berg, J.Y. Zou, A.X. Lu, A.P. Amini, Protein structure generation via folding diffusion. arXiv:2209.15611 (2022)
181. Q. Wu, C. Yang, W. Zhao, Y. He, D. Wipf, J. Yan, Difformer: scalable (graph) transformers induced by energy constrained diffusion, in *The Eleventh International Conference on Learning Representations* (2022)
182. S. Wu, Z. Shi, Itôtts and itôwave: linear stochastic differential equation is all you need for audio generation. arXiv e-prints arXiv-2105 (2021)
183. T. Wu, Z. Fan, X. Liu, Y. Gong, Y. Shen, J. Jiao, H.T. Zheng, J. Li, Z. Wei, J. Guo, et al., Ar-diffusion: auto-regressive diffusion model for text generation. arXiv preprint arXiv:2305.09515 (2023)
184. J. Wyatt, A. Leach, S.M. Schmon, C.G. Willcocks, Anoddpm: anomaly detection with denoising diffusion probabilistic models using simplex noise, in *CVPR* (2022), pp. 650–656
185. P. Xie, Q. Zhang, Z. Li, H. Tang, Y. Du, X. Hu, Vector quantized diffusion model with codeunet for text-to-sign pose sequences generation. arXiv:2208.09141 (2022)
186. T. Xie, X. Fu, O.E. Ganea, R. Barzilay, T.S. Jaakkola, Crystal diffusion variational autoencoder for periodic material generation, in *ICLR* (2021)
187. M. Xu, L. Yu, Y. Song, C. Shi, S. Ermon, J. Tang, Geodiff: a geometric diffusion model for molecular conformation generation, in *ICLR* (2021)

188. Y. Xu, M. Deng, X. Cheng, Y. Tian, Z. Liu, T. Jaakkola, Restart sampling for improving generative processes. ArXiv arXiv:2306.14878 (2023)
189. Y. Xu, Z. Liu, M. Tegmark, T. Jaakkola, Poisson flow generative models. ArXiv arXiv:2209.11178 (2022)
190. Y. Xu, Z. Liu, Y. Tian, S. Tong, M. Tegmark, T. Jaakkola, Pfgm++: unlocking the potential of physics-inspired generative models. ArXiv arXiv:2302.04265 (2023)
191. Y. Xu, S. Tong, T. Jaakkola, Stable target field for reduced variance score estimation in diffusion models. ArXiv arXiv:2302.00670 (2023)
192. T. Yan, H. Zhang, T. Zhou, Y. Zhan, Y. Xia, Scoregrad: multivariate probabilistic time series forecasting with continuous energy-based generative models. arXiv preprint arXiv:2106.10121 (2021)
193. D. Yang, J. Yu, H. Wang, W. Wang, C. Weng, Y. Zou, D. Yu, Diffsound: discrete diffusion model for text-to-sound generation. arXiv:2207.09983 (2022)
194. R. Yang, P. Srivastava, S. Mandt, Diffusion probabilistic modeling for video generation. arXiv:2203.09481 (2022)
195. Y. Yang, H. Fu, A.I. Aviles-Rivero, C.B. Schönlieb, L. Zhu, Diffmic: dual-guidance diffusion network for medical image classification, in *MICCAI* (Springer, 2023), pp. 95–105
196. J. Ye, Z. Zheng, Y. Bao, L. Qian, Q. Gu, Diffusion language models can perform many tasks with scaling and instruction-finetuning. arXiv preprint arXiv:2308.12219 (2023)
197. J. Yim, B.L. Trippe, V. De Bortoli, E. Mathieu, A. Doucet, R. Barzilay, T. Jaakkola, Se (3) diffusion model with application to protein backbone generation. arXiv preprint arXiv:2302.02277 (2023)
198. H. Yuan, Z. Yuan, C. Tan, F. Huang, S. Huang, Seqdiffuseq: text diffusion with encoder-decoder transformers. arXiv preprint arXiv:2212.10325 (2022)
199. L. Zhang, A. Rao, M. Agrawala, Adding conditional control to text-to-image diffusion models, in *ICCV* (2023), pp. 3836–3847
200. Q. Zhang, Y. Chen, Diffusion normalizing flow. NeurIPS (2021)
201. Q. Zhang, Y. Chen, Fast sampling of diffusion models with exponential integrator. arXiv:2204.13902 (2022)
202. L. Zhou, Y. Du, J. Wu, 3D shape generation and completion through point-voxel diffusion, in *ICCV* (2021), pp. 5826–5835
203. J.Y. Zhu, T. Park, P. Isola, A.A. Efros, Unpaired image-to-image translation using cycle-consistent adversarial networks, in *Proceedings of the IEEE International Conference on Computer Vision* (2017)
204. H. Zou, Z.M. Kim, D. Kang, Diffusion models in NLP: a survey. arXiv preprint arXiv:2305.14671 (2023)

Anas M. Ali is a Research Assistant within the Robotics and Internet of Things Laboratory at Prince Sultan University (PSU). He received his Master's Degree in Electrical Engineering from the Faculty of Electronic Engineering, Menoufia University, Egypt, in 2021. His primary research interests include computer vision, healthcare image analysis, deep learning, image restoration, medical image processing, and remote sensing.

Anis Koubaa is the Leader and Founder of the Robotics and Internet-of-Things Laboratory at Prince Sultan University. He is also the Executive Director of the Innovation Center at Prince Sultan University. He is a full professor of computer science and has been working on several research and development projects on robotics, unmanned systems, deep learning, and the Internet of Things. He is a Senior Fellow of the Higher Education Academy, U.K. He has presented numerous training programs on cutting-edge technologies. His research interests include developing artificial intelligence products and automated solutions for logistics, using drones and robots for delivery systems.

Bilel Benjdira is an Assistant Professor at Prince Sultan University (PSU) and a researcher within the Robotics and Internet of Things Laboratory at PSU. He received his Ph.D. degree in Electrical Engineering from ENICarthage, University of Carthage, Tunisia, in 2021. His primary research interests include computer vision, healthcare image analysis, deep learning, and remote sensing.

Enhancing Semantic Similarity Understanding in Arabic NLP with Nested Embedding Learning

Omer Nacar and Anis Koubaa

Abstract This work introduces a novel framework for training Arabic nested embedding models through Matryoshka Embedding Learning, utilizing multilingual, Arabic-specific, and English-based models to demonstrate the power of nested embeddings in various Arabic NLP downstream tasks. Our contributions include the translation of key sentence similarity datasets into Arabic, enabling a robust evaluation framework to compare these models across multiple dimensions. We trained several nested embedding models on the Arabic Natural Language Inference triplet dataset, utilizing the power of anchor-positive-negative sampling to enhance semantic differentiation, and assessed their performance using various evaluation metrics including Pearson and Spearman correlations for cosine similarity, Manhattan distance, Euclidean distance, and dot product similarity. The results show that Matryoshka embedding models significantly outperform both base models and large language models like OpenAI's text embedding models, with improvements of up to 20–25% across various similarity metrics, as evidenced by their performance on MTEB benchmarks such as STS17, STS22, and STS22-v2. These findings underscore the effectiveness of language-specific training and highlight the potential of Matryoshka models in advancing semantic textual similarity tasks for Arabic NLP.

Keywords Matryoshka learning · Nested embedding · Arabic NLP · Semantic similarity · Cross-lingual transfer learning

1 Introduction

In the rapidly evolving field of natural language processing (NLP), text embedding models have become fundamental tools for transforming textual information into structured numerical representations [1]. These embeddings capture the semantic

O. Nacar (✉)
Robotics and Internet-of-Things Lab, Prince Sultan University, Riyadh, Saudi Arabia
e-mail: onajar@psu.edu.sa

A. Koubaa
Alfaisal University, Riyadh, Saudi Arabia
e-mail: akoubaa@alfaisal.edu

A. Koubaa et al. (eds.), *Generative AI and Large Language Models: Opportunities, Challenges, and Applications*, Studies in Computational Intelligence 1214,
https://doi.org/10.1007/978-3-031-90573-5_6

and contextual nuances of words, phrases, and documents, making them essential for a variety of NLP tasks, including semantic textual similarity [2], retrieval [3], and more. The effectiveness of these models hinges on their ability to represent textual data within a dense, low-dimensional vector space that preserves meaning and context, enabling accurate and efficient downstream processing.

Arabic is the fourth most used language on the internet [4] and the fifth most spoken language globally, underscoring its significance in the digital and linguistic landscapes [5]. Despite its importance, Arabic NLP has lagged behind due to a scarcity of high-quality datasets and pre-trained models. This lack of resources has hindered the development of effective text embeddings that can capture the rich morphological and syntactic nuances of the Arabic language. Addressing these challenges is crucial for advancing Arabic NLP applications and ensuring that this widely spoken language is adequately represented in the field of NLP.

Representation learning [6] forms the backbone of modern machine learning systems, offering rich, multidimensional vectors that encode intricate information necessary for various NLP tasks [7]. These learned representations are typically static and designed to maintain high-dimensional fidelity across all applications, regardless of their specific resource and accuracy requirements. However, this rigidity can lead to inefficiencies, particularly at web-scale [8], where the deployment cost of these embeddings may exceed their initial computation cost [9]. This has prompted the exploration of more flexible and efficient approaches to representation learning.

Matryoshka Representation Learning (MRL) [10] is a new state-of-the-art text embedding models optimized to produce embeddings with increasingly higher output dimensions, representing input texts with more values. While traditional embeddings models [11] produce embeddings with fixed dimensions improvement to enhance performance, it often reduces the efficiency of downstream tasks such as search or classification. Matryoshka embedding models address this issue by training embeddings to be useful even when truncated. These models can produce effective embeddings of varying dimensions. Inspired by the concept of Matryoshka dolls-Russian nesting dolls of decreasing size placed inside one another-MRL produces embeddings that store critical information in the earlier dimensions and less important information in later ones. This structure allows for the truncation of embeddings without significant loss of information, making them useful even when reduced to lower dimensions. This adaptability is particularly valuable for large-scale classification, retrieval, and other NLP tasks, where computational efficiency and accuracy are paramount.

In this work, we make several key contributions to the field of Arabic NLP through the development and application of Matryoshka Embedding Models. First, we have translated the English Stanford Natural Language Inference (SNLI) and MultiNLI datasets into Arabic, providing essential resources for training and evaluating Arabic-specific models. Second, we have trained various English and Arabic embedding models using the Matryoshka approach, enhancing their adaptability and performance across different tasks. Finally, we have conducted comprehensive evaluations of these models, demonstrating their superior performance and making both the datasets and models publicly available on Hugging Face to support further research and application.

This chapter delves into the principles of Matryoshka Representation Learning, highlighting its potential to create efficient, adaptable embeddings tailored for Arabic NLP tasks. We explore the practical benefits of integrating MRL with established NLP models, achieving significant improvements in processing speed and accuracy. The following sections will provide a detailed overview of our methodology, the datasets used, the training process, and the results of our extensive evaluations.

2 Related Work

In the broader context of natural language processing, semantic Textual Similarity (STS) [12] is a core task that evaluates how closely two sentences align in meaning. Unlike tasks that involve binary classification, such as textual entailment or paraphrase detection, STS measures the degree of semantic equivalence on a continuum [13]. This task is crucial for various NLP applications, including machine translation [14], summarization [15], and question answering [16]. By assessing semantic similarity, STS provides a common benchmark for evaluating different techniques and models across these diverse domains.

Representation Learning has progressed significantly with the development of large-scale datasets like ImageNet [17] and JFT [9], which enable the training of versatile models applicable across both computer vision and NLP tasks. Supervised learning typically frames representation learning as a classification problem, while unsupervised and self-supervised approaches rely on proxy tasks like instance discrimination and reconstruction to achieve similar goals [18]. Recent breakthroughs in contrastive learning [19] have further enhanced the extraction of meaningful representations from large datasets, which are essential for developing cross-modal models at scale.

Building on these advancements, Matryoshka Representation Learning [10] introduces a method for encoding multiple fidelity levels within a single representation vector. This multi-fidelity approach enables adaptive deployment, optimizing resource usage without compromising accuracy. The integration of MRL into existing representation learning frameworks is straightforward, providing significant enhancements with minimal additional complexity.

Efficiency in classification [20] and retrieval [8] is paramount, particularly when handling large-scale data. Traditional methods such as dimensionality reduction, hashing, and feature selection have been employed to improve efficiency, though they often come at the cost of reduced accuracy or increased computational complexity [21]. Approximate Nearest Neighbor Search (ANNS) techniques, like Hierarchical Navigable Small World (HNSW) graphs [22], offer a balance between accuracy and efficiency but still face challenges associated with high-dimensional embeddings.

MRL addresses these challenges by reducing the dimensionality of embeddings without sacrificing the richness of the information they encode. By nesting lower-dimensional representations within higher-dimensional ones, MRL facilitates efficient and accurate classification and retrieval, particularly in adaptive systems that must operate under varying computational constraints. The hierarchical embeddings produced by MRL offer a flexible solution that scales effectively with data size and complexity.

Nested and Adaptive Neural Networks have explored the concept of nesting or packing networks of varying capacities within a larger network [23–25]. However, these approaches typically require separate forward passes for each nested network, which can be computationally expensive and inefficient for large-scale deployment. MRL differentiates itself by optimizing a logarithmic number of nesting dimensions, allowing smooth interpolation between these dimensions and enabling efficient adaptive inference.

The approach by [26], which employs nested dropout in autoencoders, shares some conceptual similarities with MRL but differs in its optimization strategy. MRL's focus on coarse-to-fine granularity and minimal overhead during inference makes it particularly suited for web-scale applications. The flexibility and efficiency of MRL open new possibilities for real-time, large-scale NLP and classification tasks, demonstrating its potential as a transformative technique in the field.

Recent advancements in large language models (LLMs) have further expanded the capabilities of text embeddings by leveraging vast parameter counts to capture complex semantic relationships. Models like *E5-mistral-7b-instruct* [27], with 7 billion parameters and 4096-dimensional embeddings, are fine-tuned on multilingual datasets, although primarily optimized for English. Similarly, *udever-bloom-1b1*, a transformer model with 1 billion parameters, fine-tuned on datasets such as MS MARCO, SNLI, and MultiNLI, offers versatile embedding capabilities [28]. OpenAI's third-generation models, including *text-embedding-3-small* and *text-embedding-3-large* [29], have demonstrated competitive performance on multilingual benchmarks like MTEB [30] by using the concept of Matryoshka embedding models. Despite their advanced capabilities, these models often impose significant resource and cost constraints, particularly in large-scale deployments.

In the context of Arabic NLP, several foundational models have played a crucial role in advancing language understanding. AraBERT [31], a BERT-based model specifically pre-trained for Arabic, addresses the language's unique morphological and syntactic characteristics. It has set new benchmarks for tasks such as sentiment analysis and named entity recognition, providing a strong foundation for Arabic NLP research. MARBERT [32], another significant model, enhances Arabic language understanding by focusing on diverse dialects, achieving state-of-the-art results across multiple Arabic NLP tasks. Its comprehensive training on a wide range of Arabic texts ensures robust dialectal coverage.

LaBSE (Language-Agnostic BERT Sentence Embedding) [33], which supports 109 languages, excels in creating multilingual sentence embeddings. By integrating techniques for both monolingual and cross-lingual representation learning, LaBSE delivers impressive performance in bi-text retrieval tasks and provides effective embeddings even for languages with limited training data. Additionally, Sentence-BERT (SBERT) [1], which modifies BERT to produce semantically meaningful sentence embeddings through a Siamese network architecture, significantly reduces computational overhead for similarity tasks while maintaining high accuracy.

Our work extends the principles of MRL to Arabic NLP, an area that has not previously benefited from such sophisticated embedding techniques. By training models specifically tailored for Arabic using sentence transformers, we introduce the first Arabic Matryoshka Embedding Models. These models address the unique challenges of the Arabic language while offering significant improvements in efficiency and adaptability for Arabic NLP systems. Our contributions include the creation and public release of translated datasets and trained models, facilitating broader research and application in the field of Arabic NLP. This pioneering work lays the groundwork for further advancements and practical implementations in this critical area.

3 Dataset Preparation

The preparation of datasets is a crucial step in developing robust sentence embeddings models, particularly for languages with fewer available resources such as Arabic. This section details the Arabic dataset used in this study, the translation process from English to Arabic, and the data preprocessing steps that were employed to ensure the datasets were ready for training nested embedding models.

3.1 Arabic Dataset

The datasets used in this study are derived from the Stanford Natural Language Inference (SNLI) [34] and MultiNLI [35] datasets, which are well-known benchmarks for evaluating models on natural language inference (NLI) tasks [36]. These datasets were originally designed to facilitate various NLP tasks by providing pairs of sentences along with labels that indicate their semantic relationships. To adapt these datasets for Arabic, we created multiple subsets to have more varieties in building different embeddings models.

Pair Subset, this subset contains pairs of sentences with the columns "anchor" and "positive." The primary purpose of this dataset is to facilitate the training of embedding models that need to learn semantic textual similarity. Table 1 shows an example of the Pair subset. The dataset includes 314 K training pairs, 6.81 K validation pairs, and 6.83 K test pairs. By using this subset, models can be trained

Table 1 Example of Arabic pair subset

Subset	Anchor	Positive
Pair	كيف أكون جيولوجياً جيداً؟ How can I be a good geologist?	ماذا علي أن أفعل لأكون جيولوجياً عظيماً؟ What should I do to be a great geologist?

to recognize and quantify the semantic similarity between two sentences, which is crucial for tasks such as paraphrase identification and duplicate question detection.

Triplet Subset, this subset extends the pair subset by including a third column, "negative," to form triplets of sentences. It contains 558K training triplets, 6.58 K validation triplets, and 6.61 K test triplets. Table 2 shows an example of the Triplet subset. The inclusion of a negative example allows the model to not only recognize similar pairs but also to distinguish between similar and dissimilar pairs. This is particularly useful for contrastive learning approaches, where the model learns to pull similar sentences closer and push dissimilar sentences apart in the embedding space.

Pair-Class Subset, this subset comprises three columns: "premise," "hypothesis," and "label," with 942K training examples, 19.7 K validation examples, and 19.7 K test examples. The label indicates the relationship between the premise and hypothesis, such as "0": "entailment", "1": "neutral", "2": "contradiction". Table 3 shows an example of the Pair-Class subset. This subset is specifically designed for natural language inference tasks, where the goal is to determine the logical relationship between two sentences. It provides a rich resource for training models that need to understand and reason about the semantic content of sentences.

Pair-Score Subset, this subset includes columns "sentence1," "sentence2," and "score," with 942 K training pairs, 19.7 K validation pairs, and 19.7 K test pairs. The score represents the degree of similarity between the two sentences on a continuous scale. Table 4 shows an example of the Pair-Score subset. This subset is ideal for tasks that require a fine-grained understanding of semantic similarity, such as ranking and retrieval systems. Models trained on this subset can learn to assign similarity scores

Table 2 Example of Arabic triplet subset

Subset	Anchor	Positive	Negative
Triplet	هناك كلب في الماء (There is a dog in the water)	الكلب يسبح في بركة (The dog is swimming in a pond)	الكلب في الرمال (The dog is in the sand)

Table 3 Example of Arabic pair-class subset

Subset	Premise	Hypothesis	Label
Pair-Class	أطفال يبتسمون و يلوحون للكاميرا (Children are smiling and waving at the camera)	إنهم يبتسمون لوالديهم (They are smiling at their parents)	1
	أطفال يبتسمون و يلوحون للكاميرا (Children are smiling and waving at the camera)	هناك أطفال حاضرون (There are children present)	0
	أطفال يبتسمون و يلوحون للكاميرا (Children are smiling and waving at the camera)	الاطفال يتجهمون (The children are frowning)	2

Table 4 Example of Arabic pair-score subset

Subset	Sentence1	Sentence2	Score
Pair-Score	رجل مسن يشرب عصير البرتقال في مطعم (An elderly man is drinking orange juice in a restaurant)	رجل يشرب العصير (A man is drinking juice)	1
Pair-Score	عائلة أجنبية تسير على طريق ترابي بجانب الماء (A foreign family is walking on a dirt road by the water)	الناس يسيرون بجانب بحيرة (People are walking by a lake)	0.5
Pair-Score	عائلة أجنبية تسير على طول طريق ترابي بجانب الماء (A foreign family is walking along a dirt road by the water)	الناس يقودون سيارة على الطريق السريع (People are driving a car on the highway)	0

to sentence pairs, which can be used to improve the performance of search engines and recommendation systems.

Arabic STSB Structure, this subset is an Arabic version of the Semantic Textual Similarity Benchmark [37]. It consists of sentence pairs drawn from diverse sources such as news headlines, video and image captions, and natural language inference data. Each pair is annotated with a similarity score normalized between 0 and 1. The dataset includes 5.75 K training pairs, 1.68 K validation pairs, and 1.38 K test pairs. Table 5 shows an example of the STSB Benchmark subset. This subset provides a benchmark for evaluating the performance of models on the task of semantic textual similarity in Arabic, offering a standardized way to measure and compare model performance. Table 6 summarizes the subsets details with their training, validation, and test splits.

These subsets cover a broad range of tasks from semantic textual similarity to classification, making them versatile for training and evaluating embedding models.

3.2 Data Preprocessing

Data preprocessing is a critical step to prepare the raw translated text for model training. The following preprocessing steps were applied to both the original Arabic and the translated datasets:

Tokenization, the text was tokenized into individual words or sub-words using SentencePiece [38]. This step breaks down the text into smaller units, making it easier for the model to learn and generalize from the data.

Table 5 Examples of Arabic STSB benchmark subset

Subset	Sentence1	Sentence2	Score
STSB Ex1	طائرة ستقلع (A plane is taking off)	طائرة جوية ستقلع (An airplane is taking off)	1
STSB Ex2	رجل يعزف على ناي كبير (A man is playing a large flute)	رجل يعزف على الناي (A man is playing the flute)	0.76
STSB Ex3	رجل ينشر الجبن المزق على البيتزا (A man is spreading shredded cheese on pizza)	رجل ينشر الجبن المزق على بيتزا غير مطبوخة (A man is spreading shredded cheese on an uncooked pizza)	0.76
STSB Ex4	ثلاثة رجال يلعبون الشطرنج (Three men are playing chess)	رجلين يلعبان الشطرنج (Two men are playing chess)	0.52

Table 6 Arabic NLI subsets details

Subset	Columns	Training Examples	Validation examples	Test examples
Pair subset	"anchor", "positive"	314 K	6.81 K	6.83 K
Triplet subset	"anchor", "positive", "negative"	558 K	6.58 K	6.61 K
Pair-class subset	"premise", "hypothesis", "label"	942 K	19.7 K	19.7 K
Pair-score subset	"sentence1", "sentence2", "score"	942 K	19.7 K	19.7 K
Arabic STSB structure	"sentence1", "sentence2", "similarity score"	5.75 K	1.68 K	1.38 K

Normalization, text normalization was performed to standardize various forms of Arabic script. This included the removal of diacritics, normalization of character forms, and handling of punctuation. Normalization ensures that different forms of the same word are treated uniformly by the model.

Data Structuring, the datasets were structured into the required format with specific columns for each task. For instance, pairs of sentences were organized into "anchor" and "positive" columns for similarity tasks, while triplets included an additional "negative" column. This structuring helps in efficiently loading and processing the data during training.

Validation and Test Splits, the datasets were split into training, validation, and test sets to ensure proper evaluation of the models. The splits were carefully maintained to ensure that the distributions of data remained consistent across these sets. This step is crucial for evaluating the performance of the models and preventing overfitting.

Saving Processed Data, the processed datasets were saved in CSV format to facilitate easy loading and use in subsequent training phases. The data was encoded in $UTF-8$ to preserve the integrity of Arabic characters. Storing the data in a structured format ensures that it can be easily shared and reused by other researchers.

By obtaining these preprocessing steps, we prepared the datasets for the machine translation process to translate these subsets into Arabic language.

3.3 Translation Process

The translation of the datasets from English to Arabic was performed using Neural Machine Translation (NMT) [39], a sophisticated technique that leverages neural networks to achieve high-quality translations. This process involved several meticulous steps to ensure the quality and accuracy of the translations, which are crucial for effective downstream NLP tasks.

Dataset Loading, the first step involved loading the English datasets using the Hugging Face datasets library.[1] This library offers a seamless way to access and manipulate a variety of NLP datasets. By leveraging this library, we ensured that the data was in a structured format, making it easier to process and translate. The datasets included various subsets such as the SNLI and MultiNLI datasets, which are benchmark datasets for natural language inference.

Translation Model Configuration, for the translation task, we used the CTranslate2 model,[2] a powerful NMT model designed for efficient translation tasks. In addition, it was essential to correctly specify the source and target languages to ensure accurate translations. We used language codes *eng_Latn* for English and *arb_Arab* for Arabic. This specification helped the NMT model to understand the linguistic characteristics of both languages, thereby improving the quality of the translations.

Batch Translation, given the large size of the datasets, the translation process was performed in batches. This approach helped manage computational resources effectively and ensured that the translation process was scalable. Each batch of sentences was first tokenized into sub-words using SentencePiece. This tokenization step is crucial because it breaks down the text into manageable units, which the NMT model can then process more effectively.

Handling Special Tokens, special tokens, such as language tags and end-of-sequence markers, were handled carefully to avoid any artifacts in the translated text. For example, the source sentences were prefixed with the source language tag,

[1] https://huggingface.co/docs/datasets/en/index.

[2] https://github.com/OpenNMT/CTranslate2.

and the target sentences were postfixed with the target language tag. After translation, any unnecessary tokens were removed to ensure that the final output was clean and ready for downstream tasks.

These steps of translation are conducted and adapted for each on the data subsets with their columns ensuring translating almost the same number of samples as in original. Moreover, the translation process steps are further controlled by the following main steps including:

SentencePiece Tokenization, each sentence was tokenized into sub-words using SentencePiece. This step ensured that the NMT model could process the text more efficiently by breaking it down into smaller, manageable units.

Translation with CTranslate2, the tokenized sentences were then fed into the CTranslate2 model for translation. The model produced translated sentences in the form of sub-word tokens.

Detokenization, the sub-word tokens were then detokenized back into full sentences using SentencePiece. This step is crucial to reconstruct the sentences in the target language accurately.

Additional steps have been taken into consideration as well focusing on maintaining data consistency and quality assurance where the translated dataset was structured to maintain consistency with the original dataset format. This consistency is vital for ensuring that the translated datasets can be easily integrated into downstream tasks without requiring significant modifications. The structure included specific columns for different types of tasks, such as sentence pairs for similarity tasks and triplets for contrastive learning tasks. Moreover, to ensure the quality of the translations, we performed several checks. A random sample of the translated sentences was manually reviewed to verify the accuracy and fluency of the translations.

The meticulous translation process described above resulted in high-quality Arabic versions of the SNLI and MultiNLI datasets. These translated datasets are structured to facilitate various NLP tasks and are made publicly available on Hugging Face[3] to enable broader research and development in Arabic NLP. This translation process not only ensures high-quality data but also sets a precedent for translating other important NLP datasets into Arabic or other low-resource languages.

The combined efforts in dataset preparation, translation, and preprocessing resulted in a robust set of Arabic NLP datasets that are versatile and ready for training state-of-the-art embedding models. The public release of these datasets on Hugging Face will enable broader research and development in Arabic NLP, fostering advancements in this critical area.

[3] https://huggingface.co/collections/Omartificial-Intelligence-Space/arabic-nli-and-semantic-similarity-datasets-6671ba0a5e4cd3f5caca50c3.

4 Methodology

The methodology section outlines the steps and processes involved in the development and evaluation of the Arabic Matryoshka Embedding Models. This includes the selection of appropriate models, training procedures, and evaluation techniques used to assess their performance on Arabic NLP tasks.

4.1 Model Selection

In this study, we aimed to train and evaluate several Matryoshka embedding models using the Arabic NLI triplet dataset. The selection of models was guided by their proven effectiveness in various NLP tasks, as well as their ability to handle Arabic text to varying extents. The models chosen for this investigation include both monolingual and multilingual sentence transformers, as well as models specifically designed for the Arabic language.

English Sentence Transformer Model, we have utilized $mpnet - base - all - nli - triplet$,[4] model that is a fine-tuned version from $microsoft/mpnet - base$[5] which maps sentences and paragraphs to a 768-dimensional dense vector space. It is designed for tasks such as semantic textual similarity, semantic search, paraphrase mining, text classification, clustering, and more. Despite being primarily trained on English data, it has been exposed to a few Arabic tokens, making it a suitable candidate for assessing cross-lingual transfer capabilities in our investigation.

Multilingual Sentence Transformer Models, we have utilized $paraphrase - multilingual - mpnet - base - v2$[6] model that maps sentences and paragraphs to a 768-dimensional dense vector space and is suitable for tasks like clustering and semantic search. It supports multiple languages, making it a strong candidate for multilingual NLP applications. Additionally, we have used $LaBSE$ (Language-Agnostic BERT Sentence Embedding),[7] model that maps 109 languages including Arabic into a shared vector space, facilitating cross-lingual tasks. It is a robust choice for evaluating multilingual capabilities in embedding models.

Arabic Sentence Transformer Models, to expand our investigation, we have used Arabic based sentence transformers including $AraBERT$,[8] an arabic pretrained language model based on Google's BERT architecture. It uses the BERT-Base configuration and is tailored for Arabic NLP tasks. AraBERT has demonstrated high performance in various Arabic language benchmarks. Moreover, $MARBERT$[9] model which is designed to handle both Modern Standard Arabic (MSA) and Dialectal

[4] https://huggingface.co/tomaarsen/mpnet-base-all-nli-triplet.

[5] https://huggingface.co/microsoft/mpnet-base.

[6] https://huggingface.co/sentence-transformers/paraphrase-multilingual-mpnet-base-v2.

[7] https://huggingface.co/sentence-transformers/LaBSE.

[8] https://huggingface.co/aubmindlab/bert-base-arabertv02.

[9] https://huggingface.co/UBC-NLP/MARBERT.

Arabic (DA). It is based on the BERT architecture and trained on a large corpus of Arabic tweets, making it particularly effective for tasks involving social media text.

The choice of these models allows for a comprehensive evaluation across different dimensions of Arabic NLP. The inclusion of an English model exposed to some Arabic tokens provides insights into the transferability of learned representations across languages. The multilingual models offer a perspective on how well models trained on diverse language data perform on Arabic specific tasks. Finally, the Arabic-specific models, $AraBERT$ and $MARBERT$, serve as benchmarks for performance in native Arabic NLP contexts.

By training these models on the Arabic NLI triplet dataset, we aim to create Matryoshka embedding versions that are not only versatile and efficient but also tailored to the unique characteristics and challenges of the Arabic language. This thorough approach ensures that our models are well-equipped to handle a variety of NLP tasks, including semantic textual similarity, semantic search, paraphrase mining, text classification, and clustering, thereby advancing the state of Arabic NLP.

4.2 Matryoshka Embedding Models

Matryoshka Embedding Models [10] represent an innovative approach to creating adaptable and efficient embeddings for natural language processing tasks. These models aim to capture multi-granularity in embeddings, allowing different levels of the embedding vector to independently serve as meaningful representations. This adaptability is crucial for efficiently managing computational resources, especially in large-scale and resource-constrained environments like Arabic NLP.

One of the core strengths of Matryoshka Embedding Models lies in their ability to create hierarchical, nested representations that are particularly well-suited to handling the linguistic complexities of Arabic. The hierarchical nature of these embeddings allows the models to store the most critical linguistic information in the initial dimensions of the embedding vector, making them highly effective in capturing the subtle morphological and syntactic variations that are characteristic of Arabic.

Unlike models trained on raw corpora, which often struggle to generalize across the rich morphological and grammatical structures of Arabic, Matryoshka Embedding Models are trained on carefully curated Arabic triplet datasets. This approach ensures that the models not only capture the nuances of the language but also excel in handling the specific challenges posed by Arabic. By focusing on triplet datasets, the models learn to differentiate between fine-grained semantic distinctions, leading to more robust and generalizable embeddings compared to those trained on raw, unstructured data.

Figure 1 shows the process of Matryoshka Representation Learning (MRL) which involves a d-dimensional representation vector $z \in R^d$ for a given datapoint x in the input domain χ. This representation vector is obtained using a deep neural network $F(.; \theta_F)$, parameterized by learnable weights θ_F. The objective of MRL is to ensure that each of the first m dimensions of the embedding vector, $z_{1:m} \in R^m$, where $m \in$

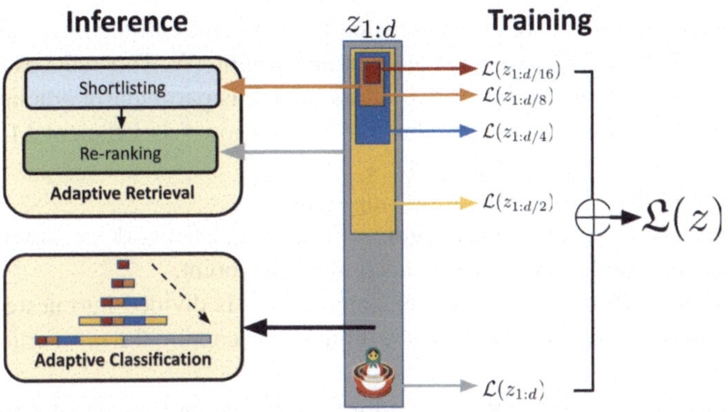

Fig. 1 Matryoshka representation learning process [10]

M, can independently serve as a transferable and general-purpose representation of the datapoint x.

The multi-granularity of these embeddings is captured through the set of chosen dimensions M, which are determined by consistent halving until the representation size reaches a minimal informative state. This nested, coarse-to-fine granularity ensures that the representations remain useful even when truncated to smaller dimensions.

Given a labeled dataset $D = \{(x_1, y_1), \ldots, (x_N, y_N)\}$, where $x_i \in \chi$ is an input point and $y_i \in [L]$ represents the label of x_i MRL optimizes the multi-class classification loss for each nested dimension $m \in M$ using standard empirical risk minimization. This is achieved by employing a separate linear classifier, parameterized by $\mathbf{W}^{(m)} \in R^{L\times m}$ for each dimension. The losses obtained from these classifiers are then aggregated, taking into account their relative importance $c_m \geq 0$ for $m \in M$. The optimization objective can be expressed as:

$$\mathcal{L}_{\text{MRL}} = \sum_{m\in M} c_m \mathcal{L}_{\text{CE}}(\mathbf{W}^{(m)}\mathbf{z}_{1:m}, y) \tag{1}$$

where:

- $\mathcal{L}_{\text{MRL}}$ is the Matryoshka Representation Learning loss.
- c_m represents the relative importance of each dimension m.
- $\mathcal{L}_{\text{CE}}$ is the multi-class softmax cross-entropy loss function.
- $\mathbf{W}^{(m)} \in \mathbb{R}^{L\times m}$ are the weights of the linear classifier for dimension m.
- $\mathbf{z}_{1:m} \in \mathbb{R}^m$ is the truncated embedding vector up to dimension m.
- y is the true label corresponding to the input x.

This formulation ensures that each subset of the embedding dimensions can independently perform well on the classification task, maintaining the flexibility and robustness of the learned representations.

To enhance efficiency, weight-tying is employed across all the linear classifiers, i.e., defining $\mathbf{W}^{(m)} = \mathbf{W}_{1:m}$ for a set of common weights $\mathbf{W}$. This reduces the memory cost associated with the linear classifiers, which is particularly crucial in cases of extremely large output spaces. This variant is known as Efficient Matryoshka Representation Learning $(MRL - E)$.

In practice, MRL involves the following steps:

Representation Vector Generation, a deep neural network generates a high-dimensional representation vector for each input datapoint.

Dimension Nesting, the high-dimensional vector is divided into nested subsets of dimensions, ensuring that the first few dimensions capture the most crucial information.

Separate Classifier Training, each subset of dimensions is used to train a separate linear classifier, optimizing the classification loss for each dimension.

Loss Aggregation, the classification losses from all subsets are aggregated based on their relative importance to form the final loss function.

By leveraging these methods, Matryoshka Embedding Models achieve flexibility and efficiency, making them suitable for adaptive deployment in various NLP tasks. Figure 2 shows the truncation step of the Matryoshka embedding where the ability to truncate embeddings is performed without significant loss of information allows for scalable and resource-efficient applications, particularly beneficial in the context of Arabic NLP where computational resources may be limited.

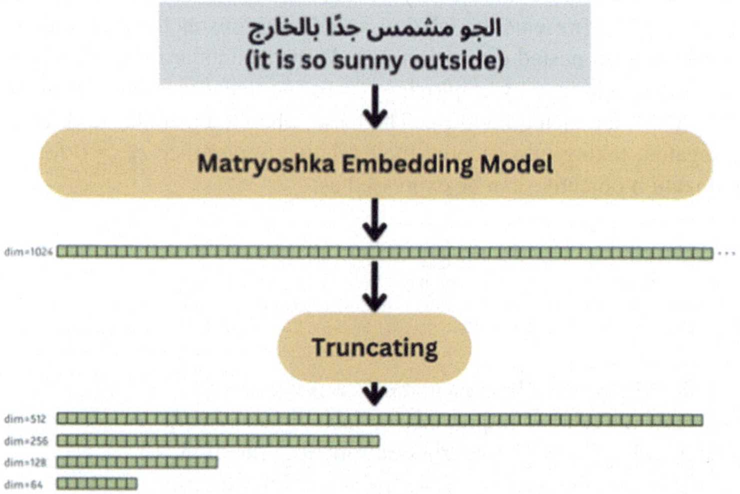

Fig. 2 Truncation step in matryoshka representation learning

4.3 Nested Embedding Training Process

The process of training nested embedding models on the translated Arabic dataset was methodically structured to ensure optimal performance. The training setup began with the initialization of several Sentence Transformer models. The primary model used was $mpnet - base - all - nli - triplet$, which maps sentences and paragraphs to a 768-dimensional dense vector space. Additionally, two multilingual sentence transformer models, $paraphrase - multilingual - mpnet - base - v2$ and $LaBSE$, and two Arabic-specific models, $AraBERT$ and $MARBERT$, were trained. These models were selected to leverage their existing capabilities while adapting them to handle the specific characteristics of the Arabic language.

The dataset used for training, the $arabic - nli - triplet$ subset, was loaded using the Hugging Face datasets library. This dataset comprises triplets of sentences, each consisting of an anchor, a positive, and a negative example with size of 558 k samples. The models were configured to handle these inputs, with settings such as using cosine similarity as the similarity function and defining the maximum sequence length as 512 tokens.

Hyper-parameter tuning was a critical aspect of the training process. The batch size was set to 128 to balance the computational load and model performance, and each model was trained for one epoch, leveraging the computational power of an A100 GPU. The Matryoshka embedding approach was implemented by specifying output dimensions of [768, 512, 256, 128, 64], allowing the models to produce embeddings at various levels of granularity. This hierarchical structure enables flexible and efficient processing of downstream tasks.

Optimization during training involved the use of specific loss functions. The primary loss function, $MultipleNegativesRankingLoss$ [40], was augmented with $MatryoshkaLoss$ to train embeddings at multiple dimensions simultaneously [10]. This combined approach ensured that the embeddings were effective across different levels of granularity, making them robust for various applications.

The training process was managed by data loading and preprocessing involving shuffling the dataset and selecting a subset of examples for training to manage computational resources effectively. Each triplet was processed to ensure consistency and quality. The $MultipleNegativesRankingLoss$ handled the ranking of the embeddings, while the $MatryoshkaLoss$ maintained their hierarchical structure. An evaluator was set up using the STS Benchmark validation dataset to monitor performance during training. This involved calculating cosine similarity between sentence pairs and tracking performance across different dimensions of the embeddings.

The training execution was conducted using $SentenceTransformerTrainer$ from the Sentence Transformers library.[10] The training arguments included specifications for the number of epochs, batch size, learning rate scheduling, and evaluation strategies. After training, the models were evaluated on the STS Benchmark test dataset to assess their performance across different dimensions. The results were tracked and recorded, highlighting the effectiveness of the Matryoshka embedding

[10] https://sbert.net/index.html.

approach. Finally, the trained models were saved locally and uploaded to the Hugging Face Hub for public access.[11]

This meticulous training ensured that the process Matryoshka embedding models were robust, efficient, and ready for deployment in various NLP tasks, particularly in the context of Arabic language processing.

5 Results and Discussion

In this section, we provide a detailed framework of the evaluation process and the metrics used to assess the performance of the trained Arabic Matryoshka embedding models on Arabic datasets. Our aim is to thoroughly examine the models' efficacy in various dimensions and discuss the outcomes in the context of Arabic natural language processing tasks.

To evaluate the performance of our models, we utilized the Arabic Semantic Textual Similarity Benchmark ($STSB$).[12] This dataset is a comprehensive collection of sentence pairs drawn from diverse sources, including news headlines, video and image captions, and natural language inference data. Each sentence pair in the dataset is human-annotated with a similarity score ranging from 1 to 5, which we normalized to a scale between 0 and 1 for this evaluation. This normalization ensures consistency and facilitates a more straightforward comparison across different models and dimensions.

The evaluation of our Arabic Matryoshka embedding models was conducted using the $EmbeddingSimilarityEvaluator$, a robust evaluation tool provided by the Sentence Transformers library. This evaluator is specifically designed to measure the similarity of embeddings by calculating the Spearman and Pearson rank correlation coefficients against the gold standard labels. These correlations provide a quantitative measure of how well the embeddings capture the semantic similarity between sentence pairs.

The $EmbeddingSimilarityEvaluator$ performs evaluation based on several metrics:

Cosine Similarity, measures the cosine of the angle between two vectors, providing a value between -1 and 1.

Manhattan Distance, calculates the absolute differences between corresponding elements of the vectors.

Euclidean Distance, computes the straight line distance between two points in the vector space.

Dot Product Similarity, computes the dot product of two vectors, which can indicate similarity in direction and magnitude.

[11] https://huggingface.co/collections/Omartificial-Intelligence-Space/arabic-matryoshka-embedding-models-666f764d3b570f44d7f77d4e.

[12] https://huggingface.co/datasets/Omartificial-Intelligence-Space/Arabic-stsb.

These metrics were computed for each dimension of the embeddings *(768, 512, 256, 128, 64)*, resulting in a comprehensive evaluation of the models' performance at various levels of granularity. For each dimension, we used the following metrics:

Pearson Correlation (Cosine, Manhattan, Euclidean, Dot, Max), measures the linear correlation between the predicted and actual similarity scores.

Spearman Correlation (Cosine, Manhattan, Euclidean, Dot, Max), Assesses the monotonic relationship between the predicted and actual similarity scores.

The evaluation process commenced with loading the pretrained Sentence Transformer models and the $STSB$ dataset using the Hugging Face datasets library. Each model was evaluated on the sentence pairs in the dataset, with the $Embedding SimilarityEvaluator$ applied to calculate the similarity scores. This process involved initializing the evaluator with the sentences and their corresponding human-annotated scores, specifying the primary similarity function (cosine similarity), and setting other relevant parameters such as batch size and truncation dimension.

The evaluator outputs a set of correlation scores for each metric and dimension, providing a detailed view of how well the embeddings align with human judgment of sentence similarity. This methodology ensures a rigorous and standardized evaluation, enabling us to compare the performance of different models and configurations objectively.

The results from this evaluation process will be discussed in the subsequent subsections, where we will delve into the performance metrics, compare the models, and interpret the findings in the context of their application to Arabic NLP tasks. This detailed analysis will highlight the strengths and potential areas for improvement in our Arabic Matryoshka embedding models, contributing to the broader understanding and development of efficient and effective NLP solutions for the Arabic language.

5.1 Comprehensive Performance Analysis For Each Trained Arabic Matryoshka Embedding Model

A deep investigation have been done to evaluate the performance of different trained Arabic Matryoshka embedding models on different dimensions. The experiments consists of evaluating Arabic embeddings models and multilingual embeddings models along with An English model subjected to a small number of Arabic tokens is included to shed light on the cross-linguistic transferability of learnt representations. Results of the first multilingual model named $Paraphrase - Multilingual - MPNet - Base - V2$ which is trained to create Arabic Matryoshka embeddings across different dimensions are summarized in Table 7.

As shown in Table 7, the model shows robust performance across all dimensions, with a slight decrease in correlation as the dimensionality reduces. The highest Pearson and Spearman correlations are observed at dimensions 768 and 512, indicating strong semantic similarity capture. Moreover, Dot product similarity metrics decrease

Table 7 Performance of trained paraphrase multilingual MPNet base V2 Matryoshka model

Dimension	Pearson cosine	Spearman cosine	Pearson manhattan	Spearman Manhattan	Pearson Euclidean	Spearman Euclidean	Pearson dot	Spearman dot	Pearson max	Spearman max
768	0.8539	0.8616	0.8497	0.8513	0.8516	0.8541	0.7281	0.7230	0.8539	0.8616
512	0.8542	0.8609	0.8487	0.8512	0.8505	0.8539	0.7076	0.7029	0.8542	0.8609
256	0.8486	0.8579	0.8405	0.8456	0.8415	0.8472	0.6669	0.6651	0.8486	0.8579
128	0.8390	0.8499	0.8287	0.8353	0.8298	0.8372	0.5856	0.5835	0.8390	0.8499
64	0.8291	0.8429	0.8101	0.8221	0.8129	0.8255	0.5067	0.5110	0.8291	0.8429

significantly with lower dimensions, highlighting its sensitivity to dimensionality reduction.

Secondly, results of the second multilingual model 'LaBSE' are detailed in Table 8.

As shown in Table 8, LaBSE exhibits high performance across all dimensions, showing minimal degradation in correlation values as dimensions decrease. The model's robustness is evident in maintaining high Pearson and Spearman correlations across various metrics.

Moving on, Tables 9 and 10 shows the performance evaluation metrics for the arabic embeddings models Arabert and MArbert respecitvely.

As shown in Table 9, Trained Arabert Matryoshka Model shows moderate performance, with correlations improving slightly as dimensions decrease from 768 to 64. While the results are lower than those for multilingual models, they are consistent across various metrics. Moreover, as shown in Table 10 for Marbert model shows strong performance at higher dimensions, with consistent Pearson and Spearman correlations. The model's performance in dot product similarity is lower compared to other metrics, suggesting an area for potential improvement.

Finally, the results of English model $MPNet - Base - All - NLI - Triplet$, which has seen a few Arabic tokens, are shown in Table 11.

As shown in Table 11, the performance of the MPNet-base-all-NLI-triplet Matryoshka model is noticeably lower compared to the multilingual model, especially in lower dimensions. Despite this, the model shows reasonable consistency across Pearson and Spearman correlations. The ability to handle Arabic tokens, though limited, provides some insights into the adaptability of English-trained models.

The evaluation results indicate that multilingual models like $Paraphrase - Multilingual - MPNet - Base - V2$ and $LaBSE$ outperform specifically Arabic-trained models in capturing semantic similarity in Arabic text. The trained $Paraphrase - Multilingual - MPNet - Base - V2$ model demonstrated the highest performance, particularly in higher dimensions. In contrast, models specifically designed for Arabic, such as $BERT - Base - AraBERTV02$and $MAR BERT$, show moderate performance but provide valuable insights into the nuances of Arabic language processing. The decrease in performance at lower dimensions for all models highlights the challenge of dimensionality reduction in maintaining semantic integrity.

Overall, these results underscore the importance of multilingual capabilities in embedding models for diverse language tasks and point to areas where Arabic-specific models can be further improved to match their multilingual counterparts.

Table 8 Performance of trained LaBSE Matryoshka model

Dimension	Pearson cosine	Spearman cosine	Pearson Manhattan	Spearman Manhattan	Pearson Euclidean	Spearman Euclidean	Pearson dot	Spearman dot	Pearson max	Spearman max
768	0.7269	0.7225	0.7259	0.721	0.726	0.7225	0.7269	0.7225	0.7269	0.7225
512	0.7268	0.7224	0.7241	0.7195	0.7248	0.7213	0.7253	0.7205	0.7268	0.7224
256	0.7283	0.7264	0.7228	0.7181	0.7251	0.7215	0.7243	0.7221	0.7283	0.7264
128	0.7102	0.7104	0.7135	0.7089	0.7172	0.713	0.6778	0.6746	0.7172	0.713
64	0.6931	0.6982	0.6971	0.6942	0.7013	0.6987	0.6377	0.6345	0.7013	0.6987

Table 9 Performance of trained AraBERT Matryoshka model

Dimension	Pearson cosine	Spearman cosine	Pearson Manhattan	Spearman Manhattan	Pearson Euclidean	Spearman Euclidean	Pearson dot	Spearman dot	Pearson max	Spearman max
768	0.595	0.616	0.6296	0.627	0.6327	0.6317	0.4282	0.4295	0.6327	0.6317
512	0.5846	0.6064	0.6288	0.6264	0.6313	0.6302	0.3789	0.3768	0.6313	0.6302
256	0.5779	0.596	0.6243	0.6217	0.6238	0.6215	0.3597	0.353	0.6243	0.6215
128	0.5831	0.6022	0.6152	0.6122	0.6162	0.6153	0.4044	0.4015	0.6162	0.6153
64	0.5725	0.5914	0.6024	0.5967	0.6069	0.6041	0.3632	0.3585	0.6069	0.6041

Table 10 Performance of trained Marbert Matryoshka model

Dimension	Pearson cosine	Spearman cosine	Pearson Manhattan	Spearman Manhattan	Pearson Euclidean	Spearman Euclidean	Pearson dot	Spearman dot	Pearson max	Spearman max
768	0.6112	0.6117	0.6444	0.6358	0.6444	0.6346	0.4724	0.4484	0.6444	0.6358
512	0.6665	0.6648	0.643	0.6335	0.6466	0.6373	0.537	0.5242	0.6665	0.6648
256	0.6601	0.6593	0.6362	0.6251	0.6408	0.63	0.5251	0.5155	0.6601	0.6593
128	0.6549	0.6523	0.6343	0.6227	0.6397	0.6281	0.4724	0.4634	0.6549	0.6523
64	0.6367	0.637	0.6264	0.6119	0.6328	0.618	0.4117	0.4044	0.6367	0.637

Table 11 Performance of trained MPNet-base-all-NLI-triplet Matryoshka model

Dimension	Pearson cosine	Spearman cosine	Pearson manhattan	Spearman Manhattan	Pearson Euclidean	Spearman Euclidean	Pearson dot	Spearman dot	Pearson max	Spearman max
768	0.6699	0.6757	0.6943	0.684	0.6973	0.6873	0.5534	0.5422	0.6973	0.6873
512	0.6628	0.6703	0.6917	0.6816	0.6949	0.6853	0.5229	0.5114	0.6949	0.6853
256	0.6368	0.6513	0.6832	0.6746	0.6844	0.676	0.4266	0.4179	0.6844	0.676
128	0.6148	0.6355	0.6731	0.6653	0.6764	0.6691	0.3513	0.3445	0.6764	0.6691
64	0.5789	0.6081	0.6579	0.6519	0.663	0.6571	0.2403	0.2331	0.663	0.6571

5.2 *Comparative Analysis of Different Arabic Trained Matryoshka Models Performance Across Metrics and Dimensions*

This subsection provides a detailed visual comparison of the performance of various Arabic Matryoshka embedding models across different dimensions. By examining the comparative plots for multiple evaluation metrics, including Pearson and Spearman correlations, we gain insights into how each model performs at different levels of granularity. Plots shown in Fig. 3 serve as a visual representation of the quantitative data, offering a clearer perspective on the relative performance of each model.

Comparison of Model Performance on Pearson Manhattan Similarity is shown in Fig. 3a where, the $paraphrase - multilingual - mpnet - base - v2$ model consistently outperformed other models across all dimensions, exhibiting the highest Pearson Manhattan similarity values. $LaBSE$ demonstrated stable performance with slight decreases at lower dimensions. The $mpnet - base - all - nli - triplet$ model showed moderate performance, surpassing the Arabic-specific models, $bert - base - arabertv02$, and MARBERT. Among these, $MARBERT$ and $bert - base - arabertv02$ had the lowest values, indicating less effectiveness in capturing semantic similarity using Pearson Manhattan similarity.

Comparison of Model Performance on Spearman Manhattan Similarity, is shown in Fig. 3b where, the $paraphrase - multilingual - mpnet - base - v2$ model again showed the highest Spearman Manhattan similarity values, reinforcing its strong performance. $LaBSE$ followed closely, maintaining stable performance across dimensions. The $mpnet - base - all - nli - triplet$ model's performance was moderate, similar to its Pearson Manhattan metric results. $MARBERT$ and $bert - base - arabertv02$ exhibited lower values, suggesting less consistency in semantic similarity capture.

Comparison of Model Performance on Pearson Euclidean Similarity is shown in Fig. 3c where, the Pearson Euclidean similarity is compared, the $paraphrase - multilingual - mpnet - base - v2$ model maintained high values across dimensions. $LaBSE$ also performed consistently well. The $mpnet - base - all - nli - triplet$ model showed moderate performance, surpassing the Arabic-specific models. $MARBERT$ and $bert - base - arabertv02$ showed lower values, with $MARBERT$ slightly outperforming $bert - base - arabertv02$.

Comparison of Model Performance on Spearman Euclidean Similarity, is shown in Fig. 3d for Spearman Euclidean similarity where, the $paraphrase - multilingual - mpnet - base - v2$ model led with the highest values. $LaBSE$ followed closely, showing stable values across dimensions. The $mpnet - base - all - nli - triplet$ model demonstrated moderate performance. The lowest values were observed in $MARBERT$ and $bert - base - arabertv02$, with $MARBERT$ slightly outperforming $bert - base - arabertv02$.

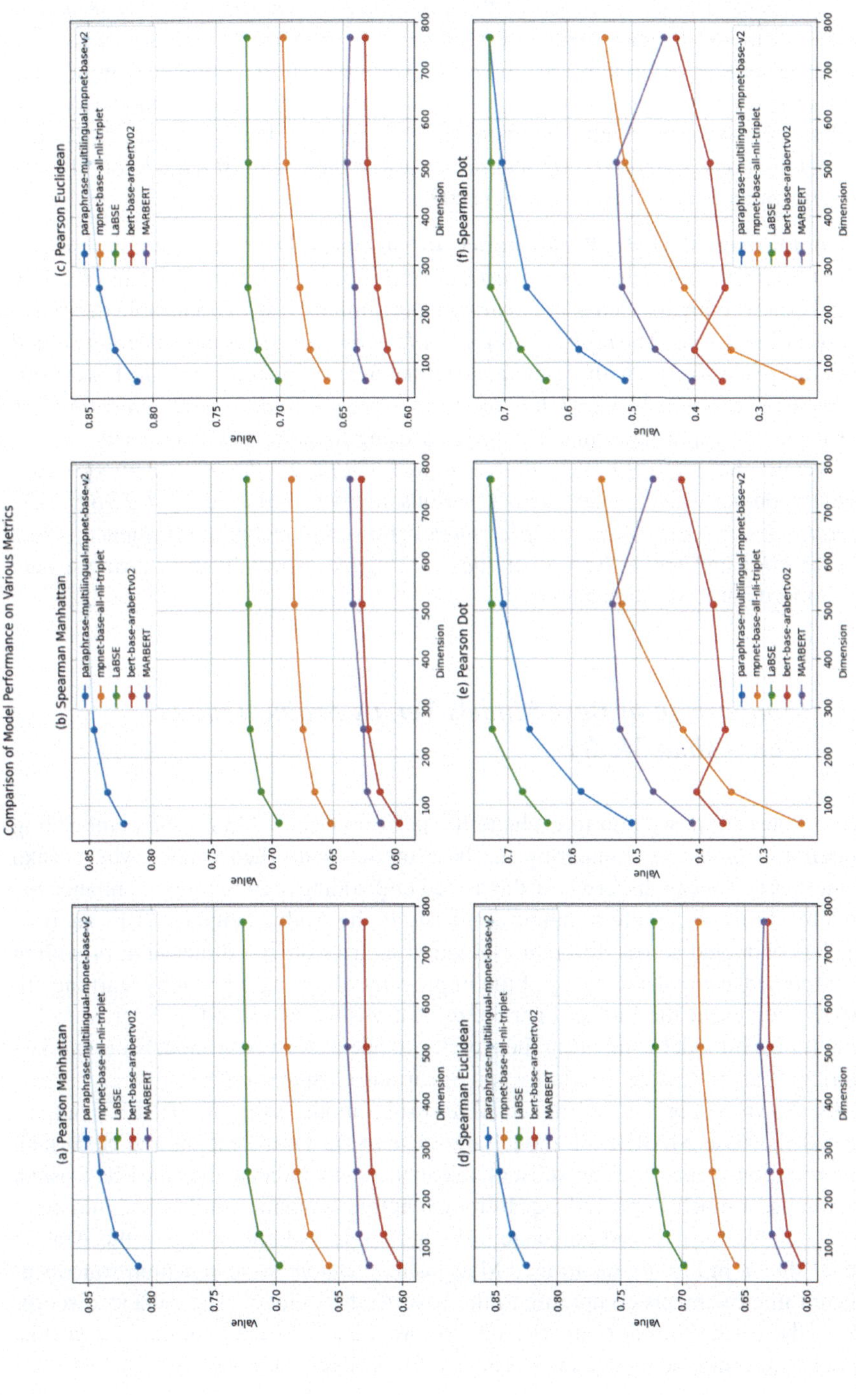

Fig. 3 Comparative analysis of model performance across different metrics and dimensions

Comparison of Model Performance on Pearson Dot Similarity, is shown in Fig. 3e where, The performance of the $paraphrase - multilingual - mpnet - base - v2$ model decreased with lower dimensions, showing the highest Pearson Dot similarity value at 768 dimensions. $LaBSE$ exhibited a similar trend, with noticeable drops at lower dimensions. The $mpnet - base - all - nli - triplet$ model showed moderate performance, with a significant decline at lower dimensions. $MARBERT$ and $bert - base - arabertv02$ had the lowest values, with $MARBERT$ slightly outperforming $bert - base - arabertv02$.

Comparison of Model Performance on Spearman Dot Similarity, is shown in Fig. 3f where, the $paraphrase - multilingual - mpnet - base - v2$ and $LaBSE$ models showed higher values in Spearman Dot similarity, with noticeable decreases at lower dimensions. The $mpnet - base - all - nli - triplet$ model demonstrated moderate performance, with significant declines at lower dimensions. The lowest values were observed in $MARBERT$ and $bert - base - arabertv02$, indicating less consistency in capturing semantic similarity using Spearman Dot similarity.

Overall, the $paraphrase - multilingual - mpnet - base - v2$ model consistently outperformed others across various metrics, followed by $LaBSE$. The $mpnet - base - all - nli - triplet$ model showed moderate performance, while $MARBERT$ and $bert - base - arabertv02$ lagged, especially in capturing semantic similarity at lower dimensions.

5.3 Comparison of Base Models Versus Arabic Trained Matryoshka Models

In this subsection, we aim to evaluate the performance of Matryoshka embedding models for Arabic by comparing the base models with their trained Matryoshka counterparts, we can discern how the nested embedding learning might enhance the models' ability to capture semantic similarity in the Arabic language. This analysis will be conducted across different evaluation metrics on 768 dimension, providing a comprehensive understanding of the improvements brought about by learning. To visually represent the impact of training Matryoshka Models, Fig. 3, provide bar plots for each model comparing their performance against base models on two key metrics: Pearson Cosine Similarity and Spearman Cosine Similarity.

As shown in Fig. 4a, the trained Matryoshka model using MARBERT outperformed the base MARBERT model significantly in both Pearson and Spearman cosine similarity metrics. This substantial enhancement indicates that the Matryoshka version has a much-improved capability to capture semantic similarities, making it more suitable for nuanced language tasks in Arabic. Additionally, For the Arabert model shown in Fig. 4b, the trained Matryoshka version has also demonstrated significant improvements comparing to the base Arabert model. The enhanced scores across the cosine similarity metrics underscore the effectiveness of the Matryoshka model in refining the model's performance for Arabic text processing.

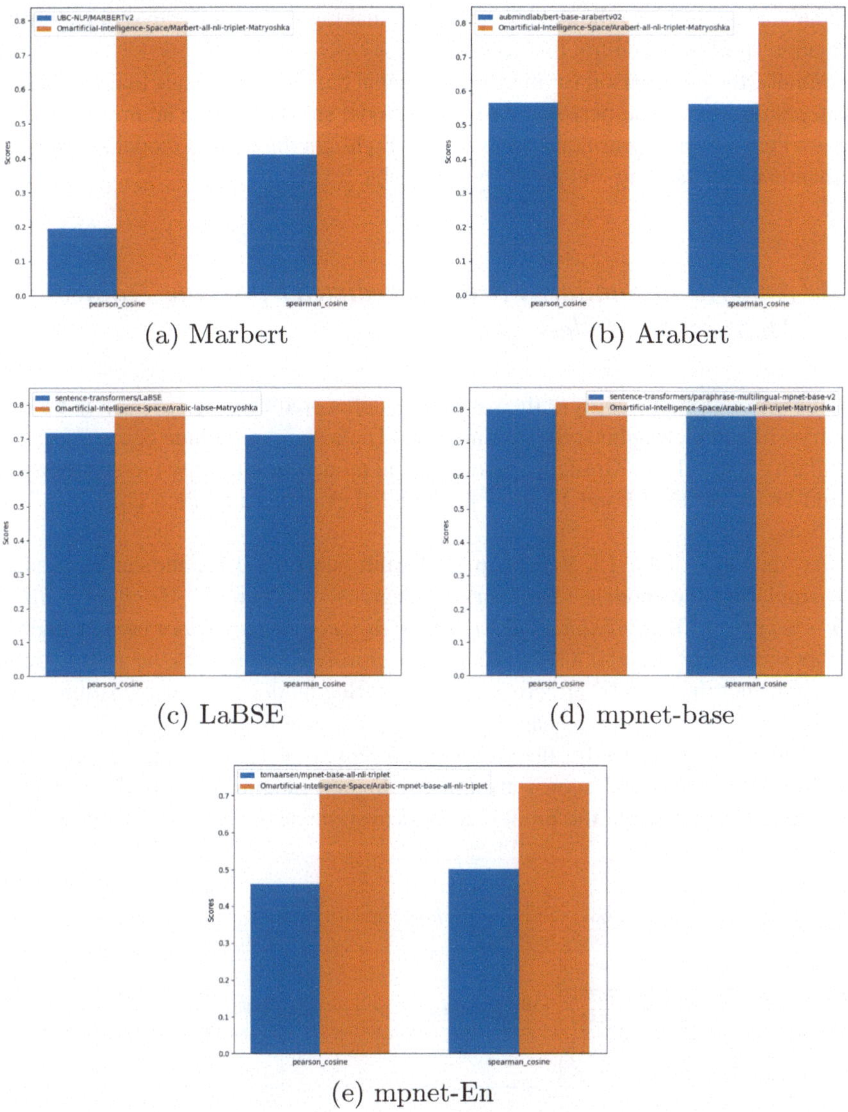

Fig. 4 Comparison of base models versus trained Matryoshka models across various metrics

For the multilingual models shown in Fig. 4c and d, the trained Matryoshka model of LaBSE model showed improved performance compared to the base LaBSE model and also the trained multilingual mpnet base model also demonstrated superior performance over its base counterpart across both Pearson and Spearman cosine similarity metrics. This consistency suggests that the Matryoshka model enhances the model's adaptability to the Arabic language, making it more effective in capturing semantic similarities within this context.

Finally, the comparison for the English mpnet base model reveals that the Arabic Matryoshka version outperformed the base model substantially. The improvements across Pearson and Spearman cosine metrics highlight the trained model's enhanced semantic understanding and accuracy for Arabic text.

5.4 Analysis of Similarity Scores Predicted by Arabic Trained Matryoshka Models

In this subsection, we analyze the similarity scores predicted by our Arabic trained Matryoshka models against the ground truth. The examples include cases of perfect similarity (score 1), no similarity (score 0), and moderate similarity (scores between 0 and 1). Tables 12, 13 and 14 show the detailed examples and their corresponding analysis.

As shown in Table 12, with a ground truth score of 0.72, indicating moderate similarity, the models show varying degrees of accuracy. $Marbert - all - nli - triplet - Matryoshka$ and $Arabic - labse - Matryoshka$ predict slightly higher scores (0.836 and 0.835, respectively), closely aligning with the ground truth. This indicates the models' effectiveness in capturing moderate semantic similarities, particularly when sentences share significant contextual overlap.

As shown in Table 13, for the ground truth score of 0.1, indicating no similarity, the Matryoshka models' scores are notably higher, ranging from 0.323 to 0.481. This suggests that while the models recognize some degree of unrelatedness, they

Table 12 Comparison of model scores for the high similarity pair

Model	Score	Sentence1	Sentence2
Ground Truth	0.72	مجموعة من الرجال يلعبون كرة القدم (A group of men playing football)	مجموعة من الأولاد يلعبون كرة القدم (A group of boys playing football)
Arabic-mpnet-base-all-nli-triplet	0.768		
Arabic-all-nli-triplet-Matryoshka	0.685		
Arabert-all-nli-triplet-Matryoshka	0.661		
Arabic-labse-Matryoshka	0.835		
Marbert-all-nli-triplet-Matryoshka	0.836		

Table 13 Comparison of model scores for the no similarity pair

Model	Score	Sentence1	Sentence2
Ground Truth	0.1	رجل يعزف على الجيتار (A man playing the guitar)	رجل يقود سيارة (A man driving a car)
Arabic-mpnet-base-all-nli-triplet	0.334		
Arabic-all-nli-triplet-Matryoshka	0.481		
Arabert-all-nli-triplet-Matryoshka	0.480		
Arabic-labse-Matryoshka	0.323		
Marbert-all-nli-triplet-Matryoshka	0.382		

Table 14 Comparison of model scores for the moderate similarity pair

Model	Score	Sentence1	Sentence2
Ground Truth	1	رجل يقوم بخدعة بالبطاقات (A man doing a card trick)	رجل يقوم بخدعة ورق (A man performing a card trick)
Arabic-mpnet-base-all-nli-triplet	0.8		
Arabic-all-nli-triplet-Matryoshka	0.91		
Arabert-all-nli-triplet-Matryoshka	0.87		
Arabic-labse-Matryoshka	0.84		
Marbert-all-nli-triplet-Matryoshka	0.85		

still infer a minimal level of semantic similarity, possibly due to shared contextual elements like " رجل" (man) in both sentences.

As shown in Table 14, the ground truth score is 1, indicating perfect similarity between the two sentences. All the Matryoshka models also predict high similarity scores, with $Arabic - all - nli - triplet - Matryoshka$ achieving the highest score of 0.906. This demonstrates the model's capability to capture near-perfect semantic similarity for sentences with minimal lexical variation.

To further analyze the performance of the trained Matryoshka models, we compare their average predicted similarity scores against the ground truth scores across three different similarity categories: low similarity, no similarity, and moderate similarity. The comparative results are illustrated in the Fig. 5.

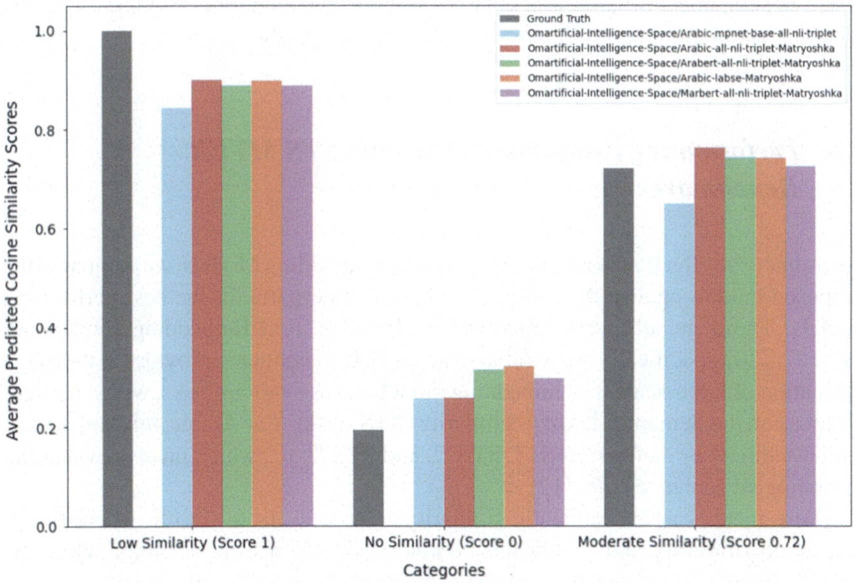

Fig. 5 Comparison of average predicted cosine similarity scores for different similarity categories

The error analysis reveals that while the Arabic trained Matryoshka models generally perform well in identifying high and moderate similarity sentence pairs, they exhibit a tendency to overestimate similarity in the no similarity category. For the low similarity category, all models predicted scores close to the ground truth, demonstrating their effectiveness in capturing high similarity relationships. Specifically, the $Marbert-all-nli-triplet-Matryoshka$ and $Arabic-labse-Matryoshka$ models showed the highest accuracy in this category. However, in the no similarity category, the predicted scores were consistently higher than the ground truth, indicating false positives.

This suggests a potential area for improvement, as the models tend to recognize some degree of similarity in dissimilar pairs. In the moderate similarity category, the models also performed well, although there was a slight tendency to overestimate similarity. Overall, these findings suggest that while the models are effective at capturing similarities, they require further refinement to accurately identify dissimilar pairs and reduce false positive rates. This enhancement can improve the robustness and accuracy of similarity models in multilingual contexts, especially for the Arabic language.

The consistent trend of improved performance across all trained Arabic Matryoshka models highlights the efficacy of the Nested learning process. The significant gains in both Pearson and Spearman cosine similarity metrics indicate that the trained models have a better grasp of the semantic nuances in the Arabic language. This enhancement makes these models more suitable for tasks requiring high precision in semantic similarity assessments, such as machine translation, information retrieval, and natural language understanding tasks specific to Arabic.

5.5 Performance Evaluation on Arabic STS MTEB Benchmarks

To evaluate the effectiveness of the Matryoshka Embedding Models, we compared the proposed models against their respective base counterparts for the best-performing models. These models were tested on the Massive Text Embedding Benchmark (MTEB) [30] specifically for Arabic. The MTEB benchmark provides a thorough evaluation of large-scale text embedding models across various tasks, with a particular focus on the Semantic Textual Similarity (STS) tasks. For Arabic, this benchmark includes three key metrics: STS17, STS22, and STS22-v2, which involve evaluating cross-lingual sentence pairs [12].

These metrics assess semantic similarity on a scale from 0 to 5, where 0 represents no similarity and 5 indicates complete equivalence in meaning. While the MTEB covers a broad spectrum of tasks, the availability of Arabic-specific datasets is somewhat limited beyond these metrics, which constrains the range of our experiments. However, the insights drawn from these STS Arabic MTEB benchmarks remain significant. The results of our experiments, presented in Table 15, illustrate a notable performance enhancement for the Matryoshka models compared to their base versions.

As demonstrated in Table 15, the Matryoshka models exhibit a significant improvement in performance over their base counterparts on the MTEB benchmark. Notably, the *Arabert-all-nli-triplet-Matryoshka* model achieved an impressive average score of 69.99, leading the STS17 metric with a score of 85.31. The *Marbert-all-nli-triplet-Matryoshka* model also performed robustly, with an average score of 67.19, showing strong results in both STS22 and STS22-v2. Similarly, the *Arabic-labse-Matryoshka* model achieved a solid average score of 66.76.

In contrast, the base models underperformed relative to the Matryoshka versions. For instance, the *paraphrase-multilingual-mpnet-base-v2* model attained an average score of 62.21, while the *bert-base-arabertv02* model lagged behind significantly with the lowest score of 50.45. These results underscore the effectiveness of Matryoshka Representation Learning (MRL) in optimizing embedding dimensions and improving performance, particularly in cross-lingual tasks like STS17.

5.6 Comparative Evaluation of Arabic Trained Matryoshka Models and LLMs

To assess the effectiveness of Arabic Matryoshka models, we performed a comparative analysis against several large language models (LLMs), including *e5-mistral-7b-instruct* (7 billion parameters), *udever-bloom-1b1* (1 billion parameters), as well as OpenAI's *text-embedding-3-small/large* and *text-embedding-ada-002* models. Despite their significantly smaller size and fewer parameters, the Matryoshka models consistently performed on par with or even exceeded the performance of these much larger models in tasks related to semantic textual similarity. Table 16 shows performance comparison between Matryoshka models and larger models on MTEB Arabic benchmarks.

As shown in Table 16, the *Arabert-all-nli-triplet-Matryoshka* model, with just 135M parameters, achieved an impressive average score of 69.99, outperforming the *e5-mistral-7b-instruct* (7B parameters, average score 68.00) and *udever-bloom-1b1* (1B parameters, average score 68.07). Similarly, OpenAI's *text-embedding-ada-002* model recorded an average score of 63.67, while the larger *text-embedding-3-large* model, despite having more parameters, scored 65.54 on average. Other Matryoshka models, such as *Marbert-all-nli-triplet-Matryoshka* and *Arabic-labse-Matryoshka*, also demonstrated competitive performance, with average scores of 67.19 and 66.76, respectively.

These findings underscore the efficiency of the Matryoshka approach, which offers strong performance in semantic similarity tasks without the need for billions of parameters. The results clearly demonstrate that smaller, well-optimized models can effectively compete with, and even surpass, much larger LLMs, making the

Table 15 Performance of Matryoshka models versus base counterparts on MTEB benchmarks

Model	Dim	# Params.	STS17	STS22	STS22-v2	Average
Arabert-all-nli-triplet-Matryoshka	768	135M	85.31	60.7	63.96	69.99
Arabic-labse-Matryoshka	768	471M	82.46	57.25	60.58	66.76
Marbert-all-nli-triplet-Matryoshka	768	163M	82.18	58.08	61.32	67.19
Arabic-all-nli-triplet-Matryoshka	768	135M	82.4	51.38	54.45	62.74
paraphrase-multilingual-mpnet-base-v2	768	135M	79.09	52.18	55.37	62.21
bert-base-arabertv02	768	135M	54.53	46.86	49.95	50.45
LaBSE	768	471M	69.07	57.66	60.98	62.57
MARBERTv2	768	163M	60.98	49.92	53.75	54.88

Table 16 Performance comparison between Matryoshka models and larger models on MTEB Arabic benchmarks

Model	Dim	# Params.	STS17	STS22	STS22-v2	Average
Arabert-all-nli-triplet-Matryoshka	768	135M	85.31	60.7	63.96	69.99
Arabic-labse-Matryoshka	768	471M	82.46	57.25	60.58	66.76
Marbert-all-nli-triplet-Matryoshka	768	163M	82.18	58.08	61.32	67.19
Arabic-all-nli-triplet-Matryoshka	768	135M	82.4	51.38	54.45	62.74
paraphrase-multilingual-mpnet-base-v2	768	135M	79.09	52.18	55.37	62.21
e5-mistral-7b-instruct	4096	7111M	81.87	54.12	–	68.00
udever-bloom-1b1	1536	1000M	81.13	55	–	68.07
text-embedding-ada-002	1536	–	70.62	58.52	61.86	63.67
text-embedding-3-small	1536	–	72.32	54.36	56.68	61.12
text-embedding-3-large	3072	–	80.59	56.92	59.11	65.54

Fig. 6 User interface of the Arabic sentence similarity application

Matryoshka technique particularly advantageous for resource-constrained applications.

6 Arabic Sentence Similarity Application

In addition to the comprehensive evaluation of the Arabic Matryoshka embedding models, we have developed an interactive *Gradio* application that leverages these models to compute semantic similarity between Arabic sentences. This tool is designed to provide users with a practical interface to utilize the advanced capabilities of *SentenceTransformer* models in real-world scenarios. Figure 6 shows the user interface of the app depolyed in hugging-face.[13]

As shown in Fig. 6, the app provides a variety of choices and modes along with the Arabic trained Matryoshka models combined together in the following Key Features.

---

[13] https://huggingface.co/spaces/Omartificial-Intelligence-Space/Arabic-Sentence-Similarity-Matryoshka-Models.

Model Selection, users can choose from a variety of Arabic Matryoshka Embedding Models. This flexibility allows for comparison and selection of the most suitable model for specific needs.

Flexible Comparison Modes, the application supports two primary comparison modes. Users can either compare two sentences directly or compare one sentence against three others, offering a versatile approach to semantic similarity evaluation.

Custom Embedding Dimensions, to cater to various computational and accuracy requirements, the application allows users to select embedding dimensions from [768, 512, 256, 128, 64]. This feature ensures that the tool can be adapted to different performance and precision needs.

Detailed Similarity Scores, the application provides detailed similarity scores between sentences, enabling users to understand the nuances of the semantic relationships captured by the models.

By providing this application, we aim to bridge the gap between advanced model development and practical usability, making state-of-the-art sentence similarity computations accessible and user-friendly.

7 Conclusion

In this chapter, we conducted a comprehensive evaluation of multiple trained Arabic nested embedding models on the Arabic Semantic Textual Similarity Benchmark. Our analysis spanned models that are both multilingual and specifically trained for Arabic, using the $EmbeddingSimilarityEvaluator$ to assess their performance across several key metrics: Pearson and Spearman correlations for cosine similarity, Manhattan distance, Euclidean distance, and dot product similarity. The results clearly demonstrate that Arabic nested embedding models, when trained on Arabic-specific data, significantly outperform their multilingual counterparts. Among the models evaluated, the $paraphrase - multilingual - mpnet - base - v2$ model consistently achieved superior results across most dimensions and metrics, underscoring its robustness in capturing semantic similarity. Additionally, $LaBSE$ exhibited stable performance, particularly in higher dimensions, making it a reliable choice for multilingual tasks. The $mpnet - base - all - nli - triplet$ model performed moderately well, surpassing the $bert - base - arabertv02$ and $MARBERT$ models, which recorded the lowest values across most metrics.

Our comparative analysis extended beyond the base models to include their Matryoshka counterparts. The results of this comparison highlight the significant benefits of language-specific training. The Matryoshka models, trained on Arabic triplet datasets, demonstrated marked improvements in capturing the semantic nuances of Arabic, as evidenced by substantial increases in Pearson and Spearman correlations. Notably, when compared to larger models like OpenAI's *text-embedding-3-large* and *text-embedding-ada-002*, as well as other prominent LLMs like *E5-mistral-7b-instruct* and *udever-bloom-1b1*, our Arabic Matryoshka models not only held their own but, in many cases, outperformed these models. Despite the vast parameter counts of these LLMs, the efficiency and targeted training of Matryoshka models on Arabic data provided a competitive edge, particularly in tasks involving semantic textual similarity. This demonstrates the efficacy of smaller, more specialized models over larger, generalized ones in certain linguistic contexts.

The insights gleaned from this study pave the way for several avenues of future research. One promising direction is the further refinement of Matryoshka models by incorporating additional Arabic-specific datasets, particularly those that capture the diversity of Arabic dialects. This could further enhance the models' ability to generalize across different forms of Arabic. Another area of interest is the application of Matryoshka embedding models to other NLP tasks beyond semantic textual similarity, such as sentiment analysis, machine translation, and question answering. By exploring these tasks, we can better understand the versatility and limitations of Matryoshka models in broader contexts.

In summary, this chapter has highlighted the critical importance of language-specific training for NLP models and demonstrated the considerable advantages of Matryoshka Representation Learning in the context of Arabic. These findings not only advance the field of Arabic NLP but also contribute to the broader discourse on the development of efficient, high-performing language models for underrepresented languages.

Acknowledgements The authors thank Prince Sultan University for their support.

References

1. N. Reimers, Sentence-BERT: sentence embeddings using Siamese BERT-Networks (2019). arXiv preprint arXiv:1908.10084
2. T. Gao, X. Yao, D. Chen, Simcse: simple contrastive learning of sentence embeddings (2021). arXiv preprint arXiv:2104.08821
3. Y. Gao, Y. Xiong, X. Gao, K. Jia, J. Pan, Y. Bi, H. Wang, Retrieval-augmented generation for large language models: a survey (2023). arXiv preprint arXiv:2312.10997
4. Y. Li, T. Yang, Word embedding for understanding natural language: a survey. Guide Big Data Appl. 83–104 (2018)
5. G. Bourahouat, M. Abourezq, N. Daoudi, Word embedding as a semantic feature extraction technique in Arabic natural language processing: an overview. Int. Arab J. Inf. Technol. **21**(2), 313–325 (2024)
6. Y. LeCun, Y. Bengio, G. Hinton, Deep learning. Nature **521**(7553), 436–444 (2015)
7. P. Nayak, Understanding searches better than ever before. Google AI Blog (2019). https://blog.google/products/search/search-language-understanding-bert/
8. J. Dean, Challenges in building large-scale information retrieval systems, in *Keynote of the 2nd ACM International Conference on Web Search and Data Mining (WSDM)*, Vol. 10 (2009)
9. C. Sun, A. Shrivastava, S. Singh, A. Gupta, Revisiting unreasonable effectiveness of data in deep learning era, in *Proceedings of the IEEE International Conference on Computer Vision* (2017), pp. 843–852
10. A. Kusupati, G. Bhatt, A. Rege, M. Wallingford, A. Sinha, V. Ramanujan, A. Farhadi, Matryoshka representation learning. Adv. Neural Inform. Process. Syst. **35**, 30233–30249 (2022)

11. N. Reimers, I. Gurevych, Sentence-bert: sentence embeddings using Siamese bert-networks, in *Proceedings of the 2019 Conference on Empirical Methods in Natural Language Processing* (Association for Computational Linguistics, 2019), pp. 3980–3990
12. D. Cer, M. Diab, E. Agirre, I. Lopez-Gazpio, L. Specia, Semeval-2017 task 1: semantic textual similarity-multilingual and cross-lingual focused evaluation (2017). arXiv preprint arXiv:1708.00055
13. Y. Zhao, T. Xia, Y. Jiang, Y. Tian, Enhancing inter-sentence attention for semantic textual similarity. Inform. Process. Manage. **61**(1), 103535 (2024)
14. A. Pathak, P. Pakray, J. Bentham, English-Mizo machine translation using neural and statistical approaches. Neural Comput. Appl. **31**(11), 7615–7631 (2019)
15. S. Liu, J. Cao, R. Yang, Z. Wen, Key phrase aware transformer for abstractive summarization. Inform. Process. Manage. **59**(3), 102913 (2022)
16. Y. Wu, S. Zhao, R. Guo, A novel community answer matching approach based on phrase fusion heterogeneous information network. Inform. Process. Manage. **58**(1), 102408 (2021)
17. J. Deng, W. Dong, R. Socher, L.-J. Li, K. Li, L. Fei-Fei, Imagenet: a large-scale hierarchical image database, in *2009 IEEE Conference on Computer Vision and Pattern Recognition* (IEEE, 2009), pp. 248–255
18. K. He, X. Zhang, S. Ren, J. Sun, Deep residual learning for image recognition, in *Proceedings of the IEEE Conference on Computer Vision and Pattern Recognition* (2016), pp. 770– 778
19. M. Gutmann, A. Hyvärinen, Noise-contrastive estimation: a new estimation principle for unnormalized statistical models, in *Proceedings of the Thirteenth International Conference on Artificial Intelligence and Statistics* (JMLR Workshop and Conference Proceedings, 2010), pp. 297–304
20. M.G. Harris, C.D. Giachritsis, Coarse-grained information dominates fine-grained information in judgments of time-to-contact from retinal flow. Vision Res. **40**(6), 601–611 (2000)
21. C. Waldburger, As search needs evolve, Microsoft makes AI tools for better search available to researchers and developers. Microsoft AI Blog (2019). https://blogs.microsoft.com/ai/bing-vector-search/
22. Y.A. Malkov, D.A. Yashunin, Efficient and robust approximate nearest neighbor search using hierarchical navigable small world graphs. IEEE Trans. Pattern Anal. Mach. Intell. **42**(4), 824–836 (2018)
23. H. Cai, C. Gan, T. Wang, Z. Zhang, S. Han, Once-for-all: yrain one network and specialize it for efficient deployment 2019. arXiv preprint arXiv:1908.09791
24. M. Wallingford, H. Li, A. Achille, A. Ravichandran, C. Fowlkes, R. Bhotika, S. Soatto, Task adaptive parameter sharing for multi-task learning (2022). arXiv preprint arXiv:2203.16708
25. J. Yu, L. Yang, N. Xu, J. Yang, T. Huang, Slimmable neural networks (2018). arXiv preprint arXiv:1812.08928
26. O. Rippel, M. Gelbart, R. Adams, Learning ordered representations with nested dropout, in *International Conference on Machine Learning* (PMLR, 2014), pp. 1746–1754
27. O. Rippel, M. Gelbart, R. Adams, Learning ordered representations with nested dropout, in *International Conference on Machine Learning* (PMLR, 2014), pp. 1746–1754
28. X. Zhang, Z. Li, Y. Zhang, D. Long, P. Xie, M. Zhang, M. Zhang, Language models are universal embedders (2023). arXiv preprint arXiv:2310.08232
29. OpenAI, OpenAI embeddings documentation (2023). Retrieved from https://platform.openai.com/docs/guides/embeddings
30. N. Muennighoff, N. Tazi, L. Magne, N. Reimers, MTEB: massive text embedding benchmark (2022). arXiv preprint arXiv:2210.07316
31. W. Antoun, F. Baly, H. Hajj, Arabert: transformer-based model for Arabic language understanding (2020). arXiv preprint arXiv:2003.00104
32. M. Abdul-Mageed, A. Elmadany, E.M.B. Nagoudi, ARBERT & MARBERT: deep bidirectional transformers for Arabic (2020). arXiv preprint arXiv:2101.01785
33. F. Feng, Y. Yang, D. Cer, N. Arivazhagan, W. Wang, Language-agnostic BERT sentence embedding (2020). arXiv preprint arXiv:2007.01852

34. S.R. Bowman, G. Angeli, C. Potts, C.D. Manning, A large annotated corpus for learning natural language inference (2015). arXiv preprint arXiv:1508.05326
35. S. Kim, I. Kang, N. Kwak, Semantic sentence matching with densely-connected recurrent and co-attentive information, in *Proceedings of the AAAI Conference on Artificial Intelligence*, Vol. 33, No. 01 (2019), pp. 6586–6593
36. B. MacCartney, C.D. Manning, Modeling semantic containment and exclusion in natural language inference, in *Proceedings of the 22nd International Conference on Computational Linguistics (Coling 2008)* (2008), pp. 521–528
37. Y. Yang, S. Yuan, D. Cer, S.Y. Kong, N. Constant, P. Pilar, R. Kurzweil, Learning semantic textual similarity from conversations (2018). arXiv preprint arXiv:1804.07754
38. T. Kudo, J. ichardson, Sentencepiece: a simple and language independent subword tokenizer and Detokenizer for neural text processing (2018). arXiv preprint arXiv:1808.06226
39. G. Klein, Y. Kim, Y. Deng, J. Senellart, A.M. Rush, Opennmt: open-source toolkit for neural machine translation (2017). arXiv preprint arXiv:1701.02810
40. M. Henderson, R. Al-Rfou, B. Strope, Y.H. Sung, L. Lukács, R. Guo, R. Kurzweil, Efficient natural language response suggestion for smart reply (2017). arXiv preprint arXiv:1705.00652

Generative AI for Code Development: Advancements, Limitations, and Prospects

Moez Krichen and Mohamed S. Abdalzaher

Abstract This chapter explores the advancements of utilizing generative AI in various aspects of code development. The chapter begins with an introduction to generative artificial intelligence (AI) and its potential applications in software design. It then delves into specific areas of code development, including code generation, optimization, refactoring, repair, annotation, and testing, highlighting how generative AI can be leveraged in each of these areas. The limitations of generative AI in code development, such as contextual understanding, data bias, and creativity, are discussed, along with potential prospects for addressing these limitations. Additionally, the chapter presents advancements in combining AI technology with code development, including hybrid approaches, explainable AI, domain-specific models, collaboration, and continuous learning. Finally, the chapter concludes with a summary of the key findings and future trends in the field of generative AI for code development. By understanding the advancements, limitations, and prospects, developers and researchers can explore the potential of generative AI in code development and pave the way for more efficient and innovative software development practices.

Keywords Generative ai · Code development advancements · Limitations · Prospects

M. Krichen (✉)
Al-Baha University, Al-Baha, Saudi Arabia
e-mail: moez.krichen@redcad.org
URL: https://www.redcad.org/members/mkrichen/

ReDCAD Laboratory, University of Sfax, Sfax, Tunisia

M. S. Abdalzaher
Department of Seismology, National Research Institute of Astronomy and Geophysics, Cairo, Egypt
e-mail: msabdalzaher@nriag.sci.eg

Department of Electrical Engineering, College of Engineering, American University of Sharjah, Sharjah, UAE

A. Koubaa et al. (eds.), *Generative AI and Large Language Models: Opportunities, Challenges, and Applications*, Studies in Computational Intelligence 1214,
https://doi.org/10.1007/978-3-031-90573-5_7

1 Introduction

A subset of artificial intelligence (AI) known as generative AI has surfaced as a technologically hopeful innovation that holds the capacity to fundamentally transform numerous facets of code development. Generative AI models have the capability to generate code, optimize pre-existing code, and aid in debugging and testing through the utilization of machine learning algorithms [1]. The aforementioned functionalities provide software developers with the potential for heightened efficiency, enhanced code quality, and increased productivity [2].

Generative AI can provide assistance in the domain of software design by facilitating the generation of initial code structures and architectures in accordance with high-level specifications. This facilitates time and effort savings for developers during the initial phases of a project, allowing them to concentrate on improving and optimizing the code that is generated [3].

Generative AI models have the capability to autonomously generate code fragments during the code generation process, utilizing patterns and specific requirements identified in extensive code repositories. This is especially beneficial for repetitive or mundane duties, as it alleviates the workload of developers and enables them to dedicate their time to more intricate and innovative approaches to problem-solving [4, 5].

An additional domain in which generative AI can exert a substantial influence is code optimization. Through the analysis of pre-existing codebases, generative AI models possess the capability to detect possible performance obstacles, propose optimizations, and autonomously implement said optimizations to augment the software's efficiency and velocity [6].

Code refactoring, which involves the restructuring of pre-existing code without modifying its external behavior, may similarly derive advantages from the implementation of generative AI techniques [7]. Codebases can be analyzed by AI models, which have the capability to suggest refactorings that enhance code maintainability, legibility, and compliance with established coding standards [8, 9]. This can aid in the streamlining of codebases and the reduction of technical debt for developers.

Generative AI models have the capability to aid in the detection and automated resolution of flaws or vulnerabilities present in code [10]. Through extensive analysis of code and issue repositories, these models possess the capability to provide automated recommendations or solutions to alleviate coding errors, thereby conserving developers' time and energy during the debugging phase [11].

Generic AI models may also provide support for code annotation [12], which involves the inclusion of descriptive remarks or metadata. These models are capable of examining the code and producing valuable remarks or annotations that improve code comprehension and maintainability, particularly in situations where there is a lack of code documentation.

Another critical facet of software development in which generative AI can be utilized is code testing [13]. Automatic test case generation by AI models, predicated on code inputs, outputs, and corner cases, furnishes developers with an initial reference for testing and guarantees a more extensive coverage of code.

Although generative AI presents significant prospects for code development, it is not devoid of constraints [14]. The task of contextually comprehending code continues to be difficult, as AI models might encounter difficulties in capturing the complete semantic significance and intention conveyed in code fragments [15]. Moreover, training dataset biases have the potential to impact the code produced by AI models, which may result in solutions that are biased or less than optimal. Generative AI is deficient in areas such as the ingenuity required to generate innovative code solutions and the difficulty of debugging complex issues [16]. Privacy concerns and the possibility of employment displacement are additional ethical issues that must be thoroughly examined in relation to the application of AI in software development [17].

However, promising are the prospects for generative AI in the context of code development [18]. Hybrid strategies that integrate the benefits of generative AI models and the knowledge of human developers can capitalize on the most advantageous aspects of both domains. By ensuring interpret ability and transparency, explainable AI techniques empower programmers to have confidence in and verify the judgments rendered by AI models [19]. Optimizing domain-specific models for particular programming languages and application domains can significantly improve the precision and pertinence of the code that is generated. A symbiotic relationship can be cultivated through collaboration between human developers and AI systems, in which AI serves to support and enhance human capabilities, rather than supplant them [20]. In addition, by continuously adapting AI models to changing programming practices and learning themselves, dynamic code development environments can be maintained in which generative AI remains effective and pertinent [21].

By analyzing the developments, constraints, and future potential of generative AI in code development, this chapter endeavors to furnish an all-encompassing comprehension of the present condition of the discipline. It provides guidance and insights to practitioners, researchers, and developers regarding the optimal utilization of generative AI to improve code development processes in terms of quality, efficiency, and innovation. By acknowledging the constraints and embracing the potential, generative AI has the capacity to facilitate the development of software systems that are more effective, dependable, and innovative. The main topics covered by this chapter are illustrated in Fig. 1.

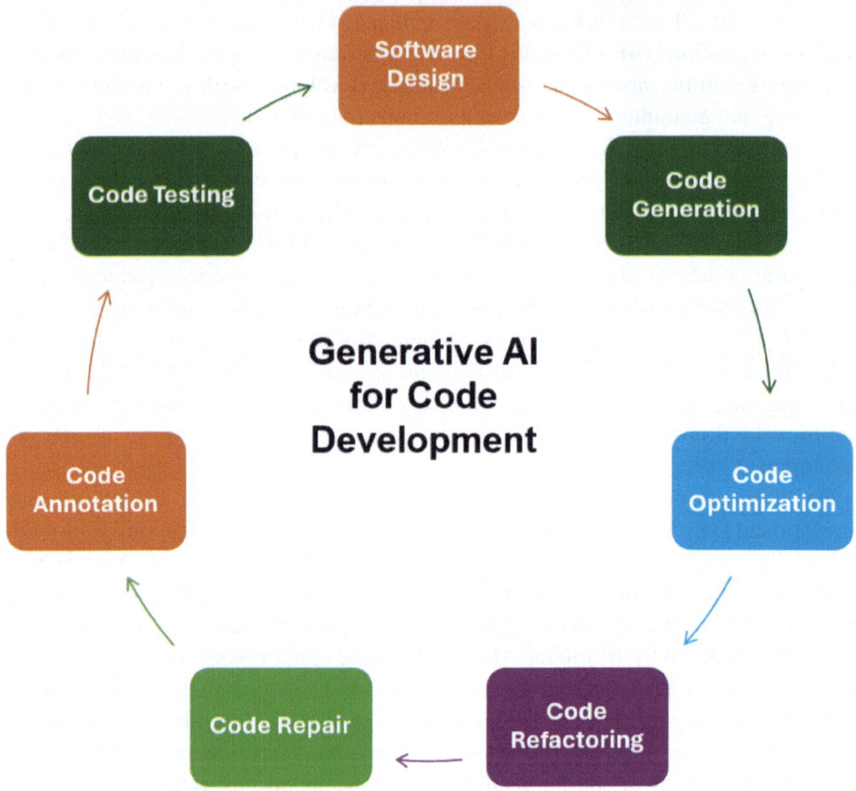

Fig. 1 Main topics covered by the chapter

2 Software Design

Generative AI has the potential to bring about a fundamental revolution in the software design process, providing significant benefits at each phase. Generic AI exhibits remarkable competence during the need identification stage through the examination of vast quantities of user data, the recognition of patterns, and the formulation of potential solutions [22, 23]. By harnessing the capabilities of sophisticated AI algorithms, designers can extract valuable insights from vast datasets, user input, and legacy software systems. This enables individuals to cultivate a more comprehensive understanding of the needs and demands of users. The application of a data-driven approach enhances the accuracy and effectiveness of the design process.

Generative AI's capability to orchestrate the generation of software specifications demonstrates its significant utility in the specification writing process [24]. By analyzing input data and high-level requirements, generative AI models possess the ability to propose exhaustive specifications encompassing functional and non-functional prerequisites, data flow diagrams, and system interactions [25]. By significantly reducing the reliance on manual labor during the process of writing specifications, this automation ensures a more efficient and consistent documentation procedure. Generative AI's capacity to produce accurate specifications and handle intricate data enables designers to devote greater focus to more sophisticated design decisions [26].

As the software development process progresses towards increasingly complex stages, generative AI continues to be an essential element. The utilization of generative AI methods in architectural design enables the independent generation or recommendation of software architectures that adhere to pre-established constraints and criteria. By conducting an analysis of various architectural alternatives and evaluating the trade-offs associated with them, generative AI assists designers in reaching informed conclusions and creating optimal designs. This procedure maximizes resource utilization by narrowing down the design space to its most feasible alternatives, resulting in software architectures that are structured and capable of expansion.

In addition to automating the generation of reusable code components, generative AI demonstrates exceptional proficiency in producing modules and components. Designers have the ability to employ generative AI models in order to generate code templates, libraries, or frameworks that adhere to specific design patterns and principles [27]. This automation not only accelerates the development process but also improves the maintainability and scalability of software systems through the promotion of code reuse and modularity. The ability of generative AI to generate code components in adherence to established best practices ensures that the software design process remains consistent and of high quality [28].

Generative AI remains a valuable asset throughout the later phases of the design process, providing ongoing support in areas such as design space investigation and optimization. By employing optimization algorithms, generative AI enables designers to explore alternative design options, optimize parameters, and determine design configurations that meet specific criteria [28]. This enables the procedure of optimizing software designs, consequently augmenting their efficiency, efficacy, or other desired characteristics. Generative AI's capacity to assess numerous design alternatives and handle intricate optimization challenges empowers designers to create software systems that not only fulfill but also exceed expected results.

To summarize (see Table 1, generative AI exhibits significant potential throughout the entirety of software design procedures, including architectural design, component generation, optimization, design space exploration, need identification, and specification writing. By harnessing the capabilities of generative AI, designers can benefit from improved design quality, streamlined decision-making, automated insights, and accelerated development processes. In conclusion, this leads to the creation of software systems that are robust, innovative, and focused on the needs of organizations and consumers.

Table 1 Summary of generative AI in software design

Key point	Explanation
Need identification	Generative AI analyzes user data, patterns, and formulates potential solutions, enhancing the accuracy and effectiveness of the design process
Specification writing	Generative AI proposes exhaustive specifications based on input data and requirements, reducing reliance on manual labor and ensuring a more efficient and consistent documentation procedure
Architectural design	Generative AI generates or recommends software architectures that adhere to constraints and criteria, helping designers make informed decisions and create optimal designs
Component generation	Generative AI produces code components, templates, libraries, or frameworks following design patterns and principles, accelerating development and improving software maintainability
Design optimization	Generative AI explores alternative options, optimizes parameters, and determines configurations that meet specific criteria, resulting in more efficient and effective software designs
Design space investigation	Generative AI supports ongoing design space investigation and optimization, enabling designers to optimize software designs and create systems that fulfill user needs and expectations

3 Code Generation

By utilizing natural language processing (NLP) and machine learning algorithms, AI-powered code generation automates the generation of code in response to a variety of inputs. These methodologies bring about a paradigm shift in the approach developers adopt when composing code, resulting in improved efficiency, code excellence, and the aggregate software development procedure. By utilizing AI technologies, developers are able to allocate their attention towards more advanced design decisions. AI models provide support in generating code that complies with functional requirements, follows established best practices, and minimizes the need for manual labor involved in repetitive coding duties [29]. The well-known techniques for code generation powered by AI are discussed as follows (see Fig. 2):

- Neural Language Models: These models have been trained on vast amounts of code repositories and programming language documentation. They can generate code snippets or complete functions based on natural language descriptions or partial code examples. Neural language models, such as transformer-based models, have demonstrated the ability to understand the semantics and syntax of programming languages, enabling them to generate syntactically correct and contextually relevant code. Developers can describe their intentions in natural language, and the models can translate those descriptions into executable code. This technique is particularly useful for automating the generation of boilerplate code, implementing common programming patterns, or providing initial code suggestions [30].

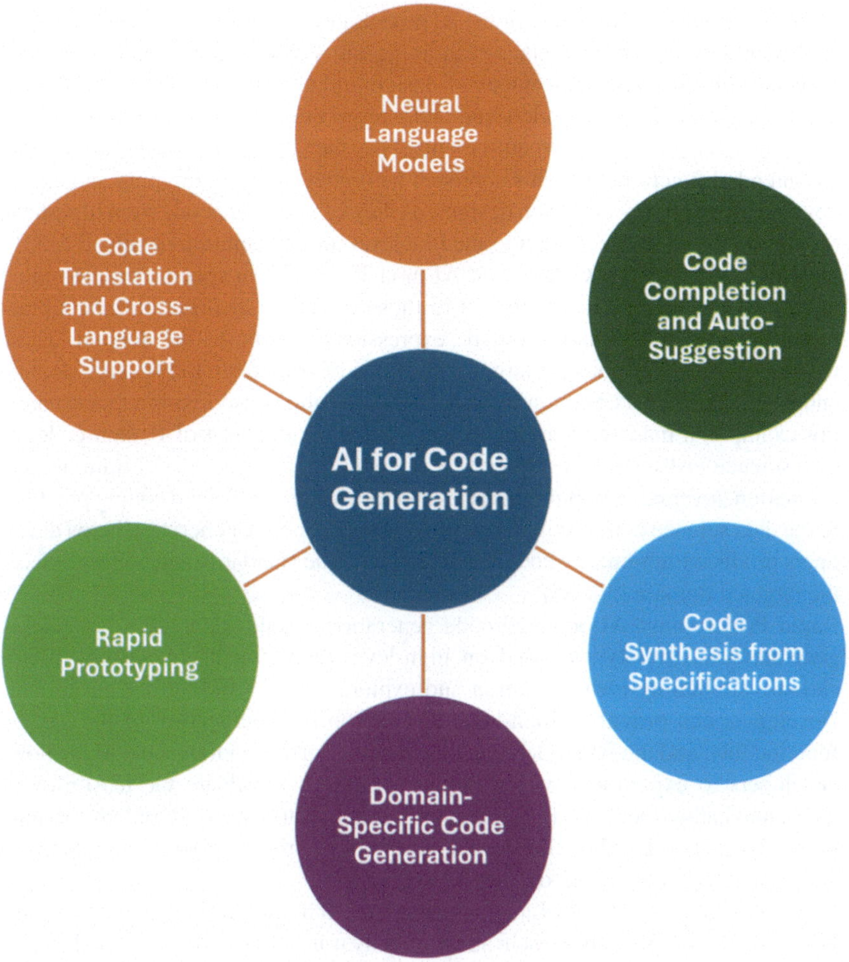

Fig. 2 AI for code generation

– Code Completion and Auto-Suggestion: AI-powered code editors and integrated development environments (IDEs) provide real-time code completion suggestions based on context, code patterns, and frequently used code snippets [31]. These suggestions enhance developers' productivity by reducing the time and effort required to write code. AI models analyze the current code being written, understand the programming language's syntax and semantics, and suggest relevant code completions. They can also provide suggestions for function parameters, method signatures, and available properties of objects. By leveraging machine learning techniques, the models continuously learn from developers' coding patterns and improve the accuracy and relevance of their suggestions over time.

- Code Synthesis from Specifications: AI models can generate code from high-level specifications or requirements, streamlining the software development process, especially during the initial stages of design and prototyping. Using NLP techniques, these models can understand and interpret textual descriptions of desired software behavior or system requirements. They then generate code that aligns with the intended functionality. This approach facilitates rapid prototyping and allows developers to quickly iterate on designs, as they can generate code from high-level requirements without getting into the low-level implementation details [32, 33].
- Domain-Specific Code Generation: AI models trained on specific programming domains can generate code tailored to those domains, incorporating best practices, design patterns, and idiomatic expressions specific to the domain. These models have been trained on large repositories of code from the specific domain and have learned the common patterns and conventions used within that domain. For example, a machine learning model trained on a dataset of Python code for data science tasks can generate code for data preprocessing, model training, and evaluation, leveraging popular libraries and following established coding practices in the data science community. This domain-specific code generation helps developers produce high-quality code that aligns with the standards and conventions of the targeted domain [34, 35].
- Rapid Prototyping: AI-powered code generation enables developers to quickly generate code prototypes based on high-level requirements or design specifications, facilitating idea validation and exploration of different design options. Developers can provide a high-level description or specification of the desired functionality, and the AI models can generate the corresponding code. This allows developers to experiment with different approaches, evaluate the feasibility of ideas, and gather feedback from stakeholders at an early stage. Rapid prototyping with AI-generated code helps accelerate the development process and supports agile and iterative software development practices.
- Code Translation and Cross-Language Support: AI models trained in multiple programming languages can assist in automatically translating code between different languages, simplifying the process of porting codes or integrating code written in different languages. These models can understand the syntax and semantics of multiple programming languages and perform automated translation between them. This is particularly useful when migrating or integrating code across different platforms or when collaborating on projects that involve multiple programming languages. The AI models can generate equivalent code snippets in the target language, preserving the functionality and logic of the original code, while adapting it to the syntax and idioms of the target language. This saves developers time and effort by automating the tedious task of manual code translation [36, 37].

AI-powered code generation techniques empower developers by automating repetitive coding tasks, providing intelligent code suggestions, enabling rapid prototyping, and facilitating code translation between languages. These advancements contribute to increased productivity, code quality, and innovation in the field of software development.

4 Code Optimization

Code optimization is the process of improving the efficiency, performance, and resource utilization of software code [38]. It aims to make programs run faster, consume less memory, and utilize system resources more effectively. Code optimization strategies and techniques are employed to enhance the execution speed, reduce the computational complexity, and optimize the memory usage of software applications.

- Algorithmic Optimization: AI techniques can analyze the behavior and performance characteristics of algorithms to identify opportunities for optimization [39]. By leveraging AI algorithms, developers can automatically analyze and optimize algorithmic choices, such as selecting more efficient sorting algorithms, improving search algorithms, or reducing time complexity through algorithmic transformations [40]. For example, AI models can analyze the input-output behavior of an algorithm and suggest modifications to reduce redundant computations or improve data access patterns. AI can provide insights and recommendations to enhance the efficiency of algorithms, leading to improved code performance.
- Machine Learning-Based Optimization: AI models can learn from large code repositories and historical performance data to make predictions and guide code optimization [41]. By training AI models on vast amounts of code, they can identify patterns and correlations between code structures and performance characteristics. These models can then suggest code transformations or optimizations that have been found to improve performance in similar situations. For instance, a machine learning model can identify code patterns that tend to be inefficient and propose alternative implementations that are more efficient based on its training data [42]. Machine learning-based optimization techniques can assist developers in making informed decisions about code changes that are likely to yield performance improvements.
- Compiler Optimization with AI: AI techniques can be integrated into compilers to enhance the effectiveness of optimization passes [43]. AI models can analyze the code and program characteristics to guide compiler optimizations. For example, AI models can help optimize loop transformations, guide register allocation decisions, or suggest code reordering to improve cache utilization. By combining AI with traditional compiler optimizations, developers can benefit from more intelligent and adaptive code optimization techniques. For instance, an AI-powered compiler can learn from a large corpus of code and performance data to make better decisions regarding optimization transformations, leading to improved code performance across a wide range of programs [44].
- Profiling with AI: Profiling tools enhanced with AI algorithms can analyze code execution patterns and identify performance bottlenecks [45]. These tools can leverage machine learning to detect patterns in code behavior and suggest optimizations based on the identified bottlenecks. For example, a profiling tool enhanced with AI can identify frequently executed code sections and recommend algorithmic or data structure changes to improve their performance. By combining profiling techniques with AI, developers can gain deeper insights into code performance

and receive intelligent recommendations for optimization. This can help developers prioritize their optimization efforts and focus on the most critical areas of the code.
- Neural Network-Based Optimization: AI models, particularly neural networks, can be trained to optimize code directly. Developers can train neural networks on large code repositories and performance data to learn patterns and transformations that lead to better performance. These models can then generate optimized versions of the code or suggest code transformations that improve performance. For example, a neural network can learn to recognize patterns in code that often result in sub-optimal performance, such as inefficient loop structures or redundant computations, and generate optimized versions of the code that eliminate these inefficiencies. Neural network-based optimization techniques can automate the process of code optimization by leveraging the learned knowledge of performance optimization patterns.
- Reinforcement Learning for Optimization: Reinforcement learning (RL) algorithms can be employed to optimize code by allowing an AI agent to learn through trial and error [46]. The agent interacts with the code and environment, applying various transformations and observing the resulting performance. Through RL, the agent can learn to make intelligent decisions about code transformations that lead to performance improvements. For example, an AI agent can learn to identify code regions that are computationally intensive and automatically parallelize them to utilize the available resources more effectively. This approach can discover novel optimizations that may not be obvious to human developers and continuously adapt to changing program and system dynamics for improved performance.

As summarized in Table 2, AI-powered code optimization techniques provide developers with intelligent insights, recommendations, and automated tools to enhance code performance. By leveraging AI algorithms and machine learning, developers can optimize their code more effectively, improving execution speed, reducing resource consumption, and enhancing overall software performance.

5 Code Refactoring

Code refactoring, leveraging AI technology, involves applying automated techniques to restructure existing code without altering its external behavior [47]. These AI-powered methods focus on improving the code's internal structure, making it more understandable, maintainable, and modifiable. By utilizing AI, developers can enhance code readability, reduce duplication, improve modularity, and ensure adherence to coding best practices.

- Do not Repeat Yourself: AI-powered tools can automatically detect and eliminate code duplication within a code. These tools use machine learning algorithms to identify repetitive code segments and suggest extracting them into reusable functions or classes. By reducing duplication, code becomes more concise and easier to

Table 2 Code optimization techniques

Optimization technique	Description
Algorithmic optimization	AI techniques analyze algorithm behavior and performance characteristics to optimize algorithmic choices, reduce time complexity, and improve code performance
Machine learning-based optimization	AI models learn from code repositories and historical performance data to suggest code transformations and optimizations for improved performance
Compiler optimization with AI	AI models integrated into compilers guide optimization passes, optimize loop transformations, register allocation, and code reordering for enhanced code performance
Profiling with AI	Profiling tools enhanced with AI analyze code execution patterns, identify bottlenecks, and suggest optimizations to improve performance
Neural network-based optimization	Neural networks trained on code repositories and performance data generate optimized code versions or suggest transformations for improved performance
Reinforcement learning for optimization	AI agents trained with reinforcement learning interact with code and environment, learning optimal code transformations through trial and error for enhanced performance

comprehend. Additionally, it minimizes the chances of introducing errors during code maintenance since changes only need to be made in one place [48].

- Single Responsibility Principle: AI-based refactoring techniques can help identify violations of the SRP. These tools analyze the code and suggest breaking down large, monolithic functions or classes into smaller, focused ones. This improves code maintainability by making it easier to understand and modify individual components. AI-powered refactoring also enhances code reusability since smaller, focused modules are more likely to be applicable in different contexts [49, 50].
- Extract Method/Function: AI-driven refactoring tools can automatically extract code segments into separate methods or functions. By doing so, developers improve code readability and promote code reuse. Extracting methods also allow for more focused unit testing since specific functionality can be tested independently. Furthermore, it facilitates easier modification of specific functionality, as changes can be made within the extracted method/function without affecting the rest of the code [49, 51].
- Rename Variables and Functions: AI-powered refactoring tools can suggest meaningful and descriptive names for variables and functions. These tools utilize NLP techniques to analyze the code and propose better names that reflect their purpose. Clear and expressive names make code self-explanatory and easier to understand for developers, reducing the cognitive load required to comprehend the code [52, 53].
- Remove Code Smells: AI-based code analysis tools can automatically detect code smells and recommend appropriate refactoring actions. These tools employ

machine learning algorithms to recognize indicators of potential issues in the code, such as long methods, excessive comments, unused variables, and complex conditionals. By addressing these code smells, code becomes more maintainable, easier to understand, and less error-prone. Removing code smells also improves code quality and reduces technical debt [54, 55].
- Simplify Conditionals: AI-driven refactoring techniques can simplify complex conditional statements. These techniques leverage machine learning algorithms to recognize patterns and suggest simplifications, such as extracting conditionals into separate methods, using guard clauses, or employing polymorphism. Simplified conditionals enhance code clarity, reduce the likelihood of introducing bugs, and make it easier to reason about the code's behavior [52, 56].
- Code Formatting and Style: AI-powered refactoring tools can automatically format code and enforce consistent coding styles. These tools analyze the code, apply indentation, spacing, and naming conventions, and ensure a cohesive and professional appearance. Consistent code formatting and style make the code easier to navigate and understand, improving readability and maintainability [57, 58].
- IDE Refactoring Tools: IDEs equipped with AI-powered refactoring tools automate common refactoring operations. These tools utilize AI techniques, such as static code analysis and pattern recognition, to suggest appropriate refactoring actions. IDEs can perform tasks like method extraction, renaming, variable extraction, and code reorganization. By using IDE refactoring tools, developers can save time and ensure the correctness of their refactoring operations [59, 60].
- Code Analysis Tools: AI-powered code analysis tools examine the code, identify code smells, and suggest refactoring opportunities. These tools use machine learning algorithms to recognize patterns and anti-patterns in the code. By analyzing the code, they provide developers with automated suggestions for refactoring, helping them identify areas of improvement and apply appropriate transformations. Code analysis tools enhance the refactoring process by providing objective insights into code quality and recommending specific refactoring actions [54, 61].
- Automated Code Restructuring: AI models trained on large code repositories can automatically refactor code by applying learned transformation rules. These models leverage patterns and transformations that improve code quality. For example, an AI model can recognize duplicated code segments and automatically extract reusable functions. Automated code restructuring techniques save developers time and effort by automating repetitive and error-prone refactoring tasks. These techniques leverage the learned knowledge from vast amounts of code to provide accurate and efficient code restructuring suggestions [62, 63].

As summarized in Table 3 by utilizing AI technology in the refactoring process, developers can significantly improve code quality, maintainability, and readability. AI-powered refactoring reduces technical debt, enhances code modularity, and facilitates future changes and feature additions. It fosters collaboration among developers by providing cleaner and more understandable code, leading to improved software development productivity and overall code health.

Table 3 Code refactoring techniques

Refactoring technique	Description
Eliminate code duplication	AI-powered tools detect and remove code duplication by suggesting the extraction of repetitive code segments into reusable functions or classes. This improves code conciseness, comprehension, and maintenance
Single responsibility principle	AI-based techniques identify violations of the SRP and suggest breaking down large functions or classes into smaller, focused ones. This improves code maintainability, understandability, and reusability
Extract method/function	AI-driven tools automatically extract code segments into separate methods or functions, enhancing code readability, reuse, and focused unit testing. It allows for easier modification of specific functionality
Rename variables and functions	AI-powered tools suggest meaningful and descriptive names for variables and functions, making the code self-explanatory and easier to understand
Remove code smells	AI-based code analysis tools detect code smells and recommend appropriate refactoring actions, improving code maintainability, understandability, and quality
Simplify conditionals	AI-driven techniques simplify complex conditional statements, enhancing code clarity, bug prevention, and reasoning
Code formatting and style	AI-powered tools automatically format code and enforce consistent coding styles, improving readability and maintainability
IDE refactoring tools	AI-equipped IDEs automate common refactoring operations, suggesting appropriate actions based on static code analysis and pattern recognition
Code analysis tools	AI-powered code analysis tools identify code smells and suggest refactoring opportunities, providing automated suggestions for improvement and transformation
Automated code restructuring	AI models trained on code repositories automatically refactor code by applying learned transformation rules, saving time and effort in repetitive and error-prone tasks

6 Code Repair

Code repair refers to the process of identifying and fixing issues or bugs in code to restore its intended functionality. It involves diagnosing and resolving problems that may arise during software development or after the deployment of an application. Code repair aims to eliminate errors, improve the reliability and performance of the code, and ensure that it functions as intended.

6.1 Common Code Issues

Code can experience various issues that require repair. Some common code issues include (see Fig. 3):

- Syntax errors arise when the code does not adhere to the rules and conventions of the programming language. These errors hinder the successful compilation or execution of the code. Fixing these errors entails detecting and correcting them, allowing the code to compile and run properly [64, 65].
- Logic Errors: Logic errors happen when the code yields unexpected or incorrect results due to faulty reasoning or improper implementation of the algorithm. These

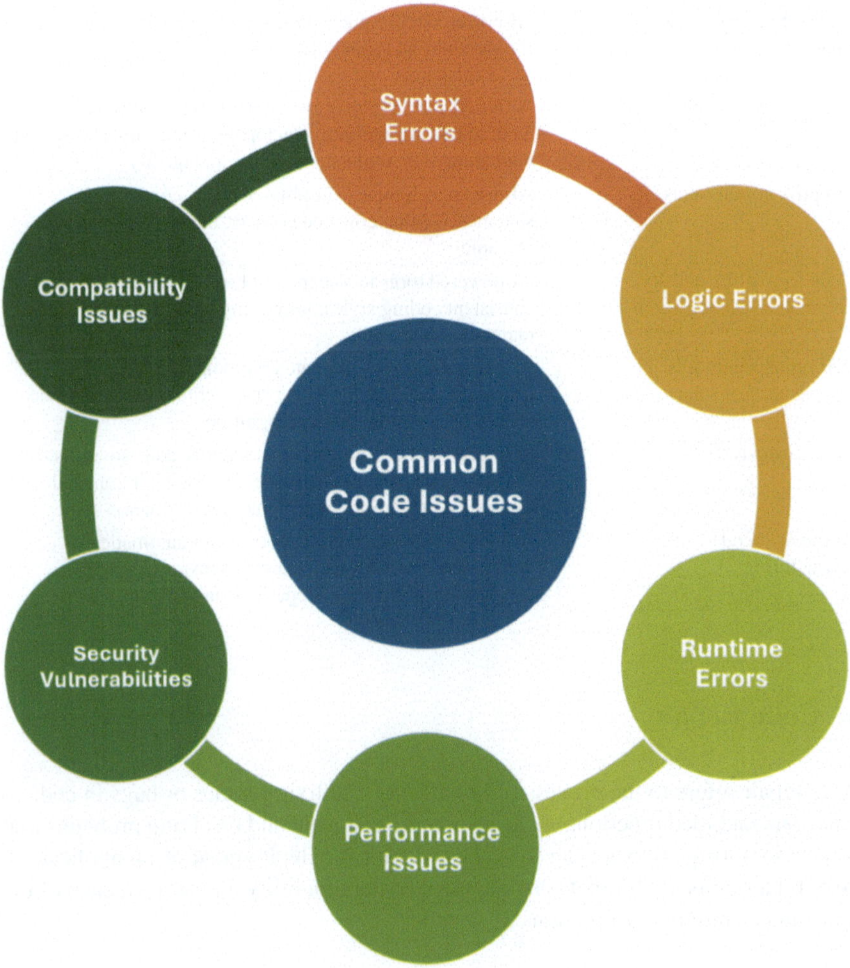

Fig. 3 Common code issues

errors can cause inaccurate calculations, unforeseen behavior, or crashes in the program. Correcting logic errors requires examining the code, pinpointing the flawed logic, and adjusting it to achieve the intended results [66].

- Runtime Errors: Runtime errors occur during the execution of the code and can cause program crashes or unexpected termination. Common runtime errors include null pointer exceptions, array out-of-bounds errors, and division by zero errors. Repairing runtime errors involves identifying the source of the error and implementing appropriate error handling mechanisms or fixing the underlying cause [67, 68].
- Performance Issues: Code may exhibit performance issues that result in slow execution or excessive resource consumption. These issues can include inefficient algorithms, memory leaks, or excessive disk or network usage. Code repair for performance issues involves analyzing the code, identifying bottlenecks, and optimizing the code to improve efficiency and resource utilization [69].
- Security Vulnerabilities: Code can contain security vulnerabilities that can be exploited by attackers. Common vulnerabilities include input validation errors, SQL injection, cross-site scripting (XSS), and insecure authentication mechanisms. Code repair for security vulnerabilities involves implementing secure coding practices, such as input validation, proper handling of user data, and using secure encryption and authentication mechanisms [70–72].
- Compatibility Issues: Compatibility issues arise when code does not function correctly on certain platforms, operating systems, or with specific dependencies or configurations. These issues can cause the code to behave differently or fail altogether. Code repair for compatibility issues involves identifying the incompatibilities and modifying the code to ensure it works correctly across different environments or configurations [73, 74].

6.2 Code Repair Techniques

Code repair often requires employing various techniques to identify and fix issues effectively. Hereafter are some commonly used code repair techniques, including AI-based approaches (see Table 4):

- Debugging: Debugging is the process of identifying and resolving issues in the code by examining its execution flow and variable values. Developers use debugging tools and techniques, such as breakpoints, stepping through code, and inspecting variable values, to locate and fix errors. Debugging helps in understanding the code's behavior and isolating the root cause of issues [75, 76].
- Unit Testing: Unit testing involves writing test cases to verify the correctness of individual units or components of the code. By designing and executing comprehensive unit tests, developers can identify bugs and ensure that repaired code functions as intended. Unit testing helps prevent regression issues and provides a safety net for code modifications [77].

Table 4 Code repair techniques

Code repair technique	Description
Debugging	The process of identifying and resolving issues in the code by examining its execution flow and variable values using debugging tools and techniques
Unit testing	Writing and executing test cases to verify the correctness of individual code units or components, helping to identify bugs and ensure repaired code functions as intended
Refactoring	Restructuring existing code to improve its internal structure without changing its external behavior, eliminating bugs and improving code maintainability and readability
Patching and hotfixes	Applying targeted changes or temporary fixes to address specific code issues, providing short-term solutions while planning a comprehensive code repair process
Code review	Having developers analyze the code for potential issues, including bugs and vulnerabilities, to identify areas requiring repair and provide suggestions for improvement
Automated bug detection and fixing	AI-powered tools analyzing code to detect bugs and propose potential fixes, leveraging static code analysis and machine learning techniques
Neural machine translation for code repair	Applying neural machine translation techniques to transform erroneous code segments into correct equivalents, particularly useful for fixing syntax errors or logic issues
Automated refactoring	AI models suggesting and applying refactoring transformations to improve code quality and fix issues, such as identifying and extracting duplicate code segments
Program synthesis	AI models generating code that repairs or extends existing code based on given input-output examples or high-level requirements, useful for fixing bugs, adding functionality, or optimizing code
Reinforcement learning for code repair	Training AI models to repair code through trial and error using reinforcement learning techniques, improving repair capabilities over time

- Refactoring: Refactoring involves restructuring existing code to improve its internal structure without changing its external behavior. Refactoring can also be used as a repair technique to fix code issues. By applying refactoring techniques, such as extracting methods, simplifying conditionals, or eliminating code smells, developers can eliminate bugs or improve the code's maintainability and readability [54, 78].
- Patching and Hotfixes: In some cases, code issues may require immediate fixes without a comprehensive code repair process. Patching and hotfixes involve applying targeted changes to address specific issues in the code. These changes may include modifying a few lines of code or adding temporary fixes to resolve critical

issues quickly. Patching and hotfixes provide short-term solutions while a more comprehensive code repair process is planned [79].

- Code Review: Code review involves having one or more developers analyze the code for potential issues, including bugs and vulnerabilities. Code review can identify areas where code repair is needed, providing insights and suggestions for improvement. By leveraging the collective knowledge and expertise of the development team, code review helps to identify and rectify issues that may have been missed during the initial development process [80].
- Automated Bug Detection and Fixing: AI-powered tools can analyze codes to detect bugs and propose potential fixes. Using techniques like static code analysis and machine learning, these tools can identify patterns and anomalies in the code that indicate the presence of bugs. They can also suggest automated fixes or generate patches to address the identified issues. By leveraging AI, developers can accelerate the bug-fixing process and reduce the manual effort required for debugging [81, 82].
- Neural Machine Translation for Code Repair: Neural machine translation (NMT) techniques, commonly used in language translation tasks, can also be applied to code repair. AI models trained on large code repositories can learn to translate buggy code segments into correct ones. By applying NMT techniques, developers can automatically repair code by transforming erroneous code segments into their correct equivalents. This approach can be particularly useful for fixing syntax errors or logic issues in the code [83].
- Automated Refactoring: AI models trained on vast codes can learn patterns and best practices for code organization and structure. These models can automatically suggest and apply refactoring transformations to improve code quality and fix issues. For example, an AI model can identify duplicate code segments and suggest their extraction into reusable functions or classes. Automated refactoring techniques powered by AI can save developers time and effort while ensuring consistent and high-quality code [62].
- Program Synthesis: Program synthesis techniques leverage AI to automatically generate code that fulfills a given specification or desired behavior. By providing input-output examples or high-level requirements, AI models can generate code that repairs or extends the existing code. Program synthesis techniques can be applied to fix bugs, add missing functionality, or optimize code. This approach can be particularly useful when faced with complex bugs or when manual repair is challenging [84].
- Reinforcement Learning for Code Repair: RL techniques can be employed to train AI models that learn to repair code through trial and error. RL models can interact with the code, explore different repair actions, and receive feedback on the correctness of their repairs. Over time, the RL model learns to make effective repairs by maximizing a reward signal. By using RL, AI models can autonomously learn repair strategies and improve their code repair capabilities [85].

These AI-based techniques for code repair offer promising avenues for automating and enhancing the code repair process. By leveraging AI's capabilities in pattern recognition, learning from large code repositories, and generating code transformations, developers can benefit from more efficient and effective code repair, leading to improved code quality and reduced development time.

7 Code Annotation

Code annotation is the process of adding explanatory comments or metadata to source code to enhance its understandability and maintainability. AI techniques can be leveraged to improve code annotation practices by automating certain aspects of the annotation process and providing intelligent suggestions. By utilizing AI in code annotation, developers can save time, improve code comprehension, and facilitate collaboration. Among the different existing code annotations in the literature, we discuss below the common AI-based techniques (see Fig. 4):

- NLP for Comment Generation: AI models trained on large code repositories and natural language data can generate meaningful comments for code segments. These models use NLP techniques to analyze the code structure, identify the purpose and functionality of the code, and generate descriptive comments that explain its behavior. AI-generated comments can provide valuable insights to developers, making the code more readable and understandable. AI-powered NLP algorithms can analyze the code syntax, identify variables, functions, and control structures, and generate comments that describe the code's purpose, expected inputs and outputs, and any potential side effects. These comments can serve as self-documenting code, enabling developers to understand the code's intent without diving into its implementation details. AI-generated comments can also assist in knowledge transfer among developers, especially in large codes or when working with legacy code [86].
- Automated Code Documentation: AI-powered tools can automatically extract relevant information from code, such as function signatures, parameter descriptions, and return types, and generate documentation. These tools employ machine learning algorithms to analyze the code structure and extract meaningful information. Automated code documentation facilitates code comprehension by providing developers with comprehensive documentation without the need for manual effort. AI-based automated code documentation techniques can process code files, extract function signatures, variable names, and inline comments, and generate API documentation or developer-friendly documentation formats such as HTML or Markdown. These automated tools can also incorporate information from external sources, such as online documentation or code repositories, to provide comprehensive documentation. By automating the documentation process, developers can focus more on coding while having up-to-date and accurate documentation at their disposal [87].
- Intelligent Code Summarization: AI models can summarize long code segments into concise descriptions. By leveraging techniques such as text summarization

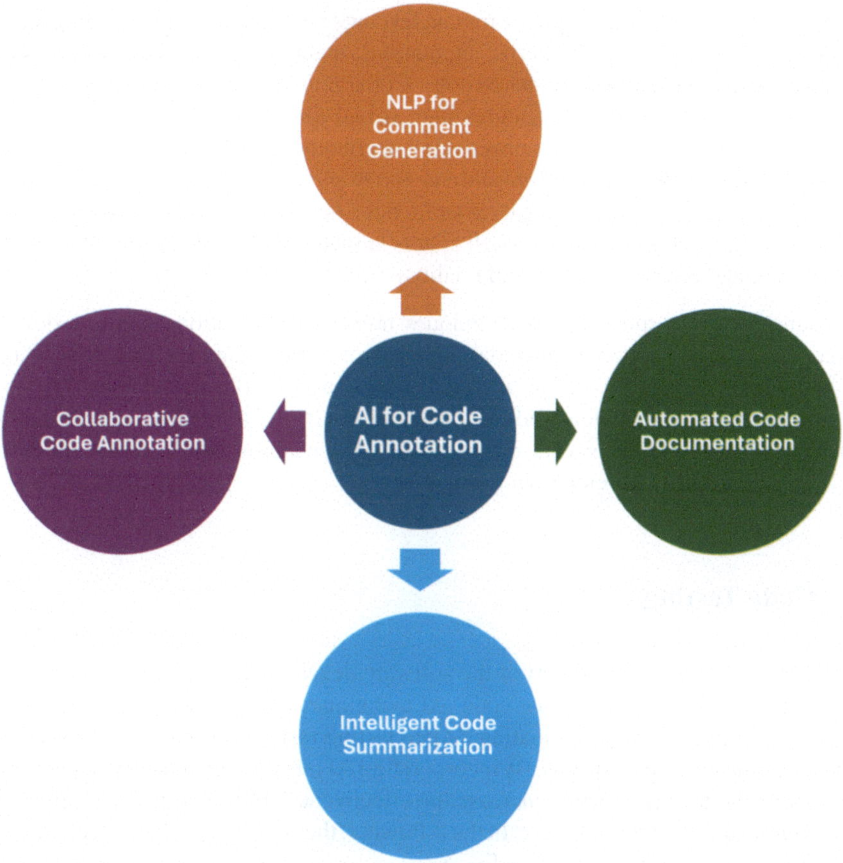

Fig. 4 AI for code annotation

and code analysis, AI-powered tools can generate short summaries that capture the essence of the code. Code summarization helps developers quickly grasp the purpose and functionality of a code segment, especially when dealing with large and complex codes. AI-based code summarization techniques analyze the code's structure, variable usage, and control flow to identify the core functionality of a code segment. These techniques generate summaries that highlight the primary purpose and behavior of the code. Code summarization can aid in code navigation, allowing developers to quickly understand the role of a code segment within a larger code. It also enables faster onboarding of new team members by providing concise explanations of existing code [88].

- Collaborative Code Annotation: AI technologies can facilitate collaborative code annotation by providing tools for shared annotations. These tools enable multiple developers to annotate code together, share comments, and provide feedback.

AI-powered collaboration platforms can leverage NLP and machine learning techniques to organize and categorize annotations, making it easier for developers to navigate and understand the shared code. Collaborative code annotation platforms powered by AI can provide features such as real-time commenting, threaded discussions, and notification systems to enable effective collaboration among developers. AI algorithms can analyze and categorize annotations based on their content, relevance, and context, making it easier for developers to find and explore the annotations. This collaborative approach fosters knowledge sharing, enables peer code reviews, and improves the overall quality of the code [89, 90].

To sum up by incorporating AI techniques into code annotation practices, developers can enhance code understandability, improve documentation, and streamline collaboration. AI-powered tools automate certain aspects of code annotation, providing intelligent suggestions and reducing manual effort. This enables developers to focus on higher-level tasks while maintaining code readability and fostering effective collaboration within development teams.

8 Code Testing

Code testing is an essential part of the software development process, ensuring the correctness and reliability of the code. AI techniques can be leveraged to enhance code testing practices by automating certain testing tasks, improving test coverage, and detecting hard-to-find bugs. By incorporating AI into code testing, developers can accelerate the testing process, increase the effectiveness of test suites, and improve the overall quality of the software. In the following, the common AI-based techniques for code testing are discussed (see Fig. 5):

- Automated Test Generation: AI-powered tools can automatically generate test cases based on code analysis and machine learning techniques. These tools can explore different code paths, identify boundary conditions, and generate test inputs that cover a wide range of scenarios. By automating test generation, AI techniques can help developers create comprehensive test suites more efficiently, reducing the manual effort required for test case creation. AI algorithms can analyze the code's structure, control flow, and data dependencies to identify potential paths and conditions that need to be covered by tests. By leveraging machine learning models, these algorithms can generate inputs that exercise different code branches and boundary cases. Automated test generation helps increase test coverage, uncover corner cases, and ensure that the code is thoroughly tested, ultimately improving the reliability of the software [91, 92].
- Fault Localization: AI techniques can assist in identifying the root causes of failures in code. By analyzing failure data, code coverage, and execution traces, AI-powered fault localization tools can pinpoint the specific lines or modules responsible for the observed failures. This information helps developers prioritize their debugging efforts and improve the efficiency of bug fixing. AI-based fault

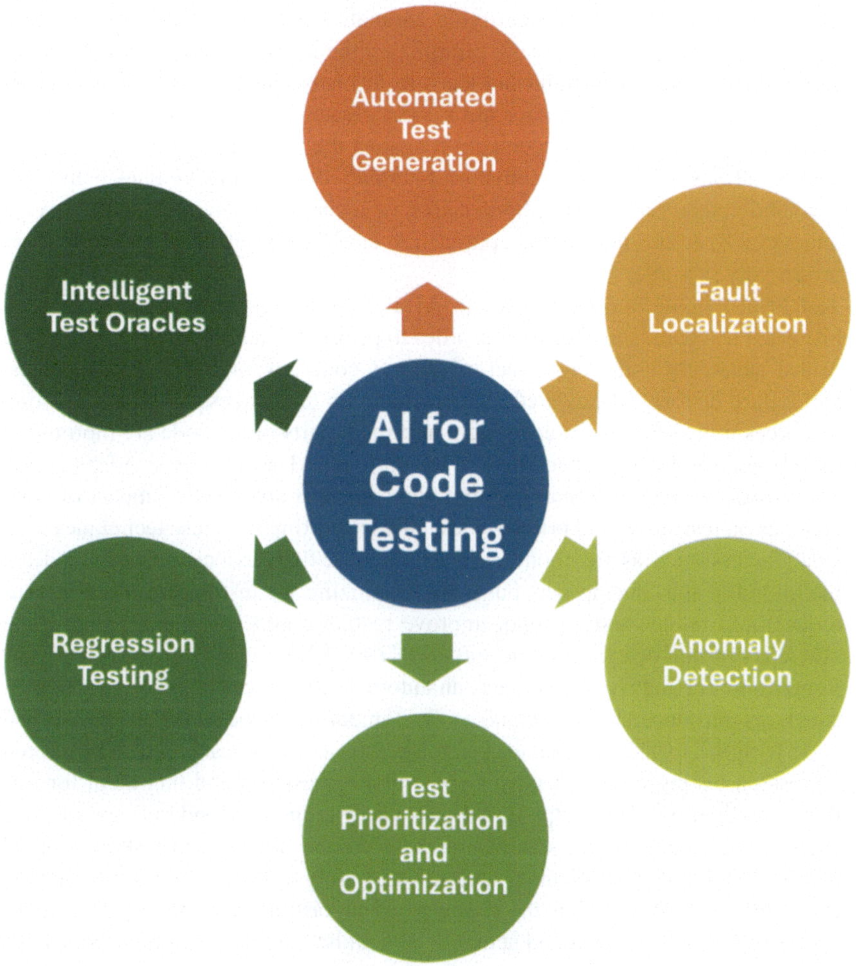

Fig. 5 AI for code testing

localization techniques leverage machine learning algorithms to analyze various factors, such as program spectra, runtime behavior, and historical failure data. These techniques can identify patterns and correlations between failures and code components, enabling developers to focus their debugging efforts on the most likely causes. By reducing the time and effort required to locate and fix bugs, AI-powered fault localization improves the overall efficiency of the development process [93, 94].

– Anomaly Detection: AI models can learn the normal behavior of a code and detect anomalies or deviations from expected patterns. By analyzing code metrics, execution logs, and historical data, AI-powered anomaly detection techniques can identify potential bugs, performance issues, or security vulnerabilities [95].

This proactive approach to identifying anomalies helps developers catch issues early in the development process, minimizing the impact on end-users. AI-based anomaly detection algorithms can learn from historical data and establish baseline patterns of normal behavior for a code. These algorithms then continuously monitor code execution and compare it against the established patterns, flagging any deviations as potential anomalies. Anomaly detection techniques help identify subtle bugs, performance bottlenecks, or security vulnerabilities that may go unnoticed by traditional testing approaches, enabling developers to address them proactively [96–99].

- Test Prioritization and Optimization: AI algorithms can analyze code changes, historical test data, and execution profiles to prioritize and optimize the execution of test cases. By considering factors such as code coverage, test dependencies, and failure history, AI-powered test prioritization techniques can allocate testing resources more effectively, ensuring that critical parts of the code are thoroughly tested while reducing redundant test executions. AI-powered test prioritization techniques leverage machine learning algorithms to analyze the impact of code changes on test suites and prioritize test cases accordingly. These techniques consider factors such as the modified code regions, their associated tests, and the likelihood of introducing new bugs. By optimizing the testing process, AI techniques help reduce testing time, improve resource allocation, and increase the efficiency of bug detection in the software [100, 101].
- Regression Testing: AI techniques can automate the process of detecting regression bugs introduced by code changes. By comparing the behavior of the modified code with the expected behavior established by existing test cases, AI-powered regression testing tools can identify potential regressions and flag them for further investigation. This helps developers catch unintended side effects of code changes and maintain the stability of the software. Regression testing with AI involves analyzing code changes, executing existing test cases, and comparing the test results before and after the changes. AI algorithms can identify discrepancies between the expected and actual results, indicating potential regressions. By automating this process, AI techniques assist in ensuring that code modifications do not unintentionally introduce bugs or affect the existing functionality of the software [102, 103].
- Intelligent Test Oracles: AI models can learn from existing test cases and code patterns to automatically generate expected outputs or predictions for new test cases. By leveraging machine learning techniques, AI-powered test oracles can provide accurate and reliable expectations for complex code behaviors, reducing the manual effort required to define test oracles and improving the accuracy of test results. AI-based test oracles can learn from historical test data and code patterns to infer the expected outputs or predictions for new test cases. Machine learning algorithms can analyze the relationships between inputs, code logic, and expected outputs, allowing the AI models to generate reliable predictions. This eliminates the need for developers to manually define test oracles for every test case, saving time and effort. The intelligent test oracles provided by AI techniques

improve the accuracy of test results and ensure that code behavior is properly validated [104, 105].

By incorporating AI techniques into code testing practices, developers can enhance the efficiency and effectiveness of their testing efforts. AI-powered tools automate various testing tasks, from test case generation to fault localization, enabling developers to focus on critical aspects of code quality. With AI's ability to analyze code patterns, historical data, and execution traces, code testing becomes more robust, improving software reliability and reducing the occurrence of hard-to-find bugs.

9 Limitations and Prospects

Generative AI, which involves the use of AI models to generate code, has shown promise in automating various aspects of code development. However, it also has its limitations and poses certain challenges. In this section, we will discuss the limitations of generative AI in code development and explore potential prospects for addressing these limitations and advancing the field. A summary is provided in Table 5.

Table 5 Limitations and prospects of generative AI in code development

Limitations	Prospects
Contextual understanding	Hybrid approaches: combining generative AI with traditional programming approaches to leverage the strengths of both AI systems and human developers
Data bias	Explainable AI: improving the interpretability and explainability of AI models to gain trust and understand their decision-making process in code development
Debugging complex issues	Domain-specific models: developing specialized AI models tailored to specific programming languages, frameworks, or application domains for more accurate and context-aware code generation
Lack of creativity	Collaborative AI development: integrating AI models into collaborative development environments to provide real-time code suggestions and assistance
Ethical considerations	Continuous learning and adaptation: Designing AI models to continually learn and adapt to evolving programming practices, languages, and frameworks

9.1 Limitations

- Contextual Understanding: Generative AI models may struggle with understanding the broader context and domain-specific intricacies of code. This limitation can lead to inaccuracies or limitations in their generated outputs. For instance, AI models might fail to account for specific programming paradigms or industry-specific coding conventions. AI developers can address this limitation by incorporating more diverse and representative training data that capture a wide range of programming contexts and by refining the model architectures to improve contextual understanding.
- Data Bias: The quality and representativeness of training data play a critical role in the performance of generative AI models. Biases in the training data, such as the underrepresentation of certain code patterns or the overrepresentation of specific programming languages, can lead to biased and suboptimal code generation. To address this issue, it is essential to identify and mitigate data biases through meticulous curation of training datasets, alongside techniques like data augmentation and fair sampling. Furthermore, continuous monitoring and evaluation of AI systems are vital for detecting and addressing any biases that may arise during deployment.
- Debugging Complex Issues: Generative AI models often struggle with complex debugging scenarios and the identification of subtle bugs that demand deep program analysis and understanding. While AI can automate certain aspects of debugging, intricate issues typically require human expertise and intuition. To address this challenge, fostering collaboration between AI systems and human developers is essential. By providing AI-generated suggestions along with clear explanations and justifications, developers can better understand the reasoning behind these suggestions, empowering them to make informed decisions throughout the debugging process.
- Lack of Creativity: AI models excel at generating code based on existing patterns and examples but may lack the creativity and innovation that human developers bring to code development. AI-generated code may lack originality, alternative approaches, or novel solutions. To address this limitation, researchers can explore techniques such as generative adversarial networks or RL to encourage AI models to generate more diverse and creative code solutions. Additionally, AI systems can be designed to facilitate collaboration between AI models and human developers, encouraging developers to iteratively refine and improve the generated code.
- Ethical Considerations: The deployment of generative AI in code development brings forth significant ethical considerations. For instance, AI models trained on extensive code repositories may inadvertently propagate vulnerabilities or reinforce biased coding practices. To ensure the ethical and responsible use of generative AI, it is crucial to focus on issues such as privacy, security, fairness, and transparency. To address these challenges, developers and researchers should actively work on creating frameworks and guidelines that encompass robust data privacy measures, bias detection and mitigation strategies, as well as transparent documentation of AI-generated code.

9.2 Prospects

- Hybrid Approaches: Combining generative AI techniques with traditional programming approaches can create hybrid models that leverage the strengths of both AI systems and human developers. This collaboration can lead to more effective and efficient code development processes. For example, AI models can be used to generate initial code drafts or automate repetitive coding tasks, while human developers can provide high-level guidance, review and refine the generated code, and inject creativity and innovation into the development process.
- Explainable AI: Improving the interpretability and explainability of AI models is crucial for gaining trust and understanding their decision-making process in code development tasks. Explainable AI techniques can provide insights into how AI models generate code, enabling developers to validate and refine the output. This can involve techniques such as attention mechanisms, visualization tools, or generating human-readable explanations alongside the generated code. By making AI models more transparent and interpretable, developers can better understand and modify the generated code to align with their requirements.
- Domain-Specific Models: Developing specialized AI models that are tailored to specific programming languages, frameworks, or application domains can lead to more accurate and context-aware code generation and development tools. These models can capture the unique characteristics and requirements of different programming contexts. By training AI models on domain-specific datasets and incorporating domain-specific knowledge, such as coding conventions and domain-specific libraries, the generated code can better match the expectations and requirements of developers working in those specific domains.
- Collaborative AI Development: Integrating AI models into collaborative development environments, such as version control systems and IDEs, can enable real-time code suggestions and assistance. AI models can provide developers with context-aware suggestions, identify potential bugs or vulnerabilities, and offer code completion options. By integrating AI into the development workflow, developers can benefit from the assistance of AI models while still retaining control and autonomy over the code they write.
- Continuous Learning and Adaptation: AI models can be designed to continually learn and adapt to evolving programming practices, languages, and frameworks. By regularly updating and retraining AI models with the latest code repositories and industry practices, the models can stay up to date and improve their code generation capabilities over time. This continuous learning approach can help address the limitations of generative AI models and ensure that they remain relevant and effective in the ever-changing landscape of code development.

In conclusion, while generative AI in code development has its limitations, such as contextual understanding, data bias, and creativity, there are prospects for addressing these limitations and advancing the field. By combining generative AI with traditional programming approaches, improving explainability, developing domain-specific models, fostering collaborative AI development, and enabling continuous learning, the potential of generative AI in code development can be realized more effectively, leading to enhanced productivity and innovation in software development. However, it is crucial to consider ethical considerations and ensure the responsible use of generative AI in code development to mitigate potential risks and biases.

10 Conclusion

Generative AI presents significant prospects for transforming code development procedures and introducing progress in diverse domains including software design, code generation, testing, annotation, refactoring, and repair. Machine learning algorithms can be utilized by generative AI models to automate repetitive tasks, improve the quality of code, and increase the productivity of developers. Nonetheless, in code development, it is critical to acknowledge and rectify the constraints associated with generative AI.

We have examined the constraints of generative AI throughout this chapter, encompassing issues such as the inability to comprehend context, bias in data, difficulty in debugging complex problems, absence of innovation, and ethical implications. The significance of maintaining a delicate equilibrium between the proficiencies of human developers and the capabilities of AI models is underscored by these constraints. By integrating the benefits of generative AI with human ingenuity, hybrid methodologies can effectively leverage the capabilities of AI while maintaining human supervision and innovative resolution of challenges. In addition, prospects and future developments in generative AI for code development have been deliberated. Explainable AI techniques empower developers to comprehend and authenticate the judgments rendered by AI models, thereby cultivating confidence and trust in the outcomes produced. By customizing models to programming languages and application domains, one can improve the precision and pertinence of the code that is generated. The synergy and interchange of knowledge fostered by collaboration between AI systems and human developers result in the development of more innovative and resilient solutions. Ongoing improvement and adjustment of AI models in response to shifting programming practices guarantee their sustained efficacy and pertinence.

It is imperative to maintain a foremost position on ethical considerations as generative AI progresses. Privacy, bias, and the possibility of employment displacement are issues that must be addressed to ensure that generative AI is utilized ethically and responsibly in code development processes.

In summary, generative AI possesses the capacity to substantially augment the efficacy, caliber, and ingenuity of code development. In regards to the integration of generative AI into their operations, developers and researchers can arrive at well-informed decisions by acquiring knowledge of its developments, constraints, and potential. By strategically integrating AI with human expertise, novel opportunities can be revealed in the realm of code development, culminating in software systems that are more streamlined, dependable, and innovative. To fully harness the benefits of generative AI in the future of code development, it is crucial to acknowledge and resolve its limitations and ethical concerns while simultaneously capitalizing on its infinite potential.

References

1. R. Schwartz, R. Schwartz, A. Vassilev, K. Greene, L. Perine, A. Burt, P. Hall, *Towards a Standard for Identifying and Managing Bias in Artificial Intelligence*. US Department of Commerce, National Institute of Standards and Technology, vol. 3 (2022)
2. M. Kurni, M.S. Mohammed, K. Srinivasa, *A Beginner's Guide to Introduce Artificial Intelligence in Teaching and Learning* (Springer Nature, 2023)
3. C.K.Y. Chan, T. Colloton, *Generative AI in Higher Education: The ChatGPT Effect* (Taylor & Francis, 2024)
4. R. Pudari, AI Supported Software Development: Moving Beyond Code Completion. Ph.D. Dissertation (2022)
5. C. Zastudil, M. Rogalska, C. Kapp, J. Vaughn, S. MacNeil, Generative AI in computing education: perspectives of students and instructors, in *IEEE Frontiers in Education Conference (FIE)* (IEEE, 2023), pp. 1–9
6. Y.K. Dwivedi, N. Kshetri, L. Hughes, E.L. Slade, A. Jeyaraj, A.K. Kar, A.M. Baabdullah, A. Koohang, V. Raghavan, M. Ahuja et al., "so what if chatgpt wrote it?" multidisciplinary perspectives on opportunities, challenges and implications of generative conversational AI for research, practice and policy. Int. J. Inform. Manage. **71**, 102642 (2023)
7. C. Seidl, S. Schuster, I. Schaefer, Generative software product line development using variability-aware design patterns. Comput. Lang. Syst. Struct. **48**, 89–111 (2017)
8. A. Peruma, S. Simmons, E.A. AlOmar, C.D. Newman, M.W. Mkaouer, A. Ouni, How do i refactor this? An empirical study on refactoring trends and topics in stack overflow. Empirical Softw. Eng. **27**(1), 11 (2022)
9. B. Van Oort, L. Cruz, M. Aniche, A. Van Deursen, The prevalence of code smells in machine learning projects, in *IEEE/ACM 1st Workshop on AI Engineering-Software Engineering for AI (WAIN)* (IEEE, 2021), pp. 1–8
10. N. Tihanyi, T. Bisztray, R. Jain, M.A. Ferrag, L.C. Cordeiro, V. Mavroeidis, The formai dataset: Generative AI in software security through the lens of formal verification, in *Proceedings of the 19th International Conference on Predictive Models and Data Analytics in Software Engineering* (2023), pp. 33–43
11. E. Dehaerne, B. Dey, S. Halder, S. De Gendt, W. Meert, Code generation using machine learning: a systematic review. IEEE Access **10**, 82434–82455 (2022)
12. J. Liu, J. Ai, M. Lu, J. Wang, H. Shi, Semantic feature learning for software defect prediction from source code and external knowledge. J. Syst. Softw. **204**, 111753 (2023)
13. C. Bull, A. Kharrufa, Generative AI assistants in software development education: a vision for integrating generative ai into educational practice, not instinctively defending against it. IEEE Softw. (2023)
14. M. Mandapuram, S.S. Gutlapalli, A. Bodepudi, M. Reddy, Investigating the prospects of generative artificial intelligence. Asian J Human. Art Lit. **5**(2), 167–174 (2018)
15. E. Wenger, *Artificial Intelligence and Tutoring Systems: Computational and Cognitive Approaches to the Communication of Knowledge* (Morgan Kaufmann, 2014)
16. M. Shahin, F. F. Chen, A. Hosseinzadeh, Harnessing customized AI to create voice of customer via gpt3. 5. Adv. Eng. Inform. **61**, 102462 (2024)

17. T.-P. Liang, L. Robert, S. Sarker, C.M. Cheung, C. Matt, M. Trenz, O. Turel, Artificial intelligence and robots in individuals' lives: how to align technological possibilities and ethical issues. Internet Res. **31**(1), 1–10 (2021)
18. S. Feuerriegel, J. Hartmann, C. Janiesch, P. Zschech, Generative AI. Bus. Inform. Syst. Eng. **66**(1), 111–126 (2024)
19. A.B. Arrieta, N. Díaz-Rodríguez, J. Del Ser, A. Bennetot, S. Tabik, A. Barbado, S. García, S. Gil-López, D. Molina, R. Benjamins et al., Explainable artificial intelligence (XAI): concepts, taxonomies, opportunities and challenges toward responsible AI. Inform. Fusion **58**, 82–115 (2020)
20. L. Zhou, S. Paul, H. Demirkan, L. Yuan, J. Spohrer, M. Zhou, J. Basu, Intelligence augmentation: towards building human-machine symbiotic relationship. AIS Trans. Human-Comput. Interact. **13**(2), 243–264 (2021)
21. C.K.Y. Chan, A comprehensive AI policy education framework for university teaching and learning. Int. J. Educ. Technol. Higher Educ. **20**(1), 38 (2023)
22. H.R. Nemati, D.M. Steiger, L.S. Iyer, R.T. Herschel, Knowledge warehouse: an architectural integration of knowledge management, decision support, artificial intelligence and data warehousing. Decis. Support Syst. **33**(2), 143–161 (2002)
23. Y.K. Dwivedi, L. Hughes, E. Ismagilova, G. Aarts, C. Coombs, T. Crick, Y. Duan, R. Dwivedi, J. Edwards, A. Eirug et al., Artificial intelligence (AI): multidisciplinary perspectives on emerging challenges, opportunities, and agenda for research, practice and policy. Int. J. Inform. Manage. **57**, 101994 (2021)
24. F. Dell'Acqua, E. McFowland, E.R. Mollick, H. Lifshitz-Assaf, K. Kellogg, S. Rajendran, L. Krayer, F. Candelon, K.R. Lakhani, Navigating the jagged technological frontier: field experimental evidence of the effects of AI on knowledge worker productivity and quality. *Harvard Business School Technology & Operations Mgt. Unit Working Paper*, no. 24-013 (2023)
25. H.-M. Heyn, E. Knauss, P. Pelliccione, A compositional approach to creating architecture frameworks with an application to distributed ai systems. J. Syst. Softw. **198**, 111604 (2023)
26. Z. Wang, P. Zheng, X. Li, C.-H. Chen, Implications of data-driven product design: from information age towards intelligence age. Adv. Eng. Inform. **54**, 101793 (2022)
27. S.L. Kumar, State of the art-intense review on artificial intelligence systems application in process planning and manufacturing. Eng. Appl. Artif. Intell. **65**, 294–329 (2017)
28. W. Liao, X. Lu, Y. Fei, Y. Gu, Y. Huang, Generative AI design for building structures. Autom. Construct. **157**, 105187 (2024)
29. E. Papagiannidis, I.M. Enholm, C. Dremel, P. Mikalef, J. Krogstie, Toward AI governance: identifying best practices and potential barriers and outcomes. Inform. Syst. Front. **25**(1), 123–141 (2023)
30. X. Du, M. Liu, K. Wang, H. Wang, J. Liu, Y. Chen, J. Feng, C. Sha, X. Peng, Y. Lou, Evaluating large language models in class-level code generation, in *2024 IEEE/ACM 46th International Conference on Software Engineering (ICSE)* (IEEE Computer Society, 2024), pp. 865–865
31. K. Jayanthi, S. Mohan, An integrated framework for emotion recognition using speech and static images with deep classifier fusion approach. Int. J. Inform. Technol. **14**(7), 3401–3411 (2022)
32. S. Bhartacharyya, R. Leupers, P. Marwedel, Software synthesis and code generation for signal processing systems. IEEE Trans. Circ. Syst. II: Anal. Digital Signal Process. **47**(9), 849–875 (2000)
33. X. Gao, T. Kim, M.D. Wong, D. Raghunathan, A.K. Varma, P.G. Kannan, A. Sivaraman, S. Narayana, A. Gupta, Switch code generation using program synthesis, in *Proceedings of the Annual Conference of the ACM Special Interest Group on Data Communication on the Applications, Technologies, Architectures, and Protocols for Computer Communication* (2020), pp. 44–61
34. S. Kelly, J.-P. Tolvanen, *Domain-Specific Modeling: Enabling Full Code Generation*. John Wiley & Sons (2008)

35. C. Schmitt, S. Kuckuk, H. Köstler, F. Hannig, J. Teich, An evaluation of domain-specific language technologies for code generation, in *2014 14th International Conference on Computational Science and Its Applications* (IEEE, 2014), pp. 18–26
36. Z. Lin, G. Li, J. Zhang, Y. Deng, X. Zeng, Y. Zhang, Y. Wan, Xcode: Towards cross-language code representation with large-scale pre-training. ACM Trans. Softw. Eng. Methodol. (TOSEM) **31**(3), 1–44 (2022)
37. G. Mathew, K.T. Stolee, Cross-language code search using static and dynamic analyses, in *Proceedings of the 29th ACM Joint Meeting on European Software Engineering Conference and Symposium on the Foundations of Software Engineering* (2021), pp. 205–217
38. J.-Y. Liou, X. Wang, S. Forrest, C.-J. Wu, Gevo: Gpu code optimization using evolutionary computation. ACM Trans. Architect. Code Optim. (TACO) **17**(4), 1–28 (2020)
39. L. Abualigah, A. Diabat, S. Mirjalili, M. Abd Elaziz, A.H. Gandomi, The arithmetic optimization algorithm. Comput. Methods Appl. Mech. Eng. **376**, 113609 (2021)
40. O. Hamdy, H. Gaber, M.S. Abdalzaher, M. Elhadidy, Identifying exposure of urban area to certain seismic hazard using machine learning and GIS: a case study of greater cairo. Sustainability **14**(17), 10722 (2022)
41. O. Issa, R. Silva-Lopez, J.W. Baker, H.V. Burton, Machine-learning-based optimization framework to support recovery-based design. Earthq. Eng. Struct. Dyn. **52**(11), 3256–3280 (2023)
42. S.S. Moustafa, M.S. Abdalzaher, F. Khan, M. Metwaly, E.A. Elawadi, N.S. Al-Arifi, A quantitative site-specific classification approach based on affinity propagation clustering. IEEE Access **9**, 155297–155313 (2021)
43. C. Cummins, B. Wasti, J. Guo, B. Cui, J. Ansel, S. Gomez, S. Jain, J. Liu, O. Teytaud, B. Steiner, Compilergym: robust, performant compiler optimization environments for AI research, in *IEEE/ACM International Symposium on Code Generation and Optimization (CGO)* (IEEE, 2022), pp. 92–105
44. M.S. Abdalzaher, S.S. Moustafa, H.A. Hafiez, W.F. Ahmed, An optimized learning model augment analyst decisions for seismic source discrimination. IEEE Trans. Geosci. Remote Sens. **60**, 1–12 (2022)
45. T. Geng, M. Amaris, S. Zuckerman, A. Goldman, G.R. Gao, J.-L. Gaudiot, A profile-based ai-assisted dynamic scheduling approach for heterogeneous architectures. Int. J. Parallel Program. **50**(1), 115–151 (2022)
46. W.-K. Hong, *Artificial Neural Network-based Optimized Design of Reinforced Concrete Structures* (CRC press, 2023)
47. M. Acher, J. Martinez, Generative AI for reengineering variants into software product lines: An experience report, in *Proceedings of the 27th ACM International Systems and Software Product Line Conference-Volume B* (2023), pp. 57–66
48. J. Pantiuchina, F. Zampetti, S. Scalabrino, V. Piantadosi, R. Oliveto, G. Bavota, M.D. Penta, Why developers refactor source code: a mining-based study. ACM Trans. Softw. Eng. Methodol. (TOSEM) **29**(4), 1–30 (2020)
49. A. Ardalani, S. Parsa, M. Zakeri-Nasrabadi, A. Chatzigeorgiou, Supporting single responsibility through automated extract method refactoring. Empirical Softw. Eng. **29**(1), 28 (2024)
50. H.C. Jiau, L.W. Mar, J.C. Chen, Obey: optimal batched refactoring plan execution for class responsibility redistribution. IEEE Trans. Softw. Eng. **39**(9), 1245–1263 (2013)
51. M. Ichii, D. Shimbara, Y. Suzuki, H. Ogawa, Refactoring verification using model transformation, in *Proceedings of the 1st International Workshop on Software Refactoring* (2016), pp. 17–24
52. M. Fowler, *Refactoring: Improving the Design of Existing Code*. (Addison-Wesley Professional, 2018)
53. P. Weißgerber, S. Diehl, Identifying refactorings from source-code changes, in *21st IEEE/ACM International Conference on Automated Software Engineering (ASE'06)* (IEEE, 2006), pp. 231–240

54. G. Lacerda, F. Petrillo, M. Pimenta, Y.G. Guéhéneuc, Code smells and refactoring: a tertiary systematic review of challenges and observations. J. Syst. Softw. **167**, 110610 (2020)
55. S. Singh, S. Kaur, A systematic literature review: refactoring for disclosing code smells in object oriented software. Ain Shams Eng. J. **9**(4), 2129–2151 (2018)
56. A.J. Mooij, J. Ketema, S. Klusener, M. Schuts, Reducing code complexity through code refactoring and model-based rejuvenation, in *2020 IEEE 27th International Conference on Software Analysis, Evolution and Reengineering (SANER)* (IEEE, 2020), pp. 617–621
57. D. Mazinanian, N. Tsantalis, A. Mesbah, Discovering refactoring opportunities in cascading style sheets, in *Proceedings of the 22nd ACM SIGSOFT International Symposium on Foundations of Software Engineering* (2014), pp. 496–506
58. M. Bosch, P. Genevès, N. Layaïida, Automated refactoring for size reduction of CSS style sheets, in *Proceedings of the 2014 ACM Symposium on Document Engineering* (2014), pp. 13–16
59. S.R. Foster, W.G. Griswold, S. Lerner, Witchdoctor: Ide support for real-time auto-completion of refactorings, in *34th International Conference on Software Engineering (ICSE)* (IEEE, 2012), pp. 222–232
60. Z. Kurbatova, V. Kovalenko, I. Savu, B. Brockbernd, D. Andreescu, M. Anton, R. Venediktov, E. Tikhomirova, T. Bryksin, Refactorinsight: enhancing ide representation of changes in git with refactorings information, in *2021 36th IEEE/ACM International Conference on Automated Software Engineering (ASE)* (IEEE, 2021), pp. 1276–1280
61. P. Mayer, A. Schroeder, Cross-language code analysis and refactoring, in *2012 IEEE 12th International Working Conference on Source Code Analysis and Manipulation* (IEEE, 2012), pp. 94–103
62. A.A.B. Baqais, M. Alshayeb, Automatic software refactoring: a systematic literature review. Softw. Qua. J. **28**(2), 459–502 (2020)
63. X. Ge, S. Sarkar, J. Witschey, E. Murphy-Hill, Refactoring-aware code review, in *IEEE Symposium on Visual Languages and Human-Centric Computing (VL/HCC)* (IEEE, 2017), pp. 71–79
64. T. Ahmed, N.R. Ledesma, P. Devanbu, Synshine: improved fixing of syntax errors. IEEE Trans. Softw. Eng. **49**(4), 2169–2181 (2022)
65. F. Medeiros, M. Ribeiro, R. Gheyi, Investigating preprocessor-based syntax errors, in *Proceedings of the 12th International Conference on Generative Programming: Concepts & Experiences* (2013), pp. 75–84
66. A. Ettles, A. Luxton-Reilly, P. Denny, Common logic errors made by novice programmers, in *Proceedings of the 20th Australasian Computing Education Conference* (2018), pp. 83–89
67. W. Weimer, G.C. Necula, Finding and preventing run-time error handling mistakes, in *Proceedings of the 19th Annual ACM SIGPLAN Conference on Object-Oriented Programming, Systems, Languages, and Applications* (2004), pp. 419–431
68. G.R. Luecke, J. Coyle, J. Hoekstra, M. Kraeva, Y. Li, O. Taborskaia, Y. Wang, A survey of systems for detecting serial run-time errors. Concurrency Comput. Pract. Exp. **18**(15), 1885–1907 (2006)
69. D. Athanasiou, A. Nugroho, J. Visser, A. Zaidman, Test code quality and its relation to issue handling performance. IEEE Trans. Softw. Eng. **40**(11), 1100–1125 (2014)
70. H. Shahriar, M. Zulkernine, Mitigating program security vulnerabilities: approaches and challenges. ACM Comput. Surv. (CSUR) **44**(3), 1–46 (2012)
71. J. Heffley, P. Meunier, Can source code auditing software identify common vulnerabilities and be used to evaluate software security?' in *Proceedings of the 37th Annual Hawaii International Conference on System Sciences* (IEEE, 2004), pp. 10
72. M.S. Abdalzaher, K. Seddik, O. Muta, A. Abdelrahman, Using stackelberg game to enhance node protection in WSNA, in *13th IEEE Annual Consumer Communications and Networking Conference (CCNC)* (IEEE, 2016), pp. 853–856
73. H. Huang, M. Wen, L. Wei, Y. Liu, S.-C. Cheung, Characterizing and detecting configuration compatibility issues in android apps, in *2021 36th IEEE/ACM International Conference on Automated Software Engineering (ASE)* (IEEE, 2021), pp. 517–528

74. L. Wei, Y. Liu, S.-C. Cheung, H. Huang, X. Lu, X. Liu, Understanding and detecting fragmentation-induced compatibility issues for android apps. IEEE Trans. Softw. Eng. **46**(11), 1176–1199 (2018)
75. M. Stitt, *Debugging: Creative Techniques and Tools for Software Repair* (John Wiley & Sons, Inc., 1992)
76. S. Benton, X. Li, Y. Lou, L. Zhang, Evaluating and improving unified debugging. IEEE Trans. Softw. Eng. **48**(11), 4692–4716 (2021)
77. E. Daka, G. Fraser, A survey on unit testing practices and problems, in *IEEE 25th International Symposium on Software Reliability Engineering* (IEEE, 2014), pp. 201–211
78. J. Xuan, B. Cornu, M. Martinez, B. Baudry, L. Seinturier, M. Monperrus, B-refactoring: Automatic test code refactoring to improve dynamic analysis. Inform. Softw. Technol. **76**, 65–80 (2016)
79. M. Tufano, C. Watson, G. Bavota, M.D. Penta, M. White, D. Poshyvanyk, An empirical study on learning bug-fixing patches in the wild via neural machine translation. ACM Trans. Softw. Eng. Methodol. (TOSEM) **28**(4), 1–29 (2019)
80. F. Huq, M. Hasan, M.M.A. Haque, S. Mahbub, A. Iqbal, T. Ahmed, Review4repair: code review aided automatic program repairing. Inform. Softw. Technol. **143**, 106765 (2022)
81. T. Mamatha, B.R.S. Reddy, C.S. Bindu, A literature review on automated code repair, in *Proceedings of the 2nd International Conference on Recent Trends in Machine Learning, IoT, Smart Cities and Applications: ICMISC 2021* (Springer, 2022), pp. 249–260
82. A. Koyuncu, K. Liu, T.F. Bissyandé, D. Kim, J. Klein, M. Monperrus, Y. Le Traon, Fixminer: mining relevant fix patterns for automated program repair. Empirical Softw. Eng. **25**, 1980–2024 (2020)
83. N. Jiang, T. Lutellier, L. Tan, Cure: code-aware neural machine translation for automatic program repair, in *2021 IEEE/ACM 43rd International Conference on Software Engineering (ICSE)* (IEEE, 2021), pp. 1161–1173
84. B. Zhou, G. Ding, Survey of intelligent program synthesis techniques, in *International Conference on Algorithms, High Performance Computing, and Artificial Intelligence (AHPCAI 2023)*, vol. 12941 (SPIE, 2023), pp. 1122–1136
85. A. Barriga, L. Mandow, J.L.P. de la Cruz, A. Rutle, R. Heldal, L. Iovino, A comparative study of reinforcement learning techniques to repair models, in *Proceedings of the 23rd ACM/IEEE International Conference on Model Driven Engineering Languages and Systems: Companion Proceedings*, (2020), pp. 1–9
86. J.S. Grosman, P.H. Furtado, A.M. Rodrigues, G.G. Schardong, S.D. Barbosa, H.C. Lopes, Eras: improving the quality control in the annotation process for natural language processing tasks. Inform. Syst. **93**, 101553 (2020)
87. S. Rai, R.C. Belwal, A. Gupta, A review on source code documentation. ACM Trans. Intell. Syst. Technol. (TIST) **13**(5), 1–44 (2022)
88. S. Iyer, I. Konstas, A. Cheung, L. Zettlemoyer, Summarizing source code using a neural attention model, in *54th Annual Meeting of the Association for Computational Linguistics. Association for Computational Linguistics* (2016), pp. 2073–2083
89. K. Bontcheva, H. Cunningham, I. Roberts, A. Roberts, V. Tablan, N. Aswani, G. Gorrell, Gate teamware: a web-based, collaborative text annotation framework. Lang. Res. Eval. **47**, 1007–1029 (2013)
90. M. Adeli, N. Nelson, S. Chattopadhyay, H. Coffey, A. Henley, A. Sarma, Supporting code comprehension via annotations: right information at the right time and place, in *IEEE Symposium on Visual Languages and Human-Centric Computing (VL/HCC)* (IEEE, 2020), pp. 1–10
91. K. Burr, W. Young, Combinatorial test techniques: table-based automation, test generation and code coverage, in *Proceedings of the International Conference on Software Testing Analysis and Review* (Citeseer, 1998)
92. S. Zhang, D. Saff, Y. Bu, M.D. Ernst, Combined static and dynamic automated test generation, in *Proceedings of the 2011 International Symposium on Software Testing and Analysis* (2011), pp. 353–363

93. W.E. Wong, Y. Qi, L. Zhao, K.-Y. Cai, Effective fault localization using code coverage, in *31st Annual International Computer Software and Applications Conference (COMPSAC)*, vol. 1 (IEEE 2007), pp. 449–456
94. L. Kampel, D. E. Simos, D.R. Kuhn, R.N. Kacker, An exploration of combinatorial testing-based approaches to fault localization for explainable AI. Ann. Math. Artif. Intell. 1–14 (2022)
95. M.S. Abdalzaher, M. Elwekeil, T. Wang, S. Zhang, A deep autoencoder trust model for mitigating jamming attack in IoT assisted by cognitive radio. IEEE Syst. J. **16**(3), 3635–3645 (2021)
96. P. Afric, L. Sikic, A.S. Kurdija, M. Silic, Repd: source code defect prediction as anomaly detection. J. Syst. Softw. **168**, 110641 (2020)
97. M.N.K. Sikder, F.A. Batarseh, Outlier detection using AI: a survey. AI Assurance 231–291 (2023)
98. R. Lima, A.M.R. da Cruz, J. Ribeiro, Artificial intelligence applied to software testing: a literature review, in *2020 15th Iberian Conference on Information Systems and Technologies (CISTI)* (IEEE, 2020), pp. 1–6
99. M.S. Abdalzaher, M.M. Fouda, A. Emran, Z.M. Fadlullah, M.I. Ibrahem, A survey on key management and authentication approaches in smart metering systems. Energies **16**(5), 2355 (2023)
100. S. Tahvili, L. Hatvani, *Artificial Intelligence Methods for Optimization of the Software Testing Process: With Practical Examples and Exercises* (Academic Press, 2022)
101. A. Swain, K. Swain, S. Swain, S. Samal, G. Palai, Automated test case prioritization using machine learning, in *International Conference on Metaheuristics in Software Engineering and its Application* (Springer, 2022), pp. 200–208
102. A.M. Memon, Automated GUI regression testing using AI planning, in *Artificial Intelligence Methods In Software Testing* (World Scientific, 2004), pp. 51–99
103. R.H. Rosero, O.S. Gómez, G. Rodríguez, 15 years of software regression testing techniques-a survey. Int. J. Softw. Eng. Knowl. Eng. **26**(05), 675–689 (2016)
104. S.R. Shahamiri, W.M.N.W. Kadir, S.Z. Mohd-Hashim, A comparative study on automated software test oracle methods, in *Fourth International Conference on Software Engineering Advances* (IEEE, 2009), pp. 140–145
105. Z. Khaliq, S.U. Farooq, D.A. Khan, Artificial intelligence in software testing: impact, problems, challenges and prospect. arXiv preprint arXiv:2201.05371 (2022)

Moez Krichen (Senior Member, IEEE) is an Associate Professor in Computer Science at AlBaha University, Saudi Arabia. With over 20 years of experience in scientific research and 15 years in teaching, he specializes in artificial intelligence, security, IoT, healthcare, and blockchain technologies. Dr. Krichen has published more than 100 journal articles and 70 conference papers, with an H-index of 53 and over 6,450 citations. He has been recognized as one of the World's Top 2% Scientists by Stanford University in 2022, 2023, and 2024. Dr. Krichen is also an IEEE Senior Member and actively contributes to various academic committees and research initiatives.

Mohamed S. Abdalzaher (Senior Member, IEEE) received the B.Sc. (Hons.) in electronics and communications engineering, in 2008, the M.Sc. degrees in electronics and communications engineering from Ain Shams University, Cairo, Egypt, in 2012, and the Ph.D. degree from the Electronics and Communications Engineering Department, Egypt-Japan University of Science and Technology, Madinet Borg Al Arab, Egypt, in 2016.From April to October 2019, he worked as a Postdoctoral Researcher at the Center for Japan-Egypt Cooperation in Science and Technology at Kyushu University. In August 2024, he took on the role of Postdoctoral Research Fellow at the American University of Sharjah. Dr. Mohamed was recognized in the list of the top 2% of highly ranked scientists. In 2022, Dr. Abdalzaher received the State Encouragement Award, followed by the Scientific Excellence Award from NRIAG in both 2023 and 2024. He serves as an

Associate Professor in the Seismology Department at the National Research Institute of Astronomy and Geophysics in Cairo. Additionally, Dr. Abdalzaher holds the position of Associate Editor for IEEE Transactions on Geoscience and Remote Sensing and has acted as a scientific editor and reviewer for numerous journals, as well as a TPC member for various conferences. He has organized and participated in multiple conferences and workshops. His research interests encompass Earthquake Engineering, Remote Sensing, Artificial Intelligence, Wireless Communications, WSN security, IoT, and data communication networks.

LLMs as a Compiler for an Arabic Programming Language

Serry Sibaee, Omer Nacar, Lahouari Ghouti, and Anis Koubaa

Abstract Programming languages force people to think logically while interacting with the computers to reach a conceptual level that makes problem solving accessible to the computer itself. Given their expressive power, natural languages outweigh their programming counterparts. In this paper, we propose a natural to programming language converter using large language models (LLMs) where the user can express the logical constructs in his own language that get converted to Python code. In this way, the converter alleviates the user fear from learning a new programming language. We identify the proposed converter as the Arabic programming language (APL). The APL model runs under the hood GPT4 application program interface (API) using an elaborate prompt engineering scheme. In addition, an intuitive, yet simple, user interface is provided to assist the APL user. As a feedback, the user will rate the APL response to the input prompt as a score ranging from 1 to 5 where 1 would indicate dissatisfactory output and 5 to indicate perfect code execution. The APL API is hosted at https://github.com/riotu-lab/apl-compiler.

Keywords Arabic programming language (APL) · Large language models (LLMS) · Python compiler · Application program interface (API) · Code execution

S. Sibaee (✉) · O. Nacar · L. Ghouti · A. Koubaa
RIOTU Labs, Prince Sultan University, Riyadh, Saudi Arabia
e-mail: ssibaee@psu.edu.sa

O. Nacar
e-mail: onajar@psu.edu.sa

L. Ghouti
e-mail: lghouti@psu.edu.sa

A. Koubaa
e-mail: akoubaa@psu.edu.sa

A. Koubaa et al. (eds.), *Generative AI and Large Language Models: Opportunities, Challenges, and Applications*, Studies in Computational Intelligence 1214,
https://doi.org/10.1007/978-3-031-90573-5_8

1 Introduction

This paper explores the synergy between programming language compilers and large language models (LLMs), proposing a novel approach where LLMs function as semi-compilers to translate Arabic structured text (APL) into Python code. We begin by outlining the architecture of this compiler, highlighting the necessary adaptations in APL's writing style. Next, we delve into the prompt engineering techniques used to effectively convert APL into executable Python. In addition, an API and user interface will be introduced, designed to facilitate the integration of this compiler into various applications.

Programming languages have undergone a significant evolution, starting with early efforts like Zuse's Plankalkül and progressing to modern languages such as Python, Java, and C++. Milestones along this journey include languages like ALGOL, COBOL, and LISP, which introduced various computing paradigms [5–7]. In parallel, Arabic-based programming languages have made their mark, with languages such as Sakhr Basic (1987) [9], ARLOGO [10], and more recent ones like AMORIA [13], Jeem [15], and Kalimat [18] contributing to programming education in Arabic-speaking communities.

The development of compilers traces back to Grace Hopper's A-0 System and continued with the introduction of Fortran, which revolutionized programming through the inclusion of key components like lexical and syntax analysis, optimization, and code generation [20–22]. This evolution parallels the rise of Pre-trained Language Models (PLMs) like BERT, GPT, and T5, which have transformed natural language processing (NLP) tasks through their rich linguistic representations [23].

Prompt engineering, a critical aspect of LLMs, has enabled the efficient use of PLMs for a range of tasks without requiring extensive training. Techniques such as zero-shot and few-shot prompting, Chain-of-Thought (CoT), and Retrieval Augmented Generation (RAG) have enhanced reasoning, logic, hallucination reduction, and code generation, making these models highly adaptable [28].

In this paper, we present the methodology for constructing the APL-to-Python compiler using LLMs, followed by a detailed look at the accompanying API and interface. We provide examples, evaluate the capabilities of APL, and conclude with a discussion of the challenges encountered in this endeavor.

2 Related Work

In the dynamic realm of contemporary programming, the advent of Large Language Models (LLMs) parallels historical shifts in programming paradigms. Analogous to the evolution from assembly language to high-level languages facilitated by optimizing compilers, LLMs signify a trans-formative leap, positioning themselves not merely as advanced autocompletion tools but as compilers. This conceptual shift is underscored by the absence of direct exploration in the literature regarding the utilization of LLMs as programming compilers.

Historically, programming proficiency resided in mastering assembly language, a skill rendered obsolete by optimizing compilers translating high-level languages into efficient machine code. A parallel transition is underway today, elevating the act of programming to a higher level of abstraction. LLMs, conceived as compilers, signify this evolution by translating natural language prompts into high-level programming languages, redefining the coding landscape.

Empirical evidence, exemplified by Copilot, accentuates this transformative potential. Copilot demonstrated efficacy in solving introductory programming problems, showcasing an ability to generate accurate solutions from problem descriptions alone. In instances of inaccuracies, developers prefer prompt manipulation over conventional debugging, reflecting a paradigm shift in code creation methodologies.

The academic discourse delves into this trans-formative trajectory, with LLM-Compiler [1] for parallel function calling and the innovative Synthesize, Execute, Debug (SED) approach [3] exemplifying engagements with evolving programming paradigms. Furthermore, critical insights from online articles [2] contribute nuanced perspectives, collectively consolidating the conceptualization of LLMs as compilers in the intricate landscape of modern programming methodologies. The absence of direct exploration in current literature underscores the nascent nature of employing LLMs as programming compilers, paving the way for future inquiries into this transformative domain.

Also there is an opposite approach of make a programming language that ease the use of LLM in python code called 'LMQL' "LMQL offers a novel way of interweaving traditional programming with the ability to call LLMs in your code." [4], there is a full survey about SQL text to code approach [27]. And a side note that no one has tried explicitly to use LLMs as a compiler to run a new programming language but the general research shows to use LLMs to automate programming or help programmers to program fast with multiple way auto complete the written code (copilot) or solving the problems by asking the chat Bot to debug and suggest solutions. And the near idea that is similar to ours is "PseuToPy" [24] which is an explicit compiler (did not use LLMs but the idea is the same) that change chinese code which follows the structure of python code then run it. Also a new Approach that shows a reverse way of using LLMs as a decompilers where the LLM was Decompile-Eval dataset improve decompilation [25]. Also another paper that used some kind of python compiler in their pipeline "PAL" [26] 'a novel approach that uses the LLM to read natural language problems and generate programs as the intermediate reasoning steps, but offloads the solution step to a runtime such as a Python interpreter (Fig. 1).

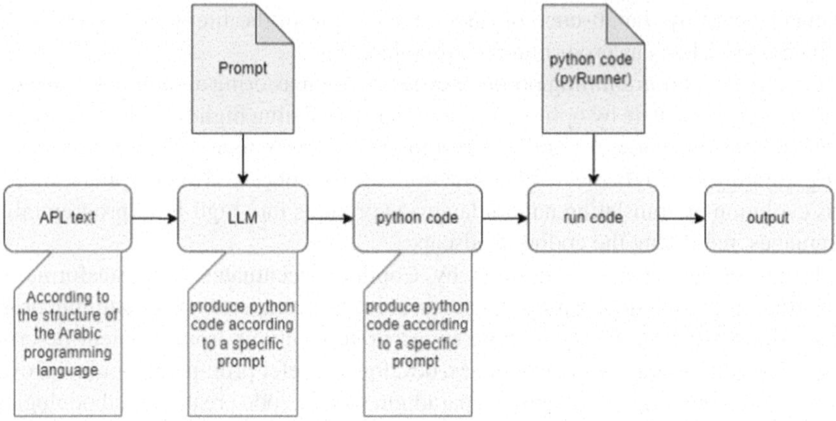

Fig. 1 The architecture of APL compiler

3 Methodology

The methodology unfolds in two key dimensions: Compiler Architecture and Prompt Engineering. Compiler Architecture examines the core components-LLM Planner, Task Fetching Unit, and Executor-offering a foundational understanding of the LLM Compiler's structure and workflow. In parallel, Prompt Engineering delves into the complexities of transforming natural language prompts into precise instructions for the LLM. This dimension not only ensures accurate code generation but also addresses edge cases, such as outliers and incorrect inputs, by returning error alerts. Together, these two dimensions provide a holistic methodology that integrates both architectural rigor and linguistic precision.

The APL code is written completely in Arabic script from right to left. And for Loops (while for example) we chose not to use indentation of python but the curly brackets because it is easier for the programmer and also easier for the LLM to compiler because copying and pasting the indented text could mess the code. By maintaining these important factors Arabic Script, Curly Brackets Usage, Line-by-Line Operations, the final condition is to make sure the programmer write the code line by line and do not use the ';' to make it clearer and easier for the reader and for the LLM.

The code input follows an Arabic script, introducing a distinctive structure influenced by a shift to the use of curly brackets "{ }" instead of indentation, departing from the traditional indentation in Python. Each operation is distinctly presented on its own line, eliminating the need for semicolons (;) in multi-line statements.

Programming Language Components For simplicity we did not include all of python properties but just these six generic ones to make it simple and also for the fragility of LLMs in the chain of combined prompts. The capabilities of the APL of generating Arabic code for Variables, Operations, File Reading (Limited to TXT Files), If Conditions, Loops, Functions and more.

Arabic Text Conventions This section details the conventions employed for Arabic text within the programming language, including variable names, operations, conditions, and loops, as well as the syntax for functions and file reading (Figs. 2, 3).

Fig. 2 Interface of APL App

vars names are with no prefix	
operations (using arabic keyboard symbols)	- multiplication * → × - division / → ÷ - power ** → ^
conditions (if)	إذا كان (كذا) {فكذا} وإن كان(كذا){فكذا} وإلا{كذا}
while loop	طالما(كذا){فافعل}
for loop	لكل _ في المدى(س،ص){افعل}
function	الدالة -اسمها-(){}
read file	اسم_النص = افتح الملف("اسم.نص")

Fig. 3 Functions in Arabic words in APL

3.1 Prompt Engineering: Dealing with the Text

Structure of the Prompt In the realm of prompt engineering, the task involves translating Arabic text into executable Python code. The process includes renaming variables to English for operational clarity and substituting Arabic functions with their Python equivalents (e.g., replacing with print). The overarching goal is to ensure the prompt results in valid and runnable Python code, facilitating seamless integration within the programming environment.

Prompt Structure Guidelines: Dealing with Keywords In the process of crafting prompts for translation into Python code, it is essential to handle specific Arabic terms and translate them into their corresponding Python keywords. The following table outlines the mapping of relevant keywords, ensuring their seamless integration into Python code while preserving Arabic texts used within strings.

The Used Prompt The used prompt in the GPT-4 is: "This GPT specializes in translating Arabic algorithmic instructions into Python code. It accurately replaces Arabic keywords with their Python equivalents, ensures variables are named in English for runnable code, and checks for syntax correctness. If a syntax error is detected, it provides a specific error message to help users correct the issue. The output code is provided as plain text, not in Python markdown, to ensure compatibility with various text editors and environments." It is worth mentioning that the only usage of GPT-4 is the most cheap and efficient way for the targeted audience because it is much faster and cheaper that downloading huge models (35B and above that can handle this task or even fine tune smaller ones) (Table 1 and Fig. 4).

In this work, we conceptualize the API similarly to a traditional compiler. But rather than translating a high-level language into machine code, our API acts as a pipe, turning Arabic input into Python code. This method makes use of Python's adaptability and widespread use in AI applications and research.

3.2 The Structure of the API

The API is designed to be seamlessly integrated into existing Python environments. This integration is achieved through a dedicated Python library, which is hosted on GitHub for accessibility and collaborative development. The library encapsulates the API's functionality into a single, user-friendly class. The main component of

Table 1 Keywords and their Arabic terms in APL

Keyword	max	min	sorted	int	abs	len	open	range	sum	replace	split	eval	round	append
Term	أعلى الأكبر	أقل	رتب	صحيح	مطلق	طول	افتح الملف	المدى	مجموع	استبدال	افصل اقطع	احسب	قرب	أضف

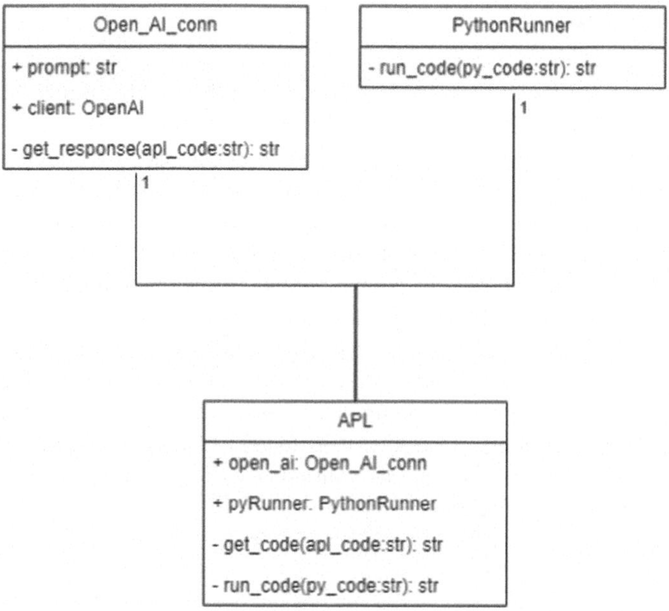

Fig. 4 Class diagram of the APL compiler

our Python library is the Compiler class. This class is initialized with an API key, enabling secure and authenticated access to the API's functionalities. The syntax is straightforward.

3.3 Sample of the Interface

To demonstrate the practical application and user interaction with our API, we introduce a sample interface designed for simplicity and effectiveness.

Design of the Interface: The interface features a dual-pane layout. The right pane serves as the input area where users can write or paste Arabic code. This design choice caters to the natural right-to-left flow of Arabic script, ensuring a familiar and comfortable user experience.

Compatibility of Arabic Text with Output Display: The output, or translated Python code, is shown in the left pane. A special focus is on making sure that Arabic text is compatible. This includes proper rendering of Arabic characters and script peculiarities in the output display in addition to left-to-right text alignment. Users' understanding and debugging abilities are improved by this dual-pane structure, which makes it simple to compare the generated Python code with the original Arabic script.

4 Illustration

In this section, we showcase the APL application of the LLM compiler in bridging linguistic barriers between Arabic and Python programming. The LLm compiler APL has been tested under the progression from foundational programming concepts to more intricate operations illustrates not only the flexibility of the model but also its potential to democratize programming for Arabic-speaking individuals.

Figure 6 demonstrates the foundational initiating with fundamental operations, such as file handling, arithmetic computations, string manipulations, and basic function definitions, this segment aims to solidify the groundwork. These examples emphasize the model's ability to comprehend and execute simple instructions, thereby laying a robust foundation for more complex tasks. Additionally, Fig. 5 shows more advanced Operations, building upon the basics, this segment delves into more sophisticated programming constructs. It includes the manipulation of data structures, control flow, and file operations with nuanced requirements. Demonstrating the model's adeptness at handling complexity.

5 Evaluation

We conducted an evaluation with 10 native Arabic speakers, who were software engineers and computer scientists from Prince Sultan University. They were asked to assess the language examples shown in the Illustration section (Sect. 5) in three aspects: readability, usefulness, and semantics (evaluating the correctness of Arabic grammar with respect to the APL texts). The results were highly positive: 90% of participants rated the examples as four or five (on a scale of 1–5) in terms of readability and semantic correctness. Additionally, 80% rated the examples similarly on the same

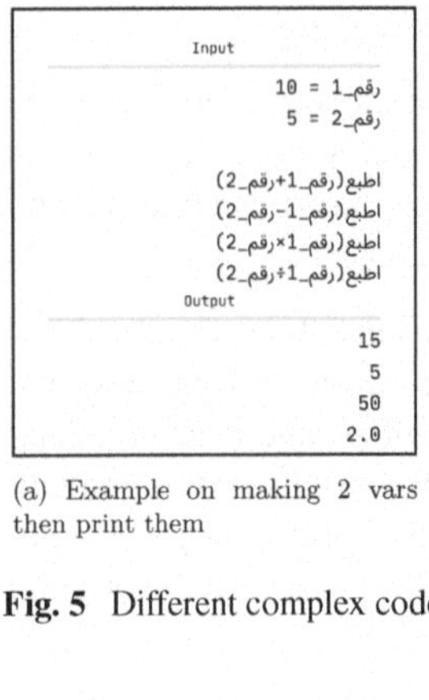

(a) Example on making 2 vars then print them

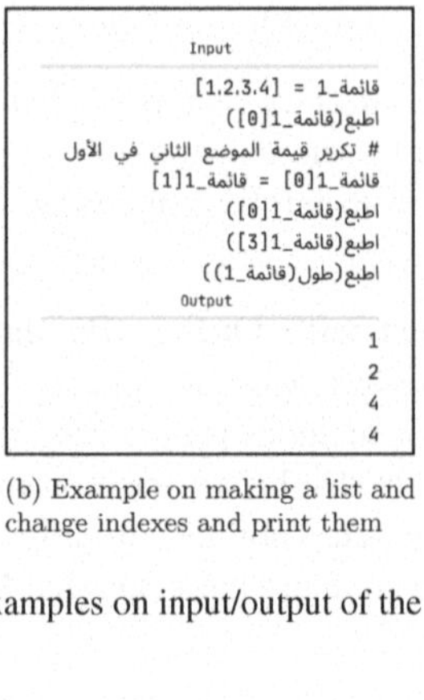

(b) Example on making a list and change indexes and print them

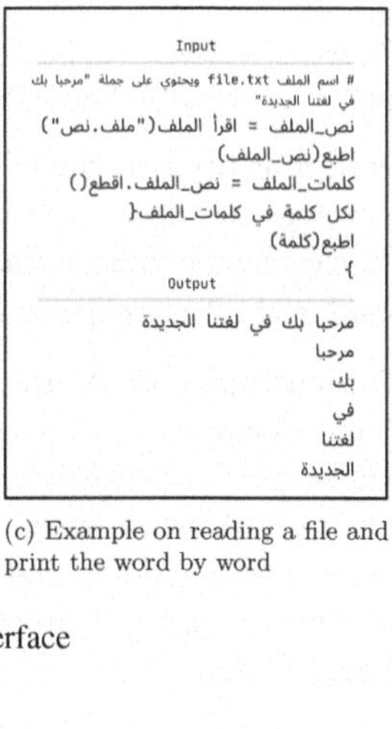

(c) Example on reading a file and print the word by word

Fig. 5 Different complex code examples on input/output of the interface

```
Input
نص١ = "مرحبا بك في لغتنا الجديدة"
اطبع(نص١)

Output
مرحبا بك في لغتنا الجديدة
```

(a) Example on print sentence

```
Input
رقم_١ = ٢
رقم_٢ = ٥
اطبع(رقم_١ ^ رقم_٢)

Output
32
```

(b) Example on simple arithmetic

```
Input
نص_الملف = اقرأ الملف("ملف.نص")
اطبع(نص_الملف)

Output
مرحبا بك في لغتنا الجديدة
```

(c) Example on reading a file

```
Input
دالة تربيع(رقم){ارجع رقم^٢ }
النتيجة = تربيع(٤)
اطبع(النتيجة)

Output
16
```

(d) Example on making a function

Fig. 6 Different code examples on input/output of the interface

scale. For the final evaluation, participants were asked to rate the usefulness of the language, with more than 70% indicating that the language is very useful for Arabic-speaking students learning programming structures and fundamentals. Main details about the survey: it was done using 'google form' where the evaluators read two sections one for the simple examples (mentioned in the paper) and then evaluate the readability and semantics, same for the complex examples and in the end the give a general value for the usefulness of the language (all from one to five).

We encountered several challenges during the development of the compiler. One major challenge was prompt engineering, particularly dealing with Arabic text in two contexts. Firstly, programming language keywords needed to be replaced with their English counterparts to avoid syntax errors in Python code. However, Arabic text within strings had to remain unchanged to ensure that printed or modified Arabic characters matched the original input. Secondly, the cost of using the GPT-4 API posed a significant challenge. While this was manageable during the development phase, the deployment phase could become prohibitively expensive.

To address these challenges in the future, we plan to explore fine-tuning open-source models. This approach could reduce the costs associated with using paid APIs and improve task-specific performance, minimizing the risk of receiving irrelevant responses from a generalized model.

6 Conclusion

In this paper, we introduced APL, an Arabic programming language that translates code into Python using large language models and prompt engineering techniques. We developed an API and a website interface for its use. An evaluation with 10 native Arabic speakers from Prince Sultan University showed that 90% rated the language highly for readability and semantic correctness, and over 70% found it very useful for teaching programming. Key challenges included prompt engineering and the high cost of using the GPT-4 API. Future work will focus on fine-tuning open-source models to reduce costs and improve task-specific performance.

Acknowledgements The author thank Prince Sultan University for their support.

References

1. S. Kim, S. Moon, R. Tabrizi, N. Lee, M.W. Mahoney, K. Keutzer, A. Gholami, An LLM compiler for parallel function calling, in *Proceedings of the Conference Name* (Springer, 2023)
2. Author, LLMs are compilers. Online article. Retrieved from https://vivekhaldar.com/articles/llms-are-compilers/ (2023)
3. V. Liventsev, A. Grishina, A. Härmä, L. Moonen, Fully autonomous programming with large language models, in *Proceedings of the Conference Name* (Springer, 2023)
4. LMQL is a programming language for large language models (LLMs) based on a superset of Python. https://github.com/eth-sri/lmql
5. R. W. Sebesta's, "Concepts of Programming Languages"
6. Computer Languages History diagram by Levenez. https://www.levenez.com/lang/
7. Wiki Article about programming languages that are not English
8. GitHub repository that summarize info about non-English programming languages. https://github.com/flightstar/Artificial-Intelligence-in-Non-English-Based-Programming-Language
9. Arabic Sakhr Basic (1987, MSX, Al Alamiah)
10. ARLOGO project homepage. http://arlogo.sourceforge.net/. Retrieved 15 Feb 2017
11. M. Al-A'Ali, M. Hamid, Design of an Arabic programming language (ARABLAN). J. Comput. Lang. **21**(3–4), 191–201 (1995)
12. E.M. Arif, *Design of an Arabic object-oriented programming language and a help system for pedagogical purposes*. Doctoral Dissertation, Illinois Institute of Technology Chicago, IL, USA (1995)
13. AMORIA homepage. http://ammoria.sourceforge.net/. Retrieved 14 Dec 2018
14. Ebda3 homepage. http://ebda3lang.blogspot.com/. Retrieved 14 Dec 2018
15. Jeem homepage. http://www.jeemlang.com/. Retrieved 14 Dec 2018
16. Y. Bassil, A. Barbar, Loughaty/MyProLang—my programming language—a template-driven automatic natural programming language, in *Proceedings of the World Congress on Engineering and Computer Science, WCECS, San Francisco, USA* (2008)
17. Qlb GitHub, https://github.com/nasser/. Retrieved 14 Dec 2018
18. Kalimat GitHub, https://github.com/lordadamson/kalimat. Retrieved 15 Feb 2017
19. Phoenix—The Arabic Object-Oriented Programming Language Youssef Bassil, LACSC—Lebanese Association for Computational Sciences, Registered under No. 957, Beirut, Lebanon (2011)
20. L. Kedward et al., The State of Fortran. Comput. Sci. Eng. https://doi.org/10.1109/MCSE.2022.3159862

21. H. Liang et al., MLSA: a static bugs analysis tool based on LLVM IR. Int. J. Netw. Distrib. Comput. **4**(3), 137–144 (2016)
22. A. Neri et al., Compiling quanta morphisms for the IBM Q Experience. arXiv:2010.10510v1 [quant-ph] 21 Oct 2020
23. Large Language Models: A Survey. arXiv:2402.06196v1 [cs.CL] 9 Feb 2024
24. PseuToPy: Towards a Non-English Natural Programming Language Patrick Wang ICER 2021, in *Proceedings of the 17th ACM Conference on International Computing Education Research* (2021), pp. 429-430. https://doi.org/10.1145/3446871.3469787
25. H. Tan, Q. Luo, J. Li, Y. Zhang, Open-access LLMs, Decompile-Eval dataset improve decompilation, enhancing program semantics. Southern University of Science and Technology The Hong Kong Polytechnic University
26. L. Gao, A. Madaan, S. Zhou, U. Alon, P. Liu, Y. Yang. J. Callan, G. Neubig, PAL: program-aided language models (2023)
27. Z. Hong, Z. Yuan, Q. Zhang, H. Chen, J. Dong, F. Huang, X. Huang, Next-generation database interfaces: a survey of LLM-based text-to-SQL
28. P. Sahoo, A.K. Singh, S. Saha, V. Jain, S. Mondal, A. Chadha, A systematic survey of prompt engineering in large language models: techniques and applications (2024)

Serry Taiseer Sibaee Serry Taiseer Sibaee (b. 2000, Homs, Syria) is a distinguished research engineer at the Robotics and Internet of Things Lab, Prince Sultan University, where he completed his Bachelor's degree in Software Engineering in 2023. Since then, he has specialized in Arabic natural language processing (NLP), contributing to the field through impactful research and innovative solutions. Over the past two years, Sari has worked on various key projects, including training Arabic language models, designing reverse dictionaries, implementing classification techniques, and addressing the complex phenomenon of language model hallucinations. His collaborative efforts with researchers have led to the development of practical tools and methodologies that enhance the usability and effectiveness of Arabic NLP systems. Serry's dedication to advancing Arabic NLP has been recognized through multiple co-authored research papers and his participation in prestigious research conferences, where he earned top awards in 2023 and 2024. His work not only showcases his technical expertise but also highlights his commitment to solving challenges unique to the Arabic language and culture. With a passion for innovation and excellence, Sari continues to push the boundaries of Arabic NLP, cementing his reputation as a key contributor to the field's growth and future development.

Omer Nacar Omer Nacar (b. December 7, 1996, Syria) is an accomplished engineer and researcher specializing in Artificial Intelligence (AI), Data Science, Signal Processing, and Deep Learning. He earned a Bachelor's degree in Communication and Network Engineering and a Master's degree in Electrical and Electronic Engineering, with a focus on the application of deep learning to signal processing. This academic foundation underscores his expertise in bridging theoretical concepts with real-world engineering applications. Throughout his career, Mr. Nacar has demonstrated a profound commitment to innovation and excellence. His contributions to the field include multiple Certificates of Excellence in Natural Language Processing, Large Language Models, Machine Learning, and Artificial Intelligence. He has also made impactful contributions to academic literature, notably authoring a pivotal chapter on "Artificial Intelligence Creativity" in the book Artificial Intelligence Ethics, addressing both technological advancements and their ethical implications. Currently, Mr. Nacar is recognized for his exceptional ability to oversee the end-to-end development of projects, from data collection and analysis to the dynamic presentation of results. His professional journey reflects a seamless integration of technical expertise and strategic insight, ensuring his work has meaningful impacts on both academic research and practical applications. Driven by a passion for advancing technology, Mr. Nacar's contributions continue to push the boundaries of AI and signal processing, combining innovation with a strong ethical framework to shape the future of intelligent systems.

Anis Koubaa Anis Koubaa is the Director of the Research and Initiatives Center and the leader of the Robotics and Internet-of-Things Lab at Prince Sultan University. He is a Full Professor in Computer Science and has contributed to numerous research and development (R&D) projects in the fields of data science and deep learning, focusing on applications such as face recognition, vehicle identification, and object classification and detection. Prof. Koubaa is an ACM Distinguished Speaker and a Senior Fellow of the Higher Education Academy (HEA) of the UK. He has delivered various training programs in areas such as data science, Python programming, Tableau for data visualization, deep learning, and other advanced technologies. His academic and professional journey reflects a strong commitment to innovation, education, and the application of cutting-edge technologies in solving real-world problems. Prof. Koubaa's work continues to inspire advancements in robotics, IoT, and AI.

Lahouri Ghouthi Dr. Lahouari Ghouti holds a PhD degree in Computer Science from Queen's University of Belfast, UK. Dr. Ghouti has been actively involved in academic research in the general field of Artificial Intelligence. He has more than 100 publications in refereed journals and conference proceedings and holds 28 US patents. His research work often involves the development of algorithms and techniques that enhance data processing and decision-making systems. Ghouti is also known for his involvement in academic publications and conferences, where he shares his insights and findings with the broader scientific community

The Silicon Oasis: IP-Enabled Generative AI Innovations for Saudi Arabia's Environmental Vision

Selma Abdelrahman Hussein and Emna Chikhaoui

Abstract Saudi Arabia's Vision 2030 sets forth an ambitious reform agenda that aims to transition from an oil-centric economy to a more diversified, knowledge-based economic structure. This transformative plan places a strong emphasis on the strategic integration of intellectual property (IP) with generative artificial intelligence (AI), which is envisioned to spur innovation across key sectors including energy, water conservation, and education. This paper explores the role of IP as a facilitator of AI-driven technological advancements that are in alignment with the kingdom's goals for sustainability and renewable energy. It examines the application of AI in various fields such as healthcare, environmental monitoring, and agricultural efficiency, highlighting the significant potential of AI to help achieve the objectives outlined in Vision 2030. Additionally, the paper discusses the challenges associated with integrating AI technologies within the existing infrastructural frameworks, emphasizing the need for evolving legal structures to keep pace with rapid technological advancements. A qualitative research approach, including interviews and observational studies, provides deep insights into how generative AI could influence Saudi Arabia's efforts towards economic diversification and environmental sustainability. The conclusion underscores the importance of continued investment in AI innovation, supported by a robust IP regime, to promote sustainable development that aligns with the nation's strategic vision.

Keywords Saudi Arabia · Vision 2030 · Intellectual property · Generative AI · Environmental sustainability

S. Abdelrahman Hussein
Research and Initiatives Center, Intellectual Property Office, Prince Sultan University, Riyadh, Riyadh, Saudi Arabia

S. Abdelrahman Hussein (✉) · E. Chikhaoui
Compliance Office, Prince Sultan University, Riyadh, Riyadh, Saudi Arabia
e-mail: sabdulrahman@psu.edu.sa

E. Chikhaoui
e-mail: echikhaoui@psu.edu.sa

A. Koubaa et al. (eds.), *Generative AI and Large Language Models: Opportunities, Challenges, and Applications*, Studies in Computational Intelligence 1214,
https://doi.org/10.1007/978-3-031-90573-5_9

Terminology, Units and Abbreviations

Terminology

- **Generative AI (GenAI)**: A subset of artificial intelligence that creates new content such as images, text, and music by learning patterns from existing data.
- **Predictive AI**: AI systems that analyze historical data to forecast future outcomes, often used in decision-making.
- **Intellectual Property (IP)**: Legal rights that protect creations of the mind, such as inventions, designs, and artistic works.
- **Sustainability**: Practices and strategies that meet the needs of the present without compromising the ability of future generations to meet their needs.
- **Vision 2030**: Saudi Arabia's strategic framework aimed at reducing dependency on oil, diversifying the economy, and developing public service sectors.

Units

- **Meter (m)**: Unit of length.
- **Kilogram (kg)**: Unit of mass.
- **Second (s)**: Unit of time.
- **Joule (J)**: Unit of energy.
- **Celsius (°C)**: Unit of temperature.

Abbreviations

- **AI**: Artificial Intelligence.
- **IP**: Intellectual Property.
- **GenAI**: Generative Artificial Intelligence.
- **SAIP**: Saudi Authority for Intellectual Property.
- **ICSDI**: International Conference on Sustainability Developments and Innovations.
- **LEAP**: Large-scale Event for Advanced Progress, a technology-focused event.
- **SI Units**: International System of Units.

Numerals

- **Decimal Points**: Use a period (.) for decimals (e.g., 3.14, 27.5).
- **Thousands Separator**: Use a comma (,) to separate thousands (e.g., 1,000; 25,000).

1 Introduction

Saudi Arabia's Vision 2030 encapsulates a transformative economic and environmental strategy aimed at diversifying the nation's economy beyond its traditional oil base [1]. This visionary framework emphasizes sustainable economic growth through incentivizing investments in non-oil industries, revolutionizing the education system, and fostering a culture of innovation and scientific research [2]. Central to achieving these objectives is the development of a dynamic, diverse, and sustainable economy characterized by enhanced economic stability, increased financial accommodation, and improved transparency. By bolstering the education sector-through the recruitment, training, and development of educators, the enhancement of curricula, and the promotion of creativity and innovation-Saudi Arabia seeks to create employment opportunities for its youth and advance its standing on the Global Competitiveness Index [3, 4]. Additionally, Vision 2030 aims to augment the share of non-oil exports within the non-oil GDP and draw foreign investment to spur competitiveness and technological advancement [5]. The interplay between intellectual property (IP) and artificial intelligence (AI) is pivotal in this context, particularly regarding AI's role in generating innovative works and their implications for IP rights and inventorship [6]. The debate on AI's legal capacity to invent and how AI-generated works should be recognized under IP law is a significant aspect of Saudi Arabia's technological and environmental vision. The document suggests that AI's capabilities to autonomously perform tasks, traditionally done by humans, necessitate a reevaluation of existing IP frameworks to accommodate the novel outputs produced by AI systems [7]. By examining different jurisdictions' approaches to AI in IP law and the economic and legal implications of recognizing AI as an inventor, Saudi Arabia's environmental vision is closely tied to navigating the challenges and opportunities presented by AI in fostering a sustainable and technologically advanced economy [8]. The role of IP-enabled generative AI in achieving Saudi Arabia's Vision 2030 is pivotal, as it intersects crucially with the nation's ambitions for technological innovation and environmental sustainability [9]. IP-enabled AI technologies are anticipated to drive significant advancements across various sectors, including energy, water conservation, and pollution control, by facilitating the development of groundbreaking solutions [10]. These AI systems, equipped with the capability to learn, adapt, and innovate, can optimize resource management and enhance the efficiency of renewable energy production, contributing to the reduction of the country's carbon footprint. Moreover, the legal and regulatory frameworks surrounding IP rights play a critical role in ensuring that innovations are protected, thereby encouraging further research and development in AI. This synergy between IP law and AI innovation is essential for fostering a conducive environment for technological advancements that align with Saudi Arabia's environmental objectives [11, 12]. This research aims to delve into the transformative potential of IP-enabled generative AI technologies in catalyzing Saudi Arabia's Vision 2030. Specifically, it focuses on leveraging AI to not only bolster economic diversification and technological innovation but also to significantly enhance environmental sustainability. By examining the intricate interplay between intellectual property (IP) and artificial intelligence (AI), this study seeks to unravel the legal, economic, and innovative landscapes shaped by AI-generated works. Central to our analysis is how AI can revolutionize sectors critical to Saudi Arabia's future-energy, water conservation, pollution control, health, food security, and higher education. In energy, AI's predictive analytics and smart grid technologies promise to optimize production and consumption patterns [9, 13, 14]. Water conservation efforts can be enhanced through AI-driven irrigation systems, reducing wastage and ensuring sustainable usage [15]. Pollution control can benefit from AI's capabilities in monitoring and managing environmental pollutants, offering real-time data to mitigate adverse effects [16]. Expanding the scope, AI's role in health encompasses improving diagnostic accuracy, personalized treatment plans, and managing health crises. In food security, AI technologies can predict crop yields, monitor soil health, and optimize agricultural practices, ensuring a stable food supply [9, 17, 18]. Higher education stands to gain through personalized learning experiences, operational efficiencies, and research and development advancements, preparing a skilled workforce for Saudi Arabia's evolving economic landscape [19]. By exploring these sectors in depth, the research will outline a roadmap for integrating AI technologies in ways that align with Saudi Arabia's environmental goals and broader Vision 2030 objectives, setting a precedent for innovative, sustainable development powered by IP-enabled AI [17]. Given these developments, this paper seeks to address the following research questions:

1. How can AI-driven innovations be effectively integrated into Saudi Arabia's energy, water, and healthcare sectors to achieve Vision 2030 goals?
2. What are the gaps in Saudi Arabia's current intellectual property (IP) frameworks for AI-generated innovations, and how can they be addressed?
3. What global best practices in AI and IP regulation can inform Saudi Arabia's policy development, considering the unique cultural and economic context of the Kingdom?
4. What are the legal and ethical challenges associated with AI-driven innovations in Saudi Arabia, and how can a tailored IP framework mitigate these issues?

These research questions frame the scope of this paper and guide the analysis of AI applications, regulatory gaps, and the potential for policy adaptation in Saudi Arabia (Fig. 1).

2 Literature Review

2.1 Introduction

The intersection of generative Artificial Intelligence (AI) and environmental sustainability represents a frontier of research with profound implications for global ecological management strategies [20]. This literature review delves into the current

Fig. 1 Transformative impact of IP-enabled generative AI technologies on Saudi Arabia's economic

state of generative AI applications in environmental sustainability, identifies gaps within the context of Saudi Arabia, and examines theoretical frameworks related to Intellectual Property (IP)-enabled AI technologies [21, 22].

2.2 How Generative AI Enhances Environmental Sustainability in Saudi Arabia

Generative AI can be used to improve environmental sustainability in Saudi Arabia by predicting and forecasting key factors related to sustainability [23]. For example, AI models such as feed-forward neural networks (FFNN), adaptive network-based fuzzy inference systems (ANFIS), and long short-term memory (LSTM) can be applied to forecast carbon dioxide (CO_2) emissions in the country [24]. Additionally, machine learning models like artificial neural networks (ANN) and support vector regression (SVR) can estimate future energy demand in the road transportation sector, which is directly linked to greenhouse gas emissions [25]. Furthermore, AI techniques such as exponential smoothing, transformers, temporal convolutional networks (TCN), and neural basis expansion analysis can be implemented to analyze and forecast CO_2 emissions, providing insights for mitigating climate change [21]. Finally, AI models like artificial neural networks (ANNs) can accurately predict crop yields based on environmental data, helping optimize agricultural production and land use planning [22].

2.2.1 Generative AI Versus Predictive AI: A Comparative Overview

Generative AI and Predictive AI serve distinct roles in various applications, each with unique strengths and limitations, and both are crucial for achieving Saudi Arabia's Vision 2030 objectives. **Generative AI** focuses on the creation of new content such as images, text, and music, leveraging advanced machine learning techniques. This technology is instrumental in industries requiring innovation and creativity, such as *healthcare*, where it is applied for medical image generation and personalized content creation to enhance diagnostic accuracy [26, 27]. However, challenges include high computational demands and ethical concerns related to copyright and data privacy [28].

On the other hand, **Predictive AI** excels at analyzing historical data to forecast outcomes, making it invaluable in decision-making processes. For example, in healthcare, predictive AI enables tailored patient care and improves disease prediction [29]. Despite its effectiveness, Predictive AI is limited by the quality and quantity of data, which can affect the accuracy of its forecasts [28].

While **Generative AI** drives innovation through content creation, **Predictive AI** supports data-driven decision-making. Together, these technologies offer complementary capabilities that are essential for advancing sectors like healthcare, energy, and agriculture, thereby contributing to Saudi Arabia's sustainability goals [30, 31].

2.2.2 Benefits of IP-Enabled Generative AI in Environmental Challenges

The use of IP-enabled generative AI can bring several benefits in addressing environmental challenges in Saudi Arabia [19]. Firstly, it can enhance energy efficiency and waste reduction, contributing to sustainability goals in the country [32]. Secondly, AI solutions can improve the efficiency and effectiveness of the healthcare system, reducing costs and enhancing the quality of care. Additionally, AI-based approaches, such as the use of drones and intelligent recognition systems, can help protect critical sectors like the oil and gas industry from adversarial drone and missile attacks [33]. Lastly, the integration of AI in higher education institutions can revolutionize teaching and learning methodologies, addressing educational challenges and accelerating progress towards the Saudi 2030 objectives [9]. Sample body text. Sample body text. Sample body text. Sample body text. Sample body text. Sample body text. Sample body text. Sample body text.

2.3 AI Applications in Energy, Water Conservation, and Healthcare

Artificial Intelligence (AI) plays a transformative role in advancing Saudi Arabia's Vision 2030, particularly within the sectors of energy, water conservation, and healthcare. These key areas are instrumental in achieving the country's goals of economic diversification and environmental sustainability.

2.3.1 AI in Energy Conservation

Artificial intelligence (AI) plays a transformative role in optimizing energy use and supporting the transition to renewable energy sources. AI applications, such as IoT-driven energy management systems and predictive modeling tools for CO_2 emissions, are helping Saudi Arabia move closer to achieving its Vision 2030 goals. These technologies enable real-time monitoring, reducing energy waste and enhancing system efficiency.

Challenges in Energy

Despite these advancements, the energy sector faces significant challenges in fully implementing AI-driven solutions:

- **Infrastructure and Cost**: Transitioning to AI-driven energy systems requires substantial infrastructure investment, especially in rural or underdeveloped areas. The cost of deploying IoT devices and creating predictive models for energy management presents a major barrier [34].
- **Data Integration and Management**: AI systems depend heavily on large volumes of real-time data from multiple sources. Poor data integration between energy systems, combined with the challenge of managing these datasets, can limit the effectiveness of AI-driven energy solutions.

2.3.2 AI in Water Conservation

In the context of Saudi Arabia's water-scarce environment, AI applications such as smart water grids and intelligent irrigation systems offer innovative solutions to optimize water usage. These technologies help reduce water wastage, improve resource management, and contribute to the country's sustainability goals.

Challenges in Water Conservation

While AI-driven technologies offer significant potential, the water sector faces persistent challenges:

- **Water Resource Scarcity**: AI technologies can optimize water usage, but they cannot resolve the underlying issue of water resource scarcity in Saudi Arabia. The nation's arid climate and limited freshwater resources remain substantial challenges [35].
- **Technological Accessibility**: Implementing advanced AI systems in rural and agricultural areas often faces hurdles due to limited technological infrastructure, such as lack of access to high-speed internet and advanced sensors, restricting the widespread use of these innovations.

2.3.3 AI in Healthcare

AI-driven innovations such as MedCore are revolutionizing healthcare in Saudi Arabia by improving diagnostic accuracy and reducing unnecessary medical tests. AI systems can support healthcare professionals in making more informed decisions, improving patient outcomes and streamlining workflows.

Challenges in Healthcare

Despite the benefits, the healthcare sector faces challenges that may slow AI adoption:

- **Public and Professional Acceptance**: While AI systems like MedCore enhance diagnostic accuracy, there remains skepticism among both healthcare professionals and patients about relying on AI-driven technologies. Many still prefer traditional human interactions, which could hinder the widespread adoption of AI in healthcare settings [36].

2.3.4 AI in Education

AI is increasingly being used in education to support personalized learning and improve student outcomes. By analyzing student data, AI systems can create tailored learning experiences, helping to bridge learning gaps and enhance academic performance.

Challenges in Education

Despite its promise, the education sector faces several challenges in implementing AI:

- **Digital Divide**: The unequal distribution of digital infrastructure across different regions of Saudi Arabia creates a barrier to implementing AI in education. Rural and underserved areas often lack the necessary technological infrastructure, such as internet access and digital devices, limiting the reach of AI-powered educational tools [?].
- **Teacher Training**: Educators may lack the skills or resources to effectively integrate AI into their teaching practices. This gap between the availability of AI technologies and teachers' ability to use them effectively in the classroom poses a significant challenge to AI adoption in education.

By integrating AI into key sectors such as energy, water, healthcare, and education, Saudi Arabia stands to make significant strides toward achieving the Vision 2030 goals. However, each sector faces distinct challenges that must be addressed for these technologies to reach their full potential. Through careful investment in infrastructure, improvements in data management, and greater emphasis on public and professional acceptance, Saudi Arabia can overcome these barriers and realize the transformative potential of AI-driven innovations (Fig. 2).

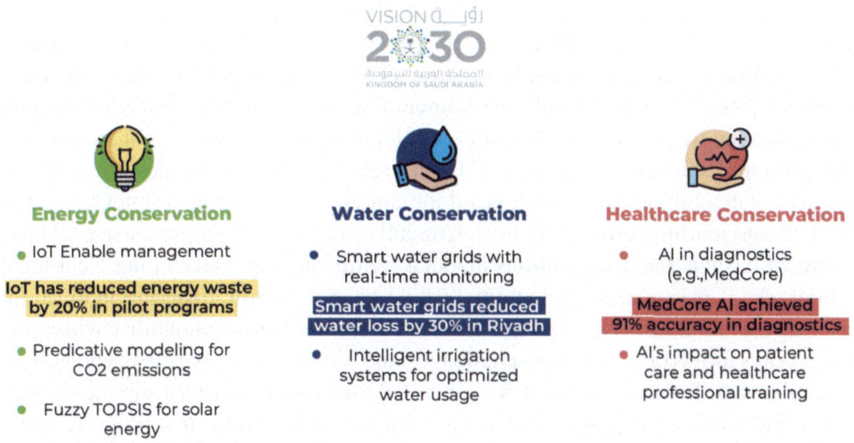

Fig. 2 AI applications across different sectors

2.4 Supporting Saudi Arabia's Environmental Vision with Generative AI

Generative AI can support Saudi Arabia's environmental vision by leveraging technology to advance sustainability goals and enhance energy efficiency [5]. The adoption of technologies such as the Internet of Things (IoT), blockchain, and artificial intelligence has played a significant role in promoting green initiatives and reducing waste in the country [37]. Saudi Arabia's Vision 2030, which prioritizes artificial intelligence (AI), aims to make the country a world leader in technology and development [9]. AI research outputs from Saudi organizations have shown a significant growth rate, driven by the government's funding and support for AI research. Additionally, machine learning techniques, such as neural networks, can be used to forecast CO_2 emissions and develop optimal models for reducing carbon emissions [38]. Furthermore, generative AI, specifically artificial neural networks (ANNs), can be utilized to predict wind speed, facilitating the integration of wind power into Saudi Arabia's electrical grid and supporting its renewable energy targets [39].

2.5 Gaps in the Literature: The Saudi Arabian Context

To address the literature gaps in IP-enabled generative AI innovations for Saudi Arabia's environmental vision, it's crucial to delve into specifics. The absence of AI applications in the Saudi judicial system suggests a need for research to evaluate AI's potential impact, particularly in streamlining processes and enhancing legal outcomes [40]. The shortage of environmental datasets hampers machine learning applications in healthcare and beyond, pointing towards a necessity for robust data collection and sharing mechanisms [41]. In higher education, the slow adoption of AI necessitates initiatives for skill development among educators to integrate AI in teaching and learning effectively [42]. Crucially, the lack of comprehensive IP laws to govern AI-generated innovations marks a significant gap. Developing clear legal frameworks is essential to protect intellectual property while fostering innovation, ensuring that AI advancements contribute to environmental sustainability within the context of Saudi Arabia's unique challenges and Vision 2030 goals [9]. The lack of comprehensive IP laws to govern AI-generated innovations marks a significant gap. Developing clear legal frameworks is essential to protect intellectual property while fostering innovation, ensuring that AI advancements contribute to environmental sustainability within the context of Saudi Arabia's unique challenges and Vision 2030 goals.

In addressing this gap, a comparative analysis with other countries offers valuable insights. Many nations, such as **Germany**, have adapted their regulations to accommodate AI-driven innovations, particularly in sectors like health technology. Germany's approach ensures that AI-generated technologies are protected, encouraging further innovation [43]. In contrast, other countries, such as the **United

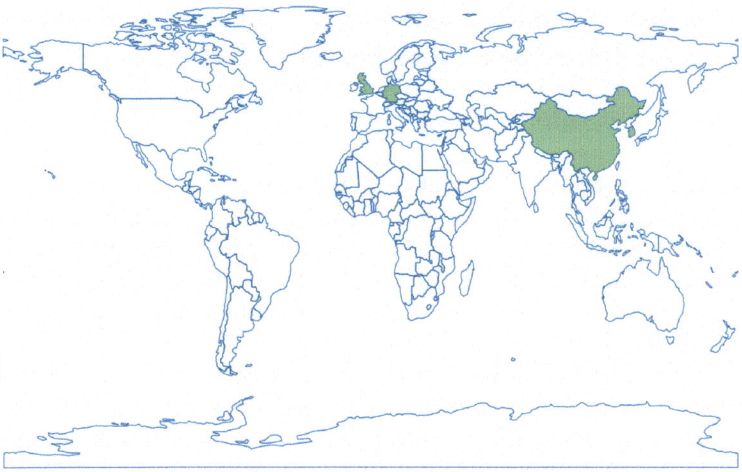

Fig. 3 Countries with significant AI and IP regulatory frameworks

States**, have initiated discussions on the complexities of AI-generated intellectual property, particularly around the issue of inventorship [44]. These global examples provide critical guidance for Saudi Arabia as it seeks to strengthen its IP framework.

However, adopting a **one-size-fits-all** approach to IP policies has proven problematic. The varying cultural and economic contexts of different countries require tailored solutions. For Saudi Arabia, integrating global best practices while considering local context will ensure that its IP framework fosters both domestic AI innovation and international collaboration [45]. Examining international strategies will help policymakers develop an adaptive, comprehensive framework that supports the Kingdom's Vision 2030 goals. This detailed exploration underscores the multi-faceted nature of the research gaps and the imperative for targeted studies to bridge these divides (Fig. 3).

2.6 Theoretical Frameworks Relevant to IP-Enabled AI Technologies

The review outlines theoretical frameworks that connect IP law with AI innovation, highlighting the role of IP rights in promoting AI technologies for environmental sustainability. It suggests exploring how Saudi Arabia's IP and policy frameworks could influence AI-driven environmental solutions [46]. Investigating the adaptation of Saudi Arabia's IP policies to the challenges and opportunities of AI technology could provide insights into sustainable development and economic diversification [8]. The conclusion calls for targeted research to develop AI solutions tailored to Saudi Arabia's unique environmental challenges and to formulate IP policies that encourage innovation in AI for sustainability.

2.7 *Comparative Analysis of AI and IP Regulations: Lessons from the United States and European Union*

2.7.1 The United States Approach to AI and IP

The United States takes a decentralized and flexible approach to AI regulation, emphasizing rapid innovation and sector-specific oversight. The U.S. framework largely allows industry autonomy and voluntary compliance, enabling AI technologies to flourish without excessive restrictions. This light-touch approach prioritizes innovation but can leave gaps in regulation, particularly regarding ethical oversight and accountability [47].

In the realm of intellectual property (IP), the U.S. provides robust protections for innovators, ensuring that patents and copyrights are strongly enforced. However, the absence of comprehensive legislation on AI-generated inventions means that only humans can be listed as inventors, limiting the scope for fully autonomous AI systems to be recognized as creators [48]. While this promotes innovation, it also leaves a regulatory gap for AI-driven inventions.

2.7.2 The European Union Approach to AI and IP

the European Union (EU) has established a more centralized and precautionary regulatory approach, exemplified by the *AI Act*, which emphasizes ethical standards, transparency, and liability frameworks. The EU focuses on regulating high-risk AI systems in sectors like healthcare and energy, where the potential for harm is significant. The EU's proposed *AI Liability Directive* aims to ensure that developers of AI technologies are held accountable for any damages caused by their innovations [49].

In terms of IP, the EU adopts a balanced approach, considering both innovation and competition. Unlike the U.S., the EU promotes a system where IP rights are weighed against broader antitrust concerns, ensuring that market monopolies do not stifle cumulative innovation [50]. Additionally, the EU's framework incorporates strict data protection standards, which align with its broader regulatory focus on privacy and security [51].

2.7.3 Lessons for Saudi Arabia

Saudi Arabia, aiming to foster AI-driven innovation under its Vision 2030, can draw several key lessons from the approaches taken by the U.S. and the EU:

- **Human-Centric Inventorship**: Both the U.S. and EU frameworks currently require human involvement in the patenting process for AI-generated works. Saudi Arabia could adopt similar regulations to align with global standards, while also preparing for future scenarios where AI may play a more significant role in IP creation.

- **Ethical and Liability Frameworks**: The EU's focus on ethical considerations, particularly through its *AI Act* and *Liability Directive*, offers a comprehensive model for Saudi Arabia. Developing similar guidelines that prioritize consumer protection and ethical AI usage in high-risk sectors could help build public trust in AI technologies.
- **Regulatory Flexibility**: The U.S.'s sector-specific and voluntary frameworks provide an example of how a flexible regulatory approach can spur rapid innovation. Saudi Arabia could adopt a hybrid model that combines the EU's stringent oversight in high-risk sectors with the U.S.'s flexible, innovation-friendly policies, allowing for regulatory experimentation through sandboxes.
- **Data Protection and Privacy**: With the EU's stringent data protection rules in mind, Saudi Arabia should prioritize safeguarding personal data and cybersecurity as AI technologies are deployed across sectors. This would align with global best practices and bolster Saudi Arabia's digital economy.

2.7.4 Comparative Table of AI and IP Policies

See Table 1.

2.7.5 Policy Recommendations for Saudi Arabia

- **Adopt a Hybrid Model**: Saudi Arabia could benefit from a hybrid regulatory approach, combining the *EU's focus on ethics and liability* with the *U.S.'s flexibility*. This would allow for innovation without compromising on consumer protection and ethical standards.

Table 1 Comparison of AI and IP policies in the U.S., EU, and Saudi Arabia

Aspect	United States	European Union	Saudi Arabia
Inventorship	Human-only inventorship for IP	Human-only inventorship for IP	Currently aligned with human-only inventorship
Regulation of AI Systems	Decentralized, sector-specific oversight	AI Act (2021), focusing on high-risk AI systems	No comprehensive AI law, but developing regulatory frameworks
Ethical Guidelines	Limited focus on ethics, more on innovation	Strong focus on ethics, liability, and consumer protection	Ethics in AI still in development in Saudi Arabia
Flexibility	Flexible, voluntary frameworks; rapid innovation	Precautionary, detailed regulation for high-risk sectors	Learning from both models to develop hybrid approach

- **Develop a Clear Ethical Framework**: Saudi Arabia should take cues from the EU in developing robust ethical guidelines, especially for AI applications in sensitive areas like healthcare and energy. Establishing clear liability rules will help build public trust in AI technologies.
- **Focus on Data Protection**: As the Kingdom continues to develop its digital economy, aligning its data protection policies with the EU's *General Data Protection Regulation (GDPR)* would ensure that personal data is safeguarded, which is crucial for long-term digital transformation.

By analyzing the regulatory frameworks of the United States and the European Union, Saudi Arabia can adapt its AI and IP regulations to support innovation while safeguarding ethical standards and data protection. A hybrid approach, emphasizing flexibility in low-risk sectors and stringent regulations in high-risk areas, would allow Saudi Arabia to navigate the complexities of AI regulation while fostering economic growth under Vision 2030.

3 Methodology

3.1 Overview of the Research Design and Approach

This study adopts a qualitative research design to explore the impact of *IP-enabled generative AI technologies* on *saudi arabia's Vision 2030*. The qualitative approach was selected to capture in-depth insights into how these technologies influence sectors such as *energy*, *environmental management*, *law*, and *policy-making*. Data collection involved both *in-depth interviews* and *focus groups*, allowing for a rich exploration of the subject matter from diverse expert perspectives (Fig. 4).

3.2 Participant Selection

The participants were selected from a diverse range of sectors to ensure comprehensive insights. These sectors included:

- **Technology**: AI engineers and software developers.
- **Law and Intellectual Property**: Legal experts specializing in intellectual property rights and AI.
- **Energy**: Professionals in renewable energy and smart grids.
- **Environmental Management**: Experts in water management, biodiversity conservation, and sustainable resource use.

The participant pool, comprising 20 experts, was selected to capture the multi-sectoral impact of generative AI technologies on *Vision 2030* (Fig. 5).

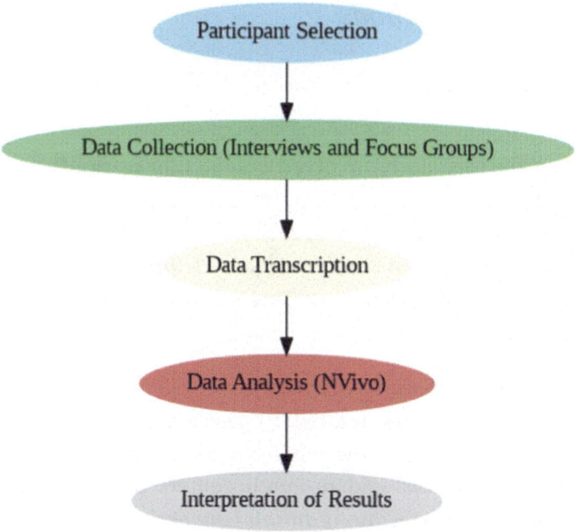

Fig. 4 Research process overview, including participant selection, data collection, transcription, analysis, and interpretation

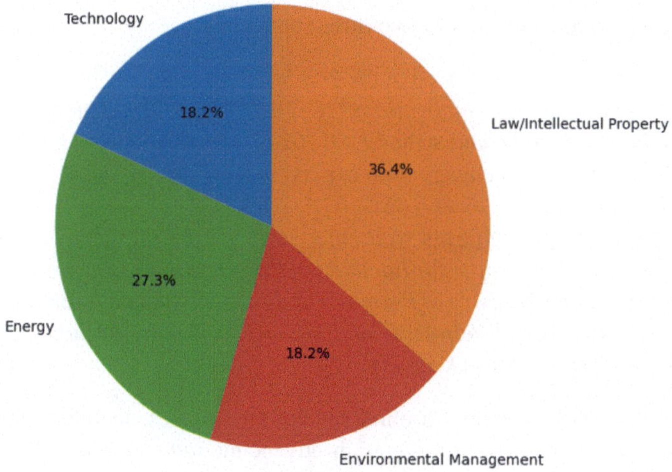

Fig. 5 Distribution of participants across different sectors

3.3 Data Collection Methods

3.4 In-Depth Interviews

The data collection for this study involved conducting **in-depth interviews** with a total of **55 participants**, selected for their expertise and involvement in sectors relevant to *AI technologies*, *intellectual property (IP)*, and *environmental management*, particularly within the context of *saudi arabia's Vision 2030.*

The participants were categorized into three primary groups based on their affiliations:

- **20 participants** from the *Saudi Authority for Intellectual Property (SAIP).*
- **20 participants** from sectors related to *energy*, *environmental management*, and *research institutions*, including environmental consulting firms, the *Ministry of Research and Innovation*, and research centers at *King Fahd University of Petroleum and Minerals* and *King Abdullah University of Science and Technology (KAUST).*
- **15 participants** from *policy-making* and *legal advisory sectors*, including legal consultancy firms and environmental policymakers.

The interviews were conducted in two different settings:

- 20 participants were interviewed at the *2nd International Conference on Sustainability Developments and Innovations (ICSDI 2024)*, held at *Prince Sultan University* in *February 2024*, and at the *LEAP 2024 conference*. These events provided an excellent platform to engage with experts in *AI*, *environmental sustainability*, and *intellectual property*.
- The remaining **35 participants** were interviewed at their respective workplaces, including the *Saudi Authority for Intellectual Property (SAIP)*, environmental consultancy offices, the *Ministry of Research and Innovation*, and research centers at *King Fahd University of Petroleum and Minerals* and *King Abdullah University of Science and Technology (KAUST).*

The interviews were structured around several key themes, including the role of *IP-enabled generative AI technologies* in advancing *saudi arabia's Vision 2030*, the challenges and opportunities in AI integration, and the impact of current *IP frameworks* on fostering innovation.

3.4.1 Key Questions Discussed

The interview discussions were guided by key questions such as:

- **"How do *IP-enabled generative AI technologies* contribute to achieving *Saudi Arabia's environmental sustainability goals* under Vision 2030?"**

- **"What challenges have you encountered in integrating *AI technologies* in sectors like *energy*, *water management*, and *environmental policy*, and how can they be addressed?"**
- **"How do you perceive the current *intellectual property (IP)* frameworks in terms of fostering or hindering AI-driven innovations in environmental projects?"**

A full set of the interview questions can be found in Appendix 1.

3.5 Focus Groups

A total of **9 focus groups** were conducted during various events held at *Prince Sultan University (PSU)* in 2024. These events provided a platform for gathering diverse insights on the role of *IP-enabled AI technologies* in achieving *saudi arabia's Vision 2030*. The focus groups were held during the following key events:

- *2nd International Conference on Sustainability Developments and Innovations (ICSDI 2024)*, held in **February 2024**.
- *7th Women in Data Science Conference*, held in **March 2024**.
- Other workshops and events organized at PSU, such as the *Sustainability Workshop Series*.

The focus groups involved participants from a variety of sectors, including *academia*, *technology*, *energy*, *environmental management*, and *policy-making*. Discussions were centered around the integration of *AI technologies* with *environmental sustainability* and the impact of *intellectual property (IP)* in fostering innovation.

3.5.1 Key Questions Discussed

The focus group discussions were guided by the following key questions:

- **"What comes to mind when you think about the intersection of artificial intelligence (AI) and environmental sustainability in Saudi Arabia?"**
- **"How do you perceive the role of intellectual property (IP) in facilitating or hindering AI-driven innovations for environmental sustainability?"**
- **"What are the major challenges in leveraging AI for environmental goals in Saudi Arabia, and how can they be overcome?"**
- **"How can IP-enabled AI technologies contribute to achieving Saudi Arabia's Vision 2030 sustainability goals?"**

These questions aimed to generate rich discussions, providing insights into participants' experiences, opinions, and knowledge regarding the role of *AI technologies* in *environmental sustainability* and the *IP frameworks* governing their implementation. A more detailed set of focus group questions can be found in Appendix 2.

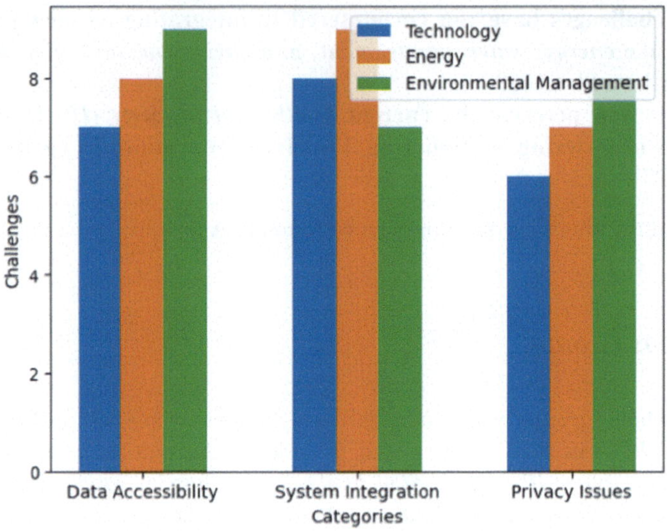

Fig. 6 Key challenges identified across sectors, including data accessibility, system integration, and privacy concerns

The insights gathered from these focus groups, particularly regarding the challenges and opportunities in *AI integration*, played a crucial role in identifying areas for *policy development* and *technological advancement* that align with *Vision 2030* (Fig. 6).

3.6 Data Analysis Using NVivo

The collected data were analyzed using *NVivo 12 software*, which facilitated detailed analysis through *content analysis* and *thematic analysis*. The steps included:

1. **Data Familiarization**: Transcripts were reviewed multiple times to identify key themes.
2. **Initial Coding**: Using NVivo's coding feature, important concepts such as *AI for energy optimization* and *IP challenges in AI adoption* were coded.
3. **Thematic Analysis**: Codes were grouped into broader themes, including *AI in Environmental Sustainability* and *Challenges in AI Implementation*.
4. **Theme Refinement**: NVivo's visualization tools (e.g., mind maps, hierarchical charts) were used to refine the themes (Fig. 7).

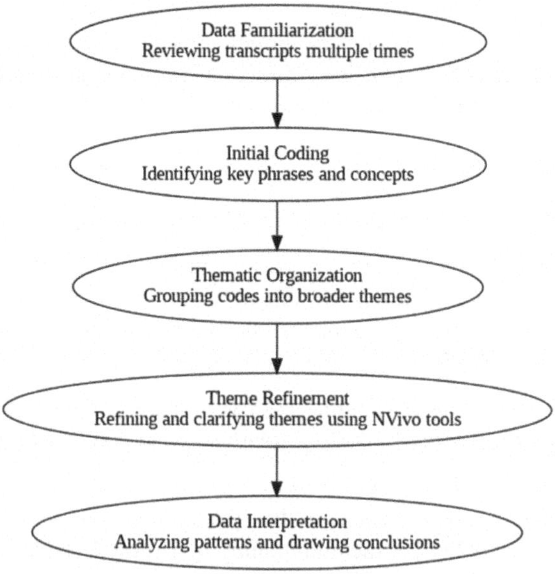

Fig. 7 Step-by-step NVivo data analysis, illustrating coding and thematic organization

3.7 *Trustworthiness and Rigor*

To ensure reliability and validity of the analysis:

- **Inter-coder reliability**: Multiple researchers independently coded the data, and discrepancies were resolved through discussion.
- **Memoing**: Memos were used to document analytical decisions and maintain transparency.
- **Data Saturation**: Data saturation was achieved when no new themes emerged from the coding process.

3.8 *Ethical Considerations*

Participants provided informed consent, and all data were handled confidentially. NVivo's privacy features were employed to ensure data protection.

4 Results

This section presents findings based on the interviews and focus group discussions, focusing on the role of *IP-enabled generative AI technologies* in achieving Saudi Arabia's *Vision 2030* goals, particularly in environmental sustainability and economic diversification. Key areas include *AI applications in environmental sustainability*, *challenges in AI implementation*, and the relationship between *intellectual property (IP)* and *AI-driven innovation*. The findings are presented with thematic analysis, participant feedback, and visual aids.

4.1 AI Applications in Environmental Sectors

Participants highlighted the growing role of AI technologies in optimizing energy production, water management, and healthcare. AI models have improved efficiency and sustainability in these sectors.

Energy Optimization: AI-driven models enhance the management of renewable energy resources, such as solar and wind power, leading to more efficient grid management.

Water Conservation: AI-based irrigation systems have reduced water consumption by up to 25%, while maintaining agricultural yields.

Healthcare Innovations: AI applications in healthcare contribute to personalized treatment and environmental monitoring within hospital systems, improving patient outcomes (Fig. 8 and Table 2).

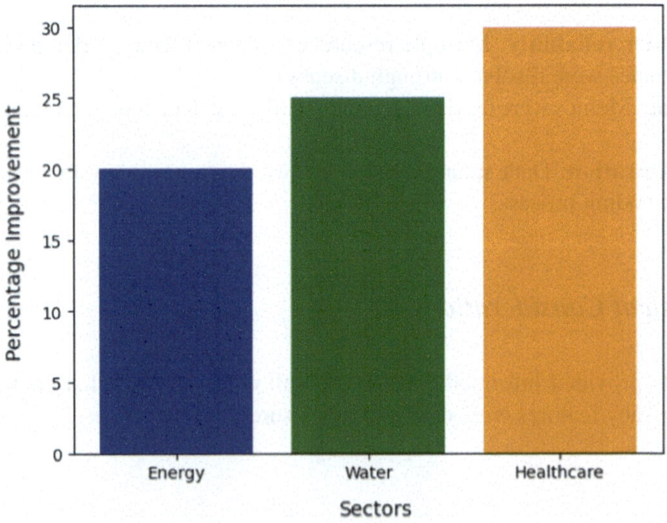

Fig. 8 AI applications in environmental sectors: percentage improvements in energy efficiency, water conservation, and healthcare optimization

Table 2 Axial coding analysis with examples

Axial code	Description	Relationship to open codes	Reason for axial coding	Examples of responses
Technological innovation in sustainability	AI's role in advancing sustainable practices across various sectors	Links AI in energy optimization, AI for sustainable agriculture	Highlights AI's transformative potential in sustainability	"AI optimization in renewable energy projects led to a 20% increase in efficiency"
AI-driven policy and conservation	AI's impact on policy making and conservation efforts, enhancing decision-making processes	Combines AI for environmental policy making, AI in biodiversity and conservation	Emphasizes AI's utility in informed policy and conservation strategies	"AI analysis informed new conservation policies, resulting in protected areas expansion"
Overcoming implementation challenges	Identifying and addressing barriers to AI integration in environmental projects	Draws from challenges in AI implementation	Showcases solutions to facilitate AI adoption	"Collaborative platforms have been established to overcome data silo issues"

4.2 *Challenges in AI Implementation*

Despite the potential of AI, several challenges were identified during the interviews and focus groups. The most significant barriers include data silos, infrastructure limitations, and legal issues related to intellectual property.

Data Siloing: Fragmented and disconnected data systems hinder the full potential of AI integration.

Infrastructure Limitations: Existing infrastructure in many areas lacks the technological capacity to integrate advanced AI systems.

Legal and IP Barriers: The slow adaptation of intellectual property laws to accommodate AI-driven innovations remains a critical obstacle (Fig. 9 and Table 3).

4.3 *Thematic Analysis from NVivo*

The analysis followed a structured thematic approach using NVivo, which identified several core themes based on participant responses.

Open Coding: The initial stage identified specific uses of AI in sectors like energy and water management, as well as challenges in integration.

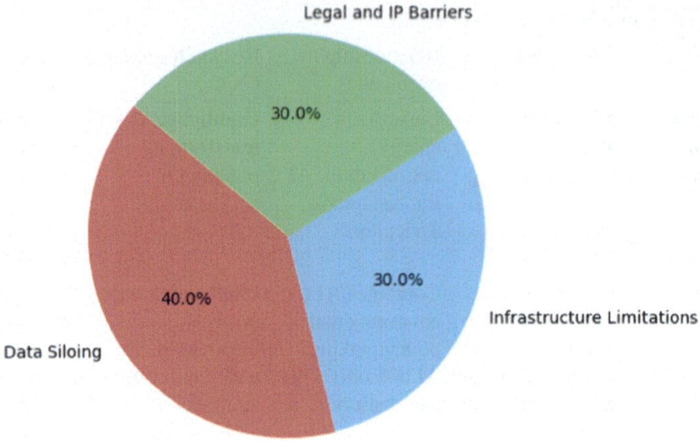

Fig. 9 Challenges in AI implementation: distribution of challenges faced across data silos, infrastructure limitations, and legal/IP barriers

Table 3 Axial coding analysis with examples

Axial code	Description	Example responses
Technological innovation	AI innovations in energy and agriculture directly support sustainability goals	"AI systems in renewable energy increased overall efficiency by 20%, reducing operational costs"
AI-driven policy insights	AI data analysis supports policy development in environmental conservation	"AI-driven insights helped us form new conservation strategies, expanding protected areas"
Challenges in AI implementation	AI faces difficulties in scaling due to legal frameworks and infrastructure issues	"The fragmented nature of our data systems makes it difficult to fully leverage AI technologies in environmental projects"

Axial Coding: Connections were made between these uses and broader environmental sustainability goals, highlighting how AI contributes to resource management.

Selective Coding: The final theme identified AI as a crucial driver for achieving *Vision 2030*'s environmental goals, while recognizing the need for improvements in policy and infrastructure (Fig. 10 and Table 4).

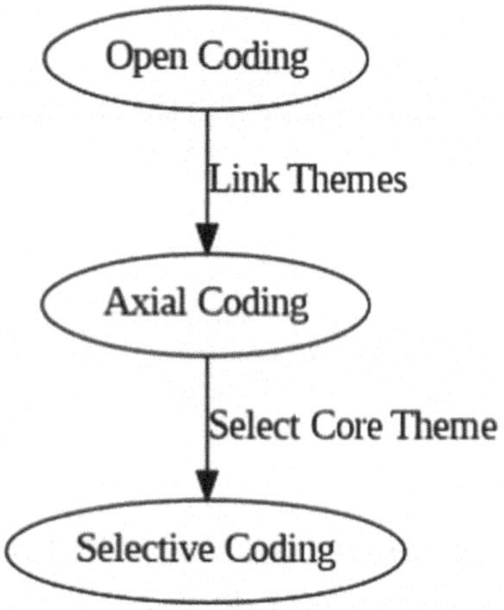

Fig. 10 Thematic analysis flowchart: process of coding from open coding to selective coding

Table 4 Selective coding analysis for AI as a catalyst for vision 2030

Selective code	Description	Integration of axial codes	Reason for selective coding	Examples
AI as a Catalyst for Vision 2030	AI technologies serve as a foundational element in realizing Saudi Arabia's Vision 2030 goals, particularly in enhancing environmental sustainability, fostering technological innovation, and overcoming implementation challenges	Combines insights from "Technological Innovation in Sustainability", "AI-Driven Policy and Conservation", and "Overcoming Implementation Challenges"	This code captures the essence of the research findings, highlighting AI's pivotal role in achieving comprehensive sustainability and innovation goals set forth by Vision 2030	Examples include AI-driven enhancements in energy efficiency, policy-making informed by AI, and overcoming data integration challenges

4.4 Focus Group Participant Responses

Focus groups provided more detailed insights into the practical challenges and opportunities related to the use of *IP-enabled AI technologies* in environmental sustainability efforts (Table 5).

Table 5 Focus group participant responses

Topic	Participant insight	Sector
AI in waste management	AI systems improved waste recycling rates by 30%	Environmental sustainability
AI for water conservation	AI models helped optimize water distribution in agriculture, reducing waste	Water management
Legal barriers	The current legal framework does not support AI innovations fast enough	Intellectual property

4.5 *Summary of Participant Insights on IP-Enabled AI for Environmental Sustainability*

Participants discussed the role of *IP-enabled AI* technologies in promoting *environmental sustainability*. The summary below aggregates the key insights from both interviews and focus groups (Table 6).

4.6 *Aggregated Insights on AI's Role in Enhancing Environmental Sustainability*

The following Table 7 aggregates insights across sectors regarding AI's role in *enhancing environmental sustainability*, based on participants' responses.

Table 6 Summary of participant insights on IP-enabled AI for environmental sustainability

Key theme	Participant insight	Supporting quotes
Energy efficiency	AI technologies improved energy efficiency in renewable projects by 20%	"AI optimization in renewable energy projects led to a 20% improvement in overall grid management efficiency"
Water management	AI-driven irrigation systems significantly reduced water usage	"The integration of AI in irrigation systems saved around 25% of water used in farming operations"
Policy making and AI	AI-driven data analytics supports more informed environmental policymaking	"AI-driven insights helped formulate new policies on renewable energy and waste management "

Table 7 Aggregated insights on AI's role in enhancing environmental sustainability

Sector	AI application	Participant example
Energy	AI improves energy efficiency and forecasting in renewable energy projects	"AI-driven forecasting models reduced energy consumption by optimizing grid distribution in solar energy systems"
Water management	AI irrigation systems improve water conservation, reducing agricultural water use	"AI systems saved 25% of water use in irrigation while maintaining high crop yields"
Waste management	AI technologies enhance waste sorting and recycling processes	"AI-driven systems increased recycling rates by 30% in urban areas"
Biodiversity conservation	AI models predict environmental impacts and assist in conservation efforts	"AI has helped track wildlife and predict habitat changes, leading to better conservation strategies"

4.7 *Themes in AI-Enabled Environmental Sustainability Efforts*

The coding analysis revealed several key themes related to *AI-enabled environmental sustainability*. The following Table 8 summarizes the major themes from the study.

Table 8 Themes in AI-enabled environmental sustainability efforts

Theme	Description	Examples
AI for resource optimization	AI technologies help optimize energy and water resources, contributing to sustainability	"AI-driven irrigation saved 25% of water usage in agricultural projects, helping meet sustainability goals"
Legal and regulatory barriers	Current IP and legal frameworks hinder the rapid deployment of AI technologies	"IP laws need to be updated to accommodate AI-driven inventions and ensure legal protection for AI-generated innovations"
Data and infrastructure challenges	Data fragmentation and outdated infrastructure pose significant barriers to AI deployment	"The lack of an integrated data system prevents full-scale AI adoption in environmental sustainability efforts"

5 Discussion

5.1 Interpretation of Results in the Context of Saudi Arabia's Vision 2030

The findings of this study illustrate the critical role of *IP-enabled AI technologies* in advancing Saudi Arabia's *Vision 2030* goals. The integration of AI in sectors such as *energy management*, *water conservation*, and *healthcare* has already led to significant improvements in *efficiency*, *resource optimization*, and *sustainability*. AI-driven technologies, when protected and incentivized by a robust intellectual property (IP) framework, can accelerate innovation, contributing to the country's economic diversification and environmental sustainability ambitions.

For example, AI's role in improving *energy efficiency* through predictive models that forecast energy needs has reduced consumption by 20%. Similarly, AI-powered *irrigation systems* have optimized water usage in agriculture, achieving up to 25% water savings. These findings underscore the importance of fostering an environment where *AI innovations* can flourish, aided by IP frameworks that protect and encourage technological advancement.

5.2 Implications for Policy, Practice, and Future Research

The results of this study carry several important implications for policymakers, practitioners, and researchers. First, there is a clear need for *policy reforms* to support AI innovations, particularly in the areas of *data integration* and *legal frameworks*. Current legal barriers, including the *slow adaptation of IP laws*, hinder the rapid deployment of AI technologies in sectors where they could significantly contribute to sustainability. By modernizing *IP regulations* to better protect *AI-generated content*, Saudi Arabia can promote both local and foreign investment in *AI-driven projects*.

For practitioners, the findings emphasize the need for *cross-sector collaboration* to overcome challenges such as *data silos* and *infrastructure limitations*. AI can only reach its full potential if integrated seamlessly into existing systems, requiring better *infrastructure* and *data-sharing mechanisms*.

Future research should focus on exploring *global best practices* in *AI and IP integration*, particularly from regions like the *European Union* and the *United States*, where AI regulations are more mature. Comparative studies would provide valuable insights into how Saudi Arabia can tailor its own policies to foster AI innovation while ensuring *ethical standards* and *data privacy*.

5.3 Recommendations for Policymakers, Researchers, and Practitioners

Based on the findings, the following recommendations are made to advance Saudi Arabia's *Vision 2030* objectives through *IP-enabled AI technologies*:

- **Policymakers** should focus on creating a *conducive regulatory environment* by modernizing *IP laws* to address the unique challenges posed by *AI technologies*. This includes developing a framework that protects AI-generated content and promotes *data-sharing* across sectors.
- **Researchers** are encouraged to conduct further studies on the *ethical implications* of AI adoption, particularly in sensitive sectors such as *healthcare* and *environmental management*. Research into the potential risks of *AI bias*, *data privacy*, and *job displacement* is crucial to ensure that AI benefits are maximized while minimizing potential harm.
- **Practitioners** should focus on *collaborative efforts* to overcome *data fragmentation* and *infrastructure limitations*. By working together, stakeholders in *technology*, *law*, and *environmental management* can ensure the successful deployment of AI in ways that contribute to both *economic growth* and *environmental sustainability*.

In conclusion, the study highlights the transformative potential of *IP-enabled AI technologies* in achieving Saudi Arabia's *Vision 2030* goals. However, realizing this potential requires comprehensive *policy reforms*, *cross-sector collaboration*, and ongoing *research* to address the challenges that currently hinder AI's full implementation.

6 Case Studies and Applications

6.1 Case Study 1: Flood Detection Secure System (FDSS) in Saudi Arabia

Introduction and Overview

The Flood Detection Secure System (FDSS) represents a significant technological advancement in Saudi Arabia's approach to environmental monitoring and disaster management [52]. Utilizing Deep Active Learning, the system is designed to enhance the accuracy and efficiency of flood detection across the country [53]. This initiative aligns with Saudi Arabia's broader environmental sustainability and disaster preparedness goals, leveraging cutting-edge technology to protect communities and landscapes from the devastating impacts of floods [54].

Technological Framework and Application

1. **Deep Active Learning and AI**: At the core of FDSS is the application of Deep Active Learning algorithms, which enable the system to process and analyze vast datasets from various sources, improving its ability to predict and detect flood events [55].
2. **Use of Images and IoT Data**: FDSS harnesses satellite imagery and data collected from an extensive network of IoT sensors. This integrated approach allows for real-time monitoring of critical environmental parameters, such as rainfall patterns, water levels, and soil moisture content, which are essential for accurate flood forecasting [56].
3. **Homomorphic Encryption for Data Privacy**: Recognizing the sensitivity of the data involved, FDSS incorporates homomorphic encryption technology. This innovative approach ensures that data can be processed and analyzed while remaining encrypted, safeguarding the privacy of potentially sensitive information without compromising the system's operational effectiveness [57, 58].

Benefits

1. Enhanced Environmental Monitoring: FDSS significantly improves Saudi Arabia's capacity for early flood detection, offering authorities a crucial lead time to implement disaster response measures and mitigate potential damages [33].
2. Increased Disaster Management Efficiency: The system enhances disaster response efforts by providing accurate and timely data, enabling more effective resource allocation and emergency planning [59].

Challenges

1. Technological Complexities: Integrating advanced AI algorithms with real-time data from diverse sources presents significant technical challenges, from data accuracy and processing to system reliability under extreme conditions [33].
2. Legal and Privacy Concerns: Ensuring compliance with national and international data protection regulations, particularly concerning the use of homomorphic encryption, requires careful legal navigation and robust security measures [60].
3. Implementation Barriers: The deployment of FDSS across Saudi Arabia's varied landscapes necessitates overcoming logistical hurdles, including the strategic placement of IoT sensors and ensuring their durability and connectivity.

Ethical and Legal Considerations from an IP Perspective

1. **Intellectual Property Compliance**: Adhering to IP laws and respecting copyright and licensing agreements, especially for software and imagery used in FDSS, is essential for the system's ethical deployment [61].
2. **Data Protection and Security**: Implementing FDSS within the framework of Saudi Arabia's data protection laws emphasizes the importance of maintaining high standards of data privacy and security, addressing potential concerns related to the collection and analysis of environmental data [62].

3. **Legal Framework Adaptation**: The dynamic nature of AI technology necessitates ongoing updates to legal and regulatory frameworks to ensure that innovations like FDSS are effectively governed, promoting both innovation and public trust [63].

Conclusion

The Flood Detection Secure System (FDSS) in Saudi Arabia exemplifies the country's commitment to leveraging advanced technology for environmental sustainability and public safety. By addressing the complex technical, legal, and ethical challenges associated with such a system, FDSS stands as a model for integrating AI and IoT technologies in environmental monitoring and disaster management efforts. As FDSS continues to evolve, it will play a pivotal role in enhancing Saudi Arabia's resilience to natural disasters, safeguarding both its people and its natural heritage.

6.2 *Forecasting CO_2 Emissions in Saudi Arabia Using AI and Statistical Modeling*

Introduction and Objectives

Saudi Arabia is advancing towards balancing economic growth with environmental sustainability. In collaboration with academic and research institutions, including King Saudi University, the country has initiated a project to forecast CO_2 emissions using Artificial Intelligence (AI) and statistical modeling. The primary aim is to inform policymaking, guide sustainable development, and support global climate change mitigation efforts [64, 65].

Methodological Framework

1. **AI and Machine Learning**: Advanced AI techniques, such as machine learning and deep learning, are employed to analyze historical CO_2 emissions data, identifying patterns and predicting future trends [38].
2. **Statistical Modeling**: Complementing AI, statistical models account for temporal trends, seasonal, and cyclical variations, ensuring forecast reliability [66].

Applications and Benefits

1. Accurate CO_2 emission forecasts enable the formulation of tailored environmental policies and initiatives, aligning with Vision 2030 and the Paris Agreement [67].
2. Insights contribute to optimizing the energy sector's performance, promoting cleaner energy sources, and efficient energy use [68].
3. The project aims to increase public awareness and encourage sustainable practices.

Challenges and Solutions

1. **Data Quality and Accessibility**: Collaboration with various stakeholders is essential to access comprehensive, high-quality data sets.
2. **Model Complexity and Interpretability**: Efforts are made to enhance the interpretability of complex AI models for policymakers and the public [69].
3. **Adapting to New Technologies**: The initiative remains flexible, incorporating advancements in AI and statistical methods to improve forecasts.

Ethical and Legal Considerations from an IP Perspective

1. Compliance with IP laws and data protection standards is crucial for ethical use and sustainability of the project [70].
2. The initiative operates within a dynamic legal environment, necessitating continuous adaptation to legal and regulatory changes [71].

Conclusion

Saudi Arabia's initiative to use AI and statistical modeling for CO_2 emission forecasting underscores its commitment to environmental sustainability and technological innovation. This project sets a precedent for responsibly leveraging technology in environmental management, contributing to national and global sustainability efforts.

6.3 Case Study 3: AI in Healthcare Environmental Monitoring in Saudi Arabia

Introduction

In Saudi Arabia, the integration of Artificial Intelligence (AI) and machine learning technologies into healthcare environmental monitoring represents a significant leap forward in improving patient care and operational efficiency. By continuously analyzing environmental parameters within healthcare facilities, AI technologies contribute to creating safer, more controlled environments conducive to patient health and recovery [72].

Technology Application

1. **Real-Time Environmental Monitoring:** Utilizing AI and machine learning algorithms, healthcare facilities in Saudi Arabia can now monitor various environmental parameters, including temperature, humidity, air quality, and the presence of airborne pathogens, in real-time [17]. This technological application ensures that any deviations from the norm can be promptly addressed, maintaining an optimal healing environment for patients.
2. **Predictive Analytics for Environmental Management:** Beyond simple monitoring, AI systems are capable of predicting potential environmental risks, enabling preemptive measures to be taken to avoid adverse impacts on patient health and healthcare facility operations [73].

Benefits

1. **Improved Healthcare Outcomes**: The primary benefit of AI-driven environmental monitoring is the direct contribution to improved patient outcomes. By ensuring environmental conditions are kept within optimal ranges, the risk of infection and disease transmission is significantly reduced [74].
2. **Enhanced Operational Efficiency**: AI technologies automate the process of environmental monitoring, reducing the burden on healthcare staff and allowing them to dedicate more time to patient care. Additionally, predictive maintenance capabilities can prevent costly and disruptive equipment failures [75].

Challenges

1. **Data Privacy Concerns**: With the collection and analysis of environmental data, ensuring patient privacy becomes a paramount challenge [76]. The healthcare sector must navigate stringent data protection regulations to safeguard sensitive information.
2. **Accuracy and Technological Integration**: Ensuring the accuracy of AI predictions and smoothly integrating new technologies with existing healthcare systems present significant challenges [77]. Any inaccuracies in environmental monitoring could have serious implications for patient safety [78].

Ethical and Legal Considerations

1. **Intellectual Property and Data Use**: The ethical use of AI in healthcare necessitates compliance with intellectual property laws, including securing appropriate licenses for data and ensuring the privacy and security of patient information [79].
2. **Legal Frameworks and Sustainability**: Updating legal frameworks to accommodate the rapid advancement of AI technologies is essential for their responsible use in healthcare. Moreover, the application of AI must align with sustainability goals, promoting environmental responsibility within healthcare facilities [80].

Conclusion

The adoption of AI in healthcare environmental monitoring in Saudi Arabia stands as a testament to the country's commitment to leveraging advanced technologies to enhance healthcare outcomes. By addressing the challenges related to data privacy, technological accuracy, and ethical use, and by ensuring legal compliance and sustainability, AI can significantly contribute to the improvement of healthcare environments. This initiative not only showcases Saudi Arabia's innovative approach to healthcare but also sets a precedent for the integration of technology in patient care and facility management on a global scale.

6.4 Case Study 5: Artificial Intelligence Framework for Modeling and Predicting Crop Yield to Enhance Food Security in Saudi Arabia

Context

Saudi Arabia faces significant challenges in achieving food security due to its arid climate, water scarcity, and reliance on food imports [81]. Recognizing the potential of technological innovation to address these challenges, the kingdom has embarked on an initiative to develop an artificial intelligence (AI) framework for modeling and predicting crop yields. This initiative aligns with Saudi Arabia's Vision 2030 goals, which emphasize sustainability, economic diversification, and technological advancement [82].

Application

The AI framework utilizes machine learning algorithms and predictive analytics to analyze a wide array of data, including satellite imagery, soil health indicators, weather patterns, and historical crop yield data [83]. By accurately predicting crop yields, the framework aims to optimize agricultural planning, resource allocation, and crop management practices.

Benefits

1. **Increased Agricultural Efficiency**: Enhanced predictive capabilities allow for better planning and resource management, reducing waste and maximizing crop yields [84].
2. **Improved Food Security**: By increasing local agricultural production and reducing dependency on imports, the framework contributes to national food security [85].
3. **Sustainability**: The efficient use of water and agricultural inputs, guided by AI predictions, supports sustainable farming practices, crucial in Saudi Arabia's arid environment [86].

Challenges

1. **Data Collection and Quality**: Gathering high-quality, comprehensive data in real-time presents logistical and technical challenges, essential for the accuracy of AI predictions [87].
2. **Technology Adoption**: Encouraging the adoption of AI-based tools among farmers and agricultural stakeholders requires overcoming barriers related to technology access, literacy, and trust [88].
3. **Environmental Variability**: The framework must account for the variability and unpredictability of environmental factors, which significantly impact crop yields [89].

Ethical and Legal Considerations from an IP Perspective

1. **Intellectual Property Rights**: Developing the AI framework involves navigating complex IP landscapes, ensuring that the use of algorithms, data, and technologies respects existing IP rights and contributes to a culture of innovation and collaboration [90].
2. **Data Privacy and Security**: Protecting the privacy and security of data used in the framework, especially when involving personal or proprietary information, is paramount, requiring adherence to strict data protection laws [91].

6.5 Case 5: Enhancing Security in the Oil and Gas Sector of Saudi Arabia

Context

The Kingdom of Saudi Arabia's oil and gas sector, vital to the global energy market and the national economy, faces increasing threats from adversarial drones and missiles [92]. In response, Saudi Arabia is leveraging the unprecedented growth in artificial intelligence (AI), alongside advancements in drones, chatbots, smart systems, and the Internet of Things (IoT) devices, to enhance sector security. This initiative illustrates the Kingdom's proactive approach to utilising technological advancements for national defence purposes [93].

Application

The proposed AI-based approach encompasses a three-zone defence strategy to safeguard the oil and gas infrastructure:

1. Drone Security Measures: Utilising drones for aerial surveillance and environmental monitoring, enabling early detection of threats [94].
2. High-Tech Power Laser Weapons: Deploying advanced laser systems in the second zone for immediate threat neutralisation [60].
3. Heavy Weapons: Stationing conventional heavy weaponry in the third zone as a final layer of defence against more significant threats [95].

An AI-based recognition system complements these zones, capable of identifying potential threats through intelligent analysis of imagery and directing appropriate responses to neutralise them effectively.

Benefits

1. Enhanced Environmental Monitoring and Management: The use of AI and drones facilitates comprehensive environmental surveillance, ensuring the safety and integrity of critical infrastructure [96].

2. Increased Security and Efficiency: The AI-driven approach allows for real-time threat detection and response, significantly reducing the risk of damage and ensuring the continuous operation of the oil and gas sector [97].
3. Cost-Effectiveness: By employing drones and AI for initial threat detection and response, Saudi Arabia can mitigate risks at a lower cost compared to traditional security measures [98].

Challenges

1. Technological Limitations: Continuous research is necessary to overcome the technical limitations of drones and AI systems, enhancing their accuracy and efficiency [99].
2. Legal and Implementation Hurdles: Navigating the legal frameworks related to drone use, AI deployment, and national defence, while ensuring effective implementation of the three-zone strategy, presents significant challenges [100].
3. Adversarial Adaptation: Keeping pace with the evolving tactics and technologies used by adversaries requires ongoing adaptation and innovation in AI and drone capabilities [101].

Ethical and Legal Considerations from an IP Perspective

1. Intellectual Property Rights: The development and deployment of AI technologies must navigate complex IP landscapes, ensuring respect for existing rights while fostering innovation.
2. Data Protection and Security: Safeguarding sensitive data collected through AI and drone surveillance is paramount, necessitating strict compliance with data protection laws [102].
3. Transparency and Accountability: Despite the operational secrecy required for national security, maintaining ethical standards in AI application and adherence to legal requirements is essential for public trust and international cooperation [103].

Conclusion

Saudi Arabia's initiative to enhance the security of its oil and gas sector through an AI-based, multifaceted defence strategy represents a pioneering approach to national defence. Addressing the dual threats of drones and missiles, the initiative underscores the Kingdom's commitment to leveraging technological advancements for the protection of its critical infrastructure. By overcoming technological, legal, and ethical challenges, Saudi Arabia aims to set a benchmark in AI-driven security measures, ensuring the resilience of its oil and gas sector against emerging threats and contributing to global energy security.

6.6 Case Study 6: Revolutionizing Higher Education in Saudi Arabia: The Integration of AI for Enhanced Learning Outcomes

Context

The higher education sector in Saudi Arabia stands on the brink of a significant transformation, driven by the adoption of artificial intelligence (AI) [104]. This shift addresses existing educational challenges and accelerates the Kingdom's progress towards its national objectives, including those outlined in Vision 2030 [105]. The initiative recognises the potential of AI to revolutionise teaching and learning methodologies, making education more accessible, personalised, and efficient [106].

Application

AI-based learning outcomes are being introduced across various levels of higher education, encompassing:

1. Personalised Learning Pathways: AI algorithms analyse students' learning styles, performance, and preferences to tailor educational content, thereby enhancing learning efficacy [107].
2. Automated Assessment and Feedback: The use of AI enables real-time assessment and feedback, allowing educators to focus on pedagogical strategies and student engagement.
3. Virtual Assistants and Chatbots: These AI tools provide students with 24/7 assistance, facilitating seamless access to information and support [108].

Benefits

1. Enhanced Educational Efficiency: AI-driven solutions streamline administrative tasks, curriculum development, and assessment processes, allowing for more efficient use of resources [109].
2. Improved Learning Outcomes: Personalised learning experiences fostered by AI contribute to higher student engagement, retention, and success rates.
3. Accessibility and Inclusivity: AI technologies make education more accessible to students with disabilities and those in remote areas, promoting inclusivity [110].

Challenges

1. Technological Adoption: Integrating AI into existing educational frameworks requires overcoming technological barriers and ensuring faculty and students are equipped with the necessary skills and resources [111].
2. Legal and Ethical Considerations: Navigating the complex landscape of IP rights, data protection laws, and ethical guidelines poses significant challenges to the ethical and legal adoption of AI in education [112].
3. Sustainability and Scalability: Ensuring the sustainable and scalable implementation of AI solutions in higher education necessitates ongoing investment, research, and policy support [113].

Ethical and Legal Considerations from an IP Perspective

1. Intellectual Property Rights: Adherence to IP laws and Creative Commons licensing is essential for the ethical use of digital content and AI technologies in education [114].
2. Data Protection and Privacy: Robust mechanisms must be in place to protect the privacy and security of student data, adhering to national and international data protection standards [115].
3. Transparency and Fairness: AI applications in education should be developed and deployed transparently, with mechanisms to address bias and ensure fairness in automated decision-making [116].

Conclusion

The integration of AI into higher education in Saudi Arabia represents a paradigm shift towards a more dynamic, responsive, and student-centred learning environment. By leveraging AI to enhance learning outcomes, the Kingdom aims to address the challenges of the modern educational landscape, aligning with its ambitious national goals. However, achieving this vision requires a balanced approach that considers technological capabilities, legal frameworks, and ethical standards. By navigating these complexities, Saudi Arabia can harness the full potential of AI to revolutionise its higher education sector, setting a benchmark for innovation and excellence on a global scale.

Appendix 1: Interview Questions and Responses

This appendix provides the full set of interview questions along with the corresponding responses from the 55 participants involved in the study. The participants represent a range of sectors, including AI technologies, intellectual property (IP), energy management, environmental sustainability, and policy-making.

Introductory Question

Can you briefly introduce yourself and describe your role in your organization?

Responses:

- **Participant 1**: "I'm a data scientist with 5 years of experience in AI, focusing on environmental projects in Saudi Arabia."
- **Participant 22**: "Environmental researcher focused on AI's role in biodiversity preservation."

- **Participant 35**: "I work on AI applications in sustainable agriculture, particularly in arid regions like Saudi Arabia."
- **Participant 48**: "I advise on environmental risk assessments, integrating AI to predict and manage risks."

Key Question 1

In what ways do you think AI technologies can contribute to achieving the goals of Saudi Arabia's Vision 2030, particularly in terms of environmental sustainability?

Responses:

- **Participant 1**: "Generative AI can significantly reduce energy consumption in industrial processes, aligning with Vision 2030s sustainability goals."
- **Participant 22**: "AI allows us to track and predict species migration and biodiversity loss in real-time, which is critical for conservation."
- **Participant 35**: "AI tools have increased crop yields while minimizing water usage, supporting sustainable agriculture."
- **Participant 48**: "AI has improved our ability to predict and manage environmental risks, which is crucial for long-term sustainability."

Key Question 2

Can you provide specific examples of AI applications in your field that have positively impacted environmental outcomes?

Responses:

- **Participant 1**: "One project I worked on optimized solar panel layouts using AI, boosting efficiency by 20%."
- **Participant 22**: "Challenges arise when dealing with fragmented data, but AI has helped unify our approach."
- **Participant 35**: "Integrating AI into farming practices was a challenge, but farmers are seeing the benefits now."
- **Participant 48**: "The integration of AI into traditional risk assessment models faced initial resistance, but the benefits have proven clear."

Probing Question

How do intellectual property rights influence the development and implementation of AI technologies in your sector?

Responses:

- **Participant 1**: "We've had to navigate complex IP frameworks, but they're crucial for protecting the innovations we create with AI."
- **Participant 22**: "The current IP laws don't fully address AI-generated content, which is a barrier to innovation in biodiversity research."
- **Participant 35**: "In agriculture, IP laws haven't caught up with AI advancements, and this slows down the adoption of new technologies."
- **Participant 48**: "There's a clear need for updated IP regulations to protect the AI models we develop for risk assessments."

Reflective Question

In your experience, what is the potential for AI to transform how Saudi Arabia addresses environmental challenges?

Responses:

- **Participant 1**: "The potential is huge; these technologies can revolutionize how we approach environmental challenges."
- **Participant 22**: "AI's ability to adapt to new data sets is one of its most powerful traits."
- **Participant 35**: "AI will be key in making agriculture more sustainable in arid regions like Saudi Arabia."
- **Participant 48**: "AI will play a significant role in managing Saudi Arabia's environmental risks."

Closing Question

What support or resources are needed to accelerate the adoption of AI in environmental management in Saudi Arabia?

Responses:

- **Participant 1**: "More public awareness and support for these initiatives are needed."
- **Participant 22**: "We need better funding for AI-based biodiversity research."

- **Participant 35**: "More investment in AI for agriculture will significantly contribute to food security."
- **Participant 48**: "More support for AI research in environmental risk management is essential."

Appendix 2: Focus Group Questions and Responses

This appendix provides a detailed overview of the focus group discussions conducted to explore the role of *IP-enabled generative AI technologies* in advancing *saudi arabia's Vision 2030* goals, particularly in relation to *environmental sustainability*. The focus groups included participants from sectors such as *energy*, *environmental conservation*, *agriculture*, and *policy-making*.

Introductory Question

What comes to mind when you think about the intersection of artificial intelligence (AI) and environmental sustainability in Saudi Arabia?

Responses:

- **Participant A**: "I've been involved in deploying AI for waste management optimization in urban areas."
- **Participant B**: "My role focuses on AI-driven conservation efforts, specifically in preserving coral reefs."
- **Participant C**: "I lead a team developing AI models for optimizing water usage in agriculture."
- **Participant D**: "As an environmental policy analyst, I assess the impact of AI on policy development."
- **Participant E**: "I'm part of a startup focusing on AI for renewable energy forecasting."

Key Question

Can you share any experiences or knowledge you have regarding current AI technologies being used in Saudi Arabia's environmental projects?

Responses:

- **Participant A**: "AI streamlines waste collection routes, reducing emissions and operational costs."

- **Participant B**: "We use AI to monitor reef health and predict areas at risk, aiding in targeted conservation efforts."
- **Participant C**: "Our AI solutions help farmers reduce water use by 25% while maintaining crop yields."
- **Participant D**: "AI provides data-driven insights for crafting more effective environmental policies."
- **Participant E**: "Our AI algorithms improve the efficiency of renewable energy distribution networks."

Probing Question

How do you perceive the role of intellectual property (IP) in facilitating or hindering AI-driven innovations for environmental sustainability?

Responses:

- **Participant A**: "Our project on smart recycling increased recycling rates by 30% in the pilot area, but IP challenges still remain in scaling the technology."
- **Participant B**: "Implementing AI for real-time monitoring of reefs was challenging due to IP concerns around data ownership."
- **Participant C**: "The main challenge was adapting AI models to local crop varieties, but IP regulations supported collaboration with developers."
- **Participant D**: "Bridging the gap between AI technology and policy implementation requires IP protections that incentivize innovation."
- **Participant E**: "The variability of renewable energy sources posed a challenge, but IP protections are necessary for proprietary AI models."

Reflective Question

What challenges have you encountered in leveraging AI for environmental goals in Saudi Arabia, and how can they be overcome?

Responses:

- **Participant A**: "AI in waste management can transform urban sustainability practices, but integrating AI into broader policies is still a challenge."
- **Participant B**: "Underwater communication limitations posed significant challenges, but advanced AI technologies are bridging the gap."
- **Participant C**: "The challenge of adapting AI models to local climates was addressed through collaboration with agricultural experts."

- **Participant D**: "Continuous dialogue between AI developers and policymakers is essential to overcoming implementation challenges."
- **Participant E**: "The variability in renewable energy sources was a challenge, but AI's predictive models helped improve efficiency."

Closing Question

What future developments in AI do you envision as crucial for Saudi Arabia's environmental vision, and what policies or IP reforms would support these innovations?

Responses:

- **Participant A**: "A holistic approach integrating AI into broader environmental policies is essential."
- **Participant B**: "Strengthening AI infrastructure and data availability for marine research is critical for future conservation efforts."
- **Participant C**: "Encouraging collaboration between AI developers and the agricultural sector is vital for sustainable agriculture."
- **Participant D**: "Enhancing the understanding of AI among policymakers is necessary for effective policy integration."
- **Participant E**: "Support for innovation and risk-taking in AI applications for renewable energy is needed to drive sustainability."

Summary of Key Themes and Insights

Energy and Resource Management: AI plays a critical role in optimizing *renewable energy distribution* and *water conservation* efforts, with significant improvements in resource efficiency (e.g., 25% reduction in water usage and 30% increase in recycling rates).

Challenges and Opportunities: *Data fragmentation*, *infrastructure limitations*, and *IP concerns* were commonly cited challenges, but *collaborative efforts* across sectors have helped address these issues.

The Role of IP in AI Adoption: Participants highlighted the importance of modernizing *IP laws* to support *AI-driven innovations*, particularly in sectors such as *energy*, *conservation*, and *agriculture*.

Policy Recommendations: Stronger collaboration between *policymakers*, *AI developers*, and *environmental experts* was emphasized as a key factor in successfully integrating AI into *Vision 2030* initiatives.

References

1. N. Kumar, Saudi Arabia's "vision 2030": structural reforms and their challenges. J. Sustain. Develop. **16**(4), 92 (2023). https://doi.org/10.5539/jsd.v16n4p92
2. F. Hasanov, N.H.A. Razek, Oil and non-oil determinants of Saudi Arabia's international competitiveness: historical analysis and policy simulations. Soc. Sci. Res. Netw. (2023). https://doi.org/10.2139/ssrn.4170277
3. S. Thaha Abdullateef, R.M. Alsheikh, B.K. Ibrahim Mohammed, Making Saudi vision 2030 a reality through educational transformation at the university level. Lab. Indus. J. Soc. Econ. Relat. Work (2023). https://doi.org/10.1080/10301763.2023.2184166
4. N. Ye, Oil and non-oil determinants of Saudi Arabia's international competitiveness: historical analysis and policy simulations. Sustainability (2023). https://doi.org/10.3390/su15119011
5. G. Tyack, Using AI-SPedia to study Saudi Universities' research outputs in the artificial intelligence field (2022). https://doi.org/10.1109/NCCC57165.2022.10067833
6. F.Y. Al Anezi, Saudi vision 2030: sustainable economic development through IoT (2021). https://doi.org/10.1109/CSNT51715.2021.9509592
7. Z.A. Memish, M.M. Altuwaijri, A.H. Almoeen, S.M. Enani, The Saudi data and artificial intelligence authority (sdaia) vision: leading the kingdom's journey toward global leadership. J. Epidemiol. Glob. Health (2021). https://doi.org/10.2991/JEGH.K.210405.001
8. N. Rauf, S. Zahoor, Artificial intelligence and the rule of law: a critical appraisal of a developing sector. Pakistan J. Soc. Res. (2023). https://doi.org/10.52567/pjsr.v5i02.1156
9. N.S. Alotaibi, Alshehri, Prospers and obstacles in using artificial intelligence in Saudi Arabia higher education institutions-the potential of AI-based learning outcomes. Sustainability **15**(13), 10723 (2023). https://doi.org/10.3390/su151310723
10. Exploring the role of information technology in supporting sustainability efforts in Saudi Arabia. Preprints **202307**(0188) (2023). https://doi.org/10.20944/preprints202307.0188.v1
11. H. Tedjini, Y. Guettaf, M.A. Hartani, Impact of artificial intelligence in renewable energy management of hybrid systems. Post Scriptum Framework (2023) https://doi.org/10.3390/psf2023006005
12. N. Altin, S.E. Eyimaya, Artificial intelligence applications for energy management in microgrid, in *IEEE International Conference on Smart Grid (icSmartGrid)* (2023). https://doi.org/10.1109/icsmartgrid58556.2023.10170860
13. S. Haleem, N. Hassan, M. Albar, H.O. AlWasem, Knowledge, awareness, and perception of Covid-19 and artificial intelligence: a cross-sectional study among the population in Saudi Arabia. Cureus (2023). https://doi.org/10.7759/cureus.40921
14. K. Alattas, Saudi Arabia corporate firms are hesitant to embrace artificial intelligence as of 2020 despite the numerous benefits. WSEAS Trans. Syst. Control (2023). https://doi.org/10.37394/23203.2023.18.4
15. F.-J. Chang, L.-C. Chang, J.-F. Chen, Artificial intelligence techniques in hydrology and water resources management. Water **15**(10), 1846 (2023). https://doi.org/10.3390/w15101846
16. A. Siddiqui, M.J. Siddiqui, N.M. Kulkarni, Artificial intelligence in water conservation: a meta-analysis study. Prabandhan: Indian J. Manage. **15**(3), 160407 (2022). https://doi.org/10.17010/pijom/2022/v15i3/160407
17. A.M. Saeed, A.B. Saeed, Saudi Arabia health systems: challenging and future transformations with artificial intelligence. Cureus Not provided 37826 (2023). https://doi.org/10.7759/cureus.37826
18. E. Chikhaoui, A. Alajmi, Artificial intelligence applications in healthcare sector: ethical and legal challenges. Emerg. Sci. J. **6**(4), 05 (2022). https://doi.org/10.28991/esj-2022-06-04-05
19. F.Y. Alenezi, Artificial intelligence versus Arab universities: an enquiry into the Saudi context. Al-Migalat al-ilmayat li-Gamiat al-Malik Faisal Not provided 220038 (2023). https://doi.org/10.37575/h/edu/220038
20. M. Yadav, G. Singh, Environmental sustainability with artificial intelligence. EPRA Int. J. Multidiscip. Res. (2023). https://doi.org/10.36713/epra13325

21. V.A. Wankhede, R. Agrawal, A. Kumar, S. Luthra, D. Pamučar, Z. Stevic, Artificial intelligence an enabler for sustainable engineering decision-making in uncertain environment: a review and future propositions. J. Glob. Oper. Strategic Sourc. (2023). https://doi.org/10.1108/JGOSS-06-2022-0057
22. S. Samoilenko, Artificial intelligence and sustainability (2023). https://doi.org/10.59287/iccar.734
23. A. Name, Exploring the role of information technology in supporting sustainability efforts in Saudi Arabia. https://doi.org/10.20944/preprints202307.0188.v1
24. A.M. Nassef, A.G. Olabi, H. Rezk, M.A. Abdelkareem, Application of artificial intelligence to predict CO_2 emissions: critical step towards sustainable environment. Sustainability **15**(9), 7648 (2023). https://doi.org/10.3390/su15097648
25. M. Shi, K.E. Currier, Z. Liu, K.W. Janowicz, N. Wiedemann, J.A. Verstegen, G. McKenzie, A. Graser, R. Zhu, G. Mai, Thinking geographically about AI sustainability. AGILE: GIScience Ser. **4**, 42 (2023). https://doi.org/10.5194/agile-giss-4-42-2023
26. J. Sun, R. Ortíz, Generative AI in healthcare. J. AI Res. (2024)
27. N.E.A. Tivari, Ai-based content creation for personalized healthcare. AI Soc. (2024)
28. P. Sidhu, Challenges of generative AI. Ethics in AI (2024)
29. A. Kaur, Predictive AI in healthcare: benefits and limitations. Healthc. Technol. J. (2024)
30. R.E.A. Alt, AI and IP integration in environmental sustainability. Int. J. AI (2024)
31. A.E.A. Sinha, Predictive AI vs generative AI: complementary technologies. AI Innov. J. (2024)
32. S.H. Khan, Artificial intelligence based modelling for predicting CO_2 emission for climate change mitigation in Saudi Arabia. Int. J. Adv. Comput. Sci. Appl. (2023). https://doi.org/10.14569/ijacsa.2023.0140421
33. A. Alsumayt, N. El-Haggar, L. Amouri, Z.M. Alfawaer, S.S. Aljameel, Smart flood detection with AI and blockchain integration in Saudi Arabia using drones. Sensors (2023). https://doi.org/10.3390/s23115148
34. R. Alghassab, Fuzzy topsis for solar energy solutions. Renew. Energy Rev. (2023)
35. A.E.A. Vani, Ai-powered smart water grids. Water Manage. J. (2024)
36. M.E.A. Alsenani, Medcore: Ai for diagnostic accuracy in healthcare. J. Healthc. AI (2024)
37. A. Nala, A. Sarah, A. Ahmed, K. Iyad, A. Aiiad, M. Rashid, Musawah: a data-driven Ai approach and tool to co-create healthcare services with a case study on cancer disease in Saudi Arabia. Sustainability **14**(6), 3313 (2022). https://doi.org/10.3390/su14063313
38. B. Kanyarat, K.C.B. Kunyanuth, Prediction of CO_2 emissions using machine learning (2023). https://doi.org/10.7250/conect.2023.099
39. B. Tayeb, Using artificial intelligence to predict wind speed for energy application in Saudi Arabia. Energies **12**(24) (2019). https://doi.org/10.3390/EN12244669
40. M. Kahia, B. Jarraya, B. Kahouli, A. Omri, Do environmental innovation and green energy matter for environmental sustainability? evidence from Saudi Arabia (1990–2018). Energies (2023). https://doi.org/10.3390/en16031376
41. A.I. Al-Alawi, A.M. A-Lmansouri, Artificial intelligence in the judiciary system of Saudi Arabia: a literature review (2023). https://doi.org/10.1109/cymaen57228.2023.10050929
42. T. Alshammari, R.A. Ramadan, A. Ahmad, Temporal variations dataset for indoor environmental parameters in northern Saudi Arabia. Appl. Sci. **13**(12), 7326 (2023). https://doi.org/10.3390/app13127326
43. A. Atabekov, Ai-driven innovations and IP challenges: a comparative analysis of international practices. J. AI Law **12**, 89–104 (2023)
44. S. Ubaydullayeva, Intellectual property and AI: a cross-national analysis. Glob. IP J. **5**, 22–37 (2023)
45. J. Smith, H. Lee, Comparative analysis of diffusion of innovative solutions in health tech. Int. J. Innov. Technol. **8**, 45–58 (2022)
46. P.G. Picht, G. Thouvenin, Ai and IP: theory to policy and back again—policy and research recommendations at the intersection of artificial intelligence and intellectual property. Int. Rev. Intellect. Property Competition Law (2023). https://doi.org/10.1007/s40319-023-01344-5

47. L. Wang, A. Haak, Ai regulation and IP law in the united states: a sector-specific approach. J. AI Policy **16**, 89–103 (2024)
48. S. Thaler, Thaler v. uspto: Ai as an inventor in the U.S. patent system. AI Law Rev. **12**, 25–34 (2019)
49. M.L. Montagnani, L. Palazzani, The Eu's AI act and liability directive: a new paradigm for regulating high-risk ai systems. Eur. J. AI Law **10**, 45–63 (2024)
50. E. Ribeiro, Balancing innovation and antitrust: the Eu's approach to AI and intellectual property rights. Competition Law J. **18**, 112–130 (2020)
51. J. Andraško, J. Habersack, Data protection and privacy in AI systems: the European perspective. Eur. Data Protect. Law Rev. **7**, 56–72 (2021)
52. A. Al-Wathinani, D.G. Barten, M. Borowska-Stefańska, P. Gołda, M.A. Alhallaf, L. Samarkandi, M. Goniewicz, K. Goniewicz, Driving sustainable disaster risk reduction: a rapid review of the policies and strategies in Saudi Arabia. Sustainability **15**(14), 10976 (2023). https://doi.org/10.3390/su151410976
53. G. A.J. Abdulnoor, S. Ahmad, A.-R. S. Tariq, A.H. Faez, Z. Maryam, A.-A. M. Ahmed, A. H. Saleh, I. Muhammad, R. Saifur, An improved flood susceptibility assessment in Jeddah, Saudi Arabia, using advanced machine learning techniques. Water **15**(14), 2511 (2023). https://doi.org/10.3390/w15142511
54. Y.E. Idrees, Iot-based system for automated floodwater detection and early warning in the East Africa region: a case study of Arusha and dar es salaam, Tanzania (2023). https://doi.org/10.58694/123456789/1340
55. K.W. Sanders, D. Li, W. Li, Z.N. Fang, Data-driven flood alert system (Fas) using extreme gradient boosting (xgboost) to forecast flood stages. Water **14**(5), 747 (2022). https://doi.org/10.3390/w14050747
56. W. Xuan, Z. Zhang, S. Xiong, W. Zhang, J.K. Tang, Z. Li, B. An, A near-real-time flood detection method based on deep learning and SAR images. Remote Sens. **15**(8), 2046 (2023). https://doi.org/10.3390/rs15082046
57. G. Kumar Mahato, S. Kumar Chakraborty, Homomorphic encryption based federated learning for financial data security. Available online at IEEE Xplore (2023). https://doi.org/10.1109/I3CS58314.2023.10127502
58. S. Dhiman, S. Nayak, G.K. Mahato, A. Ram, S.K. Chakraborty, Homomorphic encryption based federated learning for financial data security (2023). https://doi.org/10.1109/i3cs58314.2023.10127502. Available: IEEE Xplore
59. G. Sofia, Rewardljat: the operational flash-flood forecasting system for the kingdom of Saudi Arabia: a case study of the 24th November 2022 flash flood in Jeddah. EGUsphere (2023). https://doi.org/10.5194/egusphere-egu23-7434
60. A. Alqarni, D. Timko, M.L. Rahman, Saudi Arabian perspective of security, privacy, and attitude of using facial recognition technology (2023)
61. F. Garel, *Ethical Deployment* (Oxford University Press, 2022). https://doi.org/10.1093/oso/9780192847263.003.0006 . Available online at Oxford Scholarship Online
62. M. Mashaabi, R. Alnashwan, H.S. Al-Khalifa, Arabic privacy policy corpus and classification, in *Lecture Notes in Computer Science* (2023). https://doi.org/10.1007/978-3-031-35320-8_7
63. Oxford Analytica, *As it considers regulation Canberra must also boost ai* (Emerald Expert Briefings, 2023). https://doi.org/10.1108/oxan-db280017
64. M.A. Nassef, A.G. Olabi, H. Rezk, M.A. Abdelkareem, Application of artificial intelligence to predict co2 emissions: Critical step towards sustainable environment. Sustainability **15**(9), 7648 (2023). https://doi.org/10.3390/su15097648
65. H.S. Khan, Artificial intelligence based modelling for predicting CO_2 emission for climate change mitigation in Saudi Arabia. Int. J. Adv. Comput. Sci. Appl. (2023). https://doi.org/10.14569/ijacsa.2023.0140421
66. Y. Ma, H. Liu, S. Wang, Nonparametric approaches for analyzing carbon emission: from statistical and machine learning perspectives. Preprint available at arXiv:2303.14900 (2023). https://doi.org/10.48550/arxiv.2303.14900

67. Y. Gao, D. Yan, N. Liu, S. Luo, Z. Zhou, Y. Wang, Y. Chen, B. Gao, A high-frequency monitoring method for urban carbon emissions based on vertical federated deep learning and multi-source heterogeneous data fusion methods (2022). https://doi.org/10.1145/3582935.3582972
68. M. Yang, H.D. Noman, Predicting CO_2 emission footprint using AI through machine learning. Atmosphere **13**(11), 1871 (2022). https://doi.org/10.3390/atmos13111871
69. M. Yang et al., A predictive analysis on CO_2 emissions in automobiles using machine learning techniques (2023). https://doi.org/10.1109/idciot56793.2023.10053539
70. W. Qiao, *Ethical IoT: A Sustainable Way Forward* (CRC Press, 2022). https://doi.org/10.1201/9781003337423-12. Available online
71. I. Wallimann-Helmer, L. Terán, E. Portmann, H. Schubel, J. Pincay, An integrated framework for ethical and sustainable digitalization, in *Proceedings of the IEEE International Conference on e-Democracy & e-Government (ICEDEG)* (2021). https://doi.org/10.1109/ICEDEG52154.2021.9530972
72. T. Alshammari, R.A. Ramadan, A. Ahmad, Temporal variations dataset for indoor environmental parameters in northern Saudi Arabia. Appl. Sci. **13**(12), 7326 (2023). https://doi.org/10.3390/app13127326
73. N. Aljerian, M.R. Arafat, A. Aldhubib, I.K. Almohaimeed, A. Alsultan, A.A. Alhosaini, A. Aloqayli, B. Alrabiah, S. Albanyan, L.B. Ashry, R. AlGhassab, R.M. Alzahrani, G. Alhenaki, E. Alnazzawi, A. Almughaiseeb, A.A. Alanazi, Artificial intelligence in healthcare and its application in Saudi Arabia. Int. J. Innov. Res. Med. Sci. **7**(11), 1558 (2022). https://doi.org/10.23958/ijirms/vol07-i11/1558
74. A.H. Mousa, N.T. Maria, F.S. Almuntashiri, B.S. Alsaywid, M.D. Lytras, The potential of artificial intelligence in healthcare: Perceptions of healthcare practitioners and current adoption (2023). https://doi.org/10.1016/b978-0-323-98353-2.00001-0
75. F. Dabdoub, M. Colangelo, M. AlJumah, Artificial intelligence in healthcare and biotechnology: a review of the Saudi experience. J. AI Cult. (Assuming based on DOI) **1**(107) (2022). https://doi.org/10.47363/jaicc/2022(1)107
76. L.M. Al Anazi, Client satisfaction in healthcare environment in Saudi Arabia: a qualitative review. Saudi J. Nurs. Health Care **6**(7), 001 (2023). https://doi.org/10.36348/sjnhc.2023.v06i07.001
77. M.A. Alyousef, A.S.A. Enazi, A.H. Almosilhi, A.A. Alanazi, K.M. Alanazi, B.M. Aljumaily, S.M. Alanazi, A.I. Alsenaid, F.I.A. Mofeez, A. Yousef, A. Sulaiman, Impact of applying artificial intelligence: healthcare professionals insight (a qualitative survey study in Hafr-Elbatin, Saudi Arabia). Int. J. Adv. Res. (2022). https://doi.org/10.21474/ijar01/15721
78. N.A. Aljarallah, Investigating the influence of artificial intelligence on quality management in healthcare centers. F1000Research (2023). https://doi.org/10.12688/f1000research.128739.1
79. V.N. Bashkin, The impact of procedural law on artificial intelligence by improving the healthcare systems (2023). Chap. 10. https://doi.org/10.4018/978-1-6684-8337-4.ch010
80. S.P. Kennedy, Navigating the AI revolution: the case for customized regulation in healthcare. Preprint (2023). https://doi.org/10.2196/preprints.49989
81. B.A. Alotaibi, M.B. Baig, M. Najim, A.A. Shah, Y. Alamri, Water scarcity management to ensure food scarcity through sustainable water resources management in Saudi Arabia. Sustainability **15**(13), 10648 (2023). https://doi.org/10.3390/su151310648
82. N. Kumar, Saudi Arabia's "vision 2030": structural reforms and their challenges. J. Sustain. Develop. **16**(4), 92 (2023). https://doi.org/10.5539/jsd.v16n4p92
83. L.I. Licheng, W. Zhou, K. Guan, B. Peng, C. Jiang, J. Tang, S. Wang, R. Grant, S. Mezbahuddin, X. Jia, S. Xu, Knowledge-based artificial intelligence for agroecosystem carbon budget and crop yield estimation. https://doi.org/10.1002/essoar.10509206.2 (2023). https://doi.org/10.1002/essoar.10509206.2
84. P. Sharma, P. Dadheech, A.S. Senthil, AI-enabled crop recommendation system based on soil and weather patterns (2023). Chap. 10. https://doi.org/10.4018/978-1-6684-8516-3.ch010
85. V. Ishimwe, Ai based real-time weather condition prediction with optimized agricultural resources. Eur. J. Technol. (2023). https://doi.org/10.47672/ejt.1496

86. C. Lakshmi, A. Pagadala, S.M. Bandam, A. Penugonda, B. Gangaraju, T. Ganapavarapu, R. Amirtharajan, V. Thanikaiselvan, H. Mahalingam, Analysing agricultural information through machine learning and artificial intelligence for SMART IRRIGATION. https://doi.org/10.1109/vitecon58111.2023.10157366 (2023)
87. L.A. Arthur, J.W. Costello, J.E. Rea, G. Ganev, On the Challenges of Deploying Privacy-Preserving Synthetic Data in the Enterprise. Details on publication venue needed (2023)
88. P. Bo, Overcoming data challenges in AI-driven development programs. Preprint at OSF (2023). https://doi.org/10.31219/osf.io/wzms4
89. V. Charbonnier, *Governing Agricultural Data: Challenges and Recommendations* Publisher details needed (2022). https://doi.org/10.1007/978-3-031-13276-6_11
90. M. Owczarczuk, Ethical and regulatory challenges amid artificial intelligence development: an outline of the issue. Ekonomia i Prawo (2023). https://doi.org/10.12775/eip.2023.017
91. H.F. Williamson, J. Brettschneider, M. Caccamo, R.P. Davey, C. Goble, P.J. Kersey, S. May, R.J. Morris, R. Ostler, T. Pridmore, Rawlings CData management challenges for artificial intelligence in plant and agricultural research. F1000Research (2023). https://doi.org/10.12688/f1000research.52204.2
92. M.K. Alsmadi, G.M. Jaradat, S. Abahussain, M.F. Tayfour, U.A. Badawi, H. Alfagham, M. Alshabanah, D. Alrajhi, H. Alkhaldi, N. Altuwaijri, H. Shoshan, H.M. Abouelnaga, Susceptible exposed infectious recovered-machine learning for covid-19 prediction in Saudi Arabia. Int. J. Power Electron. Drive Syst. **13**(4), 4761–4776 (2023). https://doi.org/10.11591/ijece.v13i4.pp4761-4776
93. I. Mahnashi, B. Salah, A.E. Ragab, Industry 4.0 framework based on organizational diagnostics and plan-do-check-act cycle for the Saudi Arabian cement sector. Sustainability **15**(14), 11261 (2023). https://doi.org/10.3390/su151411261
94. A. Madkhali, S.T.M. Sithole, Exploring the role of information technology in supporting sustainability efforts in Saudi Arabia. Preprints (2023). https://doi.org/10.20944/preprints202307.0188.v1
95. H. El-Taj, The evolving threat landscape: an overview of attacks and response strategies against cybercrime in Saudi Arabia. Int. J. Netw. Secur. Appl. **15**(3) (2023). https://doi.org/10.5121/ijnsa.2023.15306
96. W.J. Zai, W. Fan, Y. Xie, X. Yin, Intelligent monitoring system for substation environment based on UAV inspection mode, in *Proceedings of the IEEE Conference on Power and Energy* (2023). https://doi.org/10.1109/PandaFPE57779.2023.10141068
97. T. Miller, D. Cembrowska-Lech, A. Kisiel, A. Krzemińska, I. Durlik, Harnessing AI for environmental resilience: mitigating heavy metal pollution and advancing sustainable practices in diverse spheres. Grail Sci. (2023) https://doi.org/10.36074/grail-of-science.14.04.2023.027
98. M. Cabianca, M.L. Clemente, G. Gatto, C. Impagliazzo, L. Leoni, M. Masia, R. Piras, An application of IoT in a drone inspection service for environmental control. IoT **3**(3), 20 (2022). https://doi.org/10.3390/iot3030020
99. D. Wood, J. Ahn, S. Calo, N. Greco, K. Grueneberg, T. Inoue, D. Verma, S. Wang, AI enabled processing of environmental sounds in commercial and defense environments (2022). https://doi.org/10.1002/9781119892199.ch10. Book title and editors are needed
100. E.M. Ditria, C. Buelow, M. González-Rivero, R.M. Connolly, Artificial intelligence and automated monitoring for assisting conservation of marine ecosystems: a perspective. Front. Mar. Sci. (2022). https://doi.org/10.3389/fmars.2022.918104
101. M. Hijji, Using artificial intelligence to protect, detect and mitigate oil and gas sectors of ksa from drones and missiles assaults, in **Proceedings of the International Conference on Computer and Information Technology** (2022). https://doi.org/10.1109/ICCIT52419.2022.9711599
102. P.G. Picht, G. Thouvenin, Ai and IP: theory to policy and back again—policy and research recommendations at the intersection of artificial intelligence and intellectual property. Int. Rev. Intellect. Prop. Competition Law (2023). https://doi.org/10.1007/s40319-023-01344-5
103. S.R. Asiryan, Steps of the Eu countries aimed at protecting the constitutional rights in the era of artificial intelligence. Naukovij vìsnik Užgorodskogo nacìonalnogo unìversitetu **76**, 45 (2023). https://doi.org/10.24144/2307-3322.2022.76.2.45

104. N.S. Alotaibi, A. Alshehri, Prospers and obstacles in using artificial intelligence in Saudi Arabia higher education institutions-the potential of AI-based learning outcomes. Sustainability **15**(13), 10723 (2023). https://doi.org/10.3390/su151310723
105. J. Ali, S.H.H. Madni, Iot adoption model for e-learning in higher education institutes: a case study in Saudi Arabia. Sustainability **15**(12), 9748 (2023). https://doi.org/10.3390/su15129748
106. I. Drach, O. Petroye, O. Borodiyenko, I.Y. Reheilo, O. Bazeliuk, N.V. Bazeliuk, O. Slobodianiuk, The use of artificial intelligence in higher education. Unìversiteti ì lìderstvo **15**, 66–82 (2023). https://doi.org/10.31874/2520-6702-2023-15-66-82
107. K.B. Kavitha, T.K. Pradeep Kumar, S. Nithiya, A. Suguna, Implementation of artificial intelligence in education. Int. Res. J. Comput. Sci. (2023). https://doi.org/10.26562/irjcs.2023.v1005.01
108. S.Y. Asatryan, Revolutionary changes in higher education with artificial intelligence. Mankavarzhut'yan ev hogebanut'yan himnakhndirner **10**(1) (2023). https://doi.org/10.24234/miopap.v10i1.454
109. T. Wang, B.D. Lund, A. Marengo, A. Pagano, N.R. Mannuru, Z.A. Teel, J. Pange, Exploring the potential impact of artificial intelligence (AI) on international students in higher education: generative AI, Chatbots, analytics, and international student success. Preprint at Preprints.org (2023). https://doi.org/10.20944/preprints202305.0808.v1
110. L. Chen, P. Chen, Z. Lin, Artificial intelligence in education: a review. IEEE Access (2020). https://doi.org/10.1109/ACCESS.2020.2988510
111. C.K.Y. Chan, A comprehensive AI policy education framework for university teaching and learning. Int. J. Educ. Technol. Higher Educ. (2023). https://doi.org/10.1186/s41239-023-00408-3
112. S. Gartner, M. Krasna, Artificial intelligence in education-ethical framework, in *2023 12th Mediterranean Conference on Embedded Computing (MECO)* (2023). https://doi.org/10.1109/meco58584.2023.10155012
113. X. Wang, *Introduction* (Springer, 2022). https://doi.org/10.1007/978-981-19-1496-6_1
114. M.O.F. Munshi, S. Barai, Ip rights to AI-generated works: barriers presented by existing law and reformations needed. Can. J. Bus. Inform. Stud. **22**, 037046 (2022). https://doi.org/10.34104/cjbis.022.037046
115. M. Zia-ud-Din, F.E. Elhajraoui, Role of artificial intelligence in legal education in the 21st century. FWU J. Soc. Sci. (2023). DOIurlhttps://doi.org/10.51709/19951272/summer2023/5
116. S. Nagi, *The Need for AI Ethics in Higher Education* (Springer, 2023). https://doi.org/10.1007/978-3-031-23035-6_1

Selma AbdElrahman Hussein is an accomplished Intellectual Property and Technology Transfer Specialist with over 19 years of experience in bridging academic research and industrial applications. She currently serves as the Head of the Intellectual Property Office at Prince Sultan University, Riyadh, where she fosters innovation, guides researchers in commercialization, and manages technology transfer projects. Her expertise includes intellectual property management, technology transfer, and establishing partnerships between academia and industry. Selma has held key positions, including Research Center Supervisor at Princess Nourah Bint Abdulrahman University and Chairperson of the Leather Industry Engineering Department at Sudan University of Science and Technology. She has a robust academic background with a Ph.D. in Industrial Chemical Engineering from Sudan University of Science and Technology, focusing on predictive analysis for leather manufacturing. Throughout her career, Selma has contributed to numerous intellectual property initiatives, facilitated patent applications, and led innovation competitions. She has undergone professional training in patent law, copyright, and intellectual property systems from prestigious institutions such as Harvard Law School and WIPO Academy. Selma's dedication to advancing innovation aligns with Saudi Arabia's Vision 2030 goals, emphasizing sustainable and transformative growth.

Dr. Emna Chikhaoui is an Assistant Professor of Law and serves as the Aide to the President of Prince Sultan University and Senior Manager of the Compliance, Legal, and Policies Office. She graduated from the Faculty of Law and Political Sciences of Tunisia and was seconded to Riyadh, Saudi Arabia, in 2006 to establish a law department for female students. Dr. Chikhaoui held the position of Vice Dean of the College of Law until 2021 and is the founder of the Commercial Law Master's Degree program, which launched in 2016. She also led the establishment of the Compliance, Legal, and Policies Office (CLPO) in 2021. Dr. Chikhaoui has led multiple research projects sponsored by Microsoft, with publications in reputable journals on topics such as cloud computing, data protection in the health and education sectors, and SMEs. Her recent work focuses on AI in the health sector in Saudi Arabia, addressing ethical and legal issues. She actively participates in discussions on AI-enabling ecosystems, including a recent Microsoft and Asia House roundtable in Riyadh. Dr. Chikhaoui's research interests include law and technology, intellectual property issues, and policy development.

The manufacturer's authorised representative in the EU is Springer Nature Customer Service Centre GmbH, Europaplatz 3, 69115 Heidelberg, Germany. If you have any concerns regarding our products, please contact ProductSafety@springernature.com

Printed and bound by CPI Group (UK) Ltd, Croydon, CR0 4YY
30/09/2026
02225584-0009